DNA AND RNA: PROPERTIES AND MODIFICATIONS, FUNCTIONS AND INTERACTIONS, RECOMBINATION AND APPLICATIONS

BIOINFORMATICS

GENOME BIOINFORMATICS AND COMPUTATIONAL BIOLOGY

DNA and RNA: Properties and Modifications, Functions and Interactions, Recombination and Applications

Additional books in this series can be found on Nova's website
under the Series tab.

Additional e-books in this series can be found on Nova's website
under the e-book tab.

Genetics - Research and Issues

Additional books in this series can be found on Nova's website
under the Series tab.

Additional e-books in this series can be found on Nova's website
under the e-book tab.

DNA AND RNA: PROPERTIES AND MODIFICATIONS, FUNCTIONS AND INTERACTIONS, RECOMBINATION AND APPLICATIONS

BIOINFORMATICS

GENOME BIOINFORMATICS AND COMPUTATIONAL BIOLOGY

RENU TUTEJA
EDITOR

Nova Science Publishers, Inc.
New York

For permission to use material from this book please contact us:
Telephone 631-231-7269; Fax 631-231-8175
Web Site: http://www.novapublishers.com

Library of Congress Cataloging-in-Publication Data

Bioinformatics : genome bioinformatics and computational biology / editor, Renu Tuteja.
p. ; cm.
Includes bibliographical references and index.
ISBN 978-1-62100-913-9 (hardcover)
I. Tuteja, Renu.
[DNLM: 1. Computational Biology--methods. 2. Genomics--methods. QU 26.5]
570.285--dc23
2011038582

Published by Nova Science Publishers, Inc. † New York

Contents

Preface

This book describes current research from around the world on studies involving bioinformatics and computational biology tools. Bioinformatics is the field of science in which computer science, information technology and biology, merge to form a single discipline. It is the application of information technology and computer science to the field of biology and biological systems. It is a new science which is very helpful in understanding most of the biological processes. In the last few years the use of bioinformatics has increased tremendously and it has aided in developing new methods to understand the complex biological systems. Bioinformatics, which is the melding of molecular biology with computer science, is essential to the use of genomic information in understanding human diseases and in the identification of new molecular targets for drug discovery. Bioinformatics has wide applications and its major uses in this field are gene finding, genome assembly, drug design and discovery and genome wide association studies. The field of bioinformatics has developed in a way that the critical tasks involve the detailed analysis and interpretation of the variety of data, including nucleotide and amino acid sequences, protein domains, and protein structures. This real process of analyzing and interpreting data is commonly known as computational biology. This book describes the use of bioinformatics and computational biology in a variety of processes. Some of these are biobanking, application of next-generation DNA sequencing in medical discovery, microRNA profiling involved in human tumorigenesis and cancer phylogenetics.

Renu Tuteja
Staff Research Scientist
International Centre for Genetic Engineering & Biotechnology (ICGEB)
PO Box 10504, Aruna Asaf Ali Marg, New Delhi 110 067, INDIA.
Tel: 91-11-26741358/60/61(O); 91-11-26890983(R)
Fax: 91-11-26742316

In: Bioinformatics
Editor: Renu Tuteja
ISBN: 978-1-62100-913-9

Chapter I

Repetitive DNA in *Phyllostachys pubescens* Genome and its Utilization for Genomic Evolution Analysis and Marker-Assisted Selection

Ding-Qin Tang** and Ming-Bing Zhou
The Nurturing Station for the State Key Laboratory of Subtropical Silviculture,
Zhejiang A & F University, Zhejiang, China

1. Introduction

Bamboo are members of the subfamily Bambusoideae within the grass family Poaceae, distribute all over the world, but major species richness is found in Asia Pacific and South America (Bystriakova et al., 2003a, b). More than 70 genera and 1000 species of bamboo worldwide have been described, and share some common features that include rhizomatous habit, hollow segmented culms, petiolate blade with tessellate venation, flowers with three or more lodicules, usually with six stamens, and fruit possess small embryo and linear hilum (Soderstrom, 1981). Bamboos are multipurpose useful and is in high demand in the rural life of many Asian countries. Thin culms with narrow cavities are popularly used as umbrella handles, fishing rods and flutes, while mature culms are widely used in constructing mud huts, mats, baskets, fences and the production of high quality charcoal (Park and Kwon, 1998) along with the fibers which are ideal for paper and pulp production. Highly nutritious leaves as well as young shoots are often used for preparing delicious soups and pickles (Khatta and Katoch, 1983). Characterized with high growth rate (typically matures within 5-7 years) plus a number of important fuel features such as low ash content, alkali index or heating value, bamboo is a promising energy crop for future (Scurlock et al., 2000). *Phyllostachys pubescens* (synonym: *P. edulis*) is the most commercially important species with rapid culm elongation (more than 100 cm/d) (Li et al., 1997), providing the third largest source of timber

* E-mail: tang@zafu.edu.cn.

after Chinese red pine (*Pinus massoniana*) and China fir (*Cunninghamia lanceolata*) in China. There are 4.2 million ha of bamboo forest in China, and *P. pubescens* accounts for 3 million ha (approximately 2% of the total forest area), an area that has doubled over the last 30 years (Fu, 2001).

The genome sizes of bamboo species are relative large among the grass. The basic chromosome number for most woody bamboo species is χ=12, the same as rice, while for some herbaceous bamboo (*Olyra* spp.) it is synapomorphic (χ=11) (Hunziker et al., 1982). Occurrence of poly-ploidization has been reported in the woody bamboos (Soderstrom, 1981; Pohl and Clark, 1992). Cytological studies indicated the existence of two distinct sections, tropical (hexaploids, 2n=6χ= 72 and aneuploidy with more than 100 in some species) and temperate (tetraploids, 2n= 4χ= 48) within the woody group (Ghorai and Sharma, 1980; Kellogg and Watson, 1993; Clark et al., 1995). It was later supported by the flow cytometric estimation of the genomic DNA content, in which temperate bamboos had higher DNA content (4.17–5.3 pg) than tropical ones (2.34–3.23 pg) (Gielis et al., 1997a). According to the most recent estimation (Gui et al., 2007), the genome size of tetraploid *Phyllostachys pubescens* is 2034 Mb, which is 5.4 fold larger than that of the diploid cultivated rice and 1.9 fold larger than that of tetraploid wild rice, while little less (86.92%) than that of maize genome. The genome size of *Olyra latiflia*, another herbaceous bamboo species, is approximate 9.5 pg and close to two times that of maize (unpublished data).

Many studies show that large genomes contain abundant repetitive DNA sequences which are largely responsible for the huge differences in plant genome size (Bennetzen, 2002; Feschotte et al., 2002). Therefore, it is reasonable to assume that large and diverse families of repetitive DNA sequences will be found in bamboo genomes. Repetitive DNA sequences are with high copy numbers in genomes and mainly comprise of two categories, e.g. tandem repeats and interspersed repeats (Ussery et al., 2008). Tandem repeat sequence motifs lie adjacent to each other in a block, or an array, one typical family of which are microsatellites, also known as simple-sequence repeats (SSRs), which have repeat units typically 1-6 bp with repeat length rarely exceeding hundreds of repetitions in order. The interspersed repeat sequences disperse throughout the genome as single units flanked by unique sequence, most of which are derived from transposable elements (TEs), "jumping" of a DNA segment to another place of the genome. Repetitive sequences make up a significant part of many genomes (Wickstead et al., 2003), are dynamic elements that contribute to plasticity, evolve faster than coding regions and can be used in species identification and phylogenetic inference (Fu et al., 1998). For example, TEs account for at least 45% of the human genome (Lander et al., 2001), 35% of the rice genome (International Rice Genome Sequencing Project, 2005) and 85% of the maize genome (Schnable et al., 2009). TEs are one of the propulsors of genome evolution. They cause both large-scale rearrangements and changes in the structure and expression of individual genes, through activities such as excision, integration, chromosome breakage and ectopic recombination (Naito et al., 2009; Sinzelle et al., 2009). Many genes may have been assembled or amplified through the action of TEs, which can lead to infertility in heterozygous progeny. Therefore, TEs may be responsible for the rate at which such incompatibility is generated in separated populations (Bennetzen, 2002). Microsatellites are abundant and distributed throughout the genome (Richard et al., 2008), and also thought to play a significant role in genome evolution by creating and maintaining quantitative genetic variation because of their high mutability (Tautz et al., 1986; Kashi et al., 1997). Microsatellites preserve for chromosome organization (highly clustered in the centromere), affect DNA replication and gene duplication, induce genetic disorder and regulate gene expression (reviewed by Li, 2001). For example, the length of SSRs may influence transcriptional activity in promoter regions (Kashi et al., 1997). Length of poly-glutamine or poly-proline tracts encoded by SSRs may affect protein–protein interactions involving transcription factors (Gerber et al., 1994; Perutz et al., 1994).

The repetitive sequences is an important part of the genome composition, in particular for complex genomes such as that of *P. pubescens*. Analyses of repetitive sequences can deliver important information on the structure and evolution of genomes and on the biology of repetitive sequences. Here is first comprehensive analysis of the repetitive sequences in *P. pubescens* genome with the emphasis of TEs and SSRs. Our data provide a glimpse of the structure and evolution of *P. pubescens* genome, and enrich the current resource of molecular markers used for marker-assisted selection, genetic diversity studies for *P. pubescens* subfamily.

TEs in *P. pubescens* Genome and its Utilization for Genomic Evolution Analysis

Whole genomic sequencing reveals that TEs constitute a large fraction even a majority of the DNA in plants (such agriculturally important cereal as corn and wheat) and animals (mice and humans). Only a small fraction of the *P. pubescens* genome has been sequenced thus far, it is not clear of the content of TEs and their compact on the *P. pubescens* genome evolution. We investigated the distribution, diversity and evolution of TEs in *P. pubescens* genome, and emphasis on four super families of TEs including *mariner*-like, *PIF* (*P instability factor*)-like and *Pong*-like elements and Ty1-*copia* retro-elements.

2.1. Distribution and Evolution of TEs in *P. pubescens (Poaceae) Genome* Survey Sequences and the Full-Length cDNA Sequences

Compared to other members of the grass family, genomic data for bamboo is comparatively scarce, and no bamboo species genome has yet been sequenced on a large scale.

Most of the available genome data is for *P. pubescens* (Tang, 2009). Presently, there are 2.12 Mb of *P. pubescens* genome survey sequences (GSSs) (Gui et al., 2007), including two BAC clones of 113.2-kb length GQ252886 and 139.3-kb length GQ252887 (Gui et al., 2007, 2010) and 11.4 Mb of 10608 full-length cDNA sequences (FL-cDNAs) (Peng et al., 2010) deposited in GenBank. In this sector, we 1) compared the distribution of TEs between GSSs and FL-cDNAs, 2) analyzed the phylogeny of Ty1-*copia* and Ty3-*gypsy* with the most prevalence, and 3) characterized the transcript profiles of cDNA-TEs. This data provides important information on the structure and evolution of genomes and on the biology of TEs for *P. pubescens*.

2.1.1. Distribution of TEs in P. pubescens GSSs and FL-cDNAs

P. pubescens GSS and FL-cDNA *sequence data* were downloaded from NCBI GenBank (http://www.ncbi.nlm.nih.gov/). Redundant sequences were eliminated and overlapping sequences were spliced together using CAP3 software (http://seq.cs.iastate.edu/cap3.html) (Huang and Madan, 1999). After removal of redundant sequences, there remained 1.47 Mb non-redundant GSSs and 5.90 Mb non-redundant FL-cDNA sequences or contigs. They were analyzed by RepeatMasker (http://www.repeatmasker.org/cgi-bin/WEBRepeatMasker) to identify TEs, revealing 674 GSS-TEs (total 0.29 Mb, representing 20.42% of the GSS data) and 95 cDNA-TEs (total 13.52 kb, representing 0.32% of the FL-cDNA data) (Table 1). The 674 GSS-TEs comprised 146 DNA transposons (total 0.04 Mb, representing 2.43% of the GSS data), 518 retrotransposons (total 0.25 Mb, 17.23% of the GSS data) and 10 uncharacterized TEs (total 5.31 kb, 0.36% GSS data).

Table 1. Transposable elements (TEs) in *Phyllostachys pubescens* genome survey sequences (GSSs) and the full-length cDNA sequences (FL-cDNAs) (Zhou et al., 2011)

Names of TE families	Number of TEs in GSSs/ FL-cDNAs		
	No.[a]	Length (bp)[b]	Proportion (%)[c]
RNA transposon	518/31	252790/5868	17.23/0.12
SINEs	0/0	0/0	0/0
Penelope	0/0	0/0	0/0
LINEs	11/1	3741/98	0.25 /0
CRE/SLACS	0/0	0/0	0/0
L2/CR1/Rex	0/0	0/0	0/0
R1/LOA/Jockey	0/0	0/0	0/0
R2/R4/NeSL	0/0	0/0	0/0
RTE/Bov-B	0/0	0/0	0/0
L1/CIN4	11/1	3741/98	0.25/0
LTR elements	507/30	249049/5770	16.98/0.12
BEL/Pao	0	0	0
Ty1-*copia*	179/15	117876/3396	8.03/0.07
Ty3-*gypsy*	293/15	127712/2374	8.70/0.05
Retroviral	0	0	0
DNA transposons	146/54	35656/6014	2.43/0.12
Ac/Ds	45/6	12070/713	0.82/0.01
Tc1/marier	15/11	2157/1705	0.15/0.03
En/Spm	32/8	10606/620	0.72/0.01
Mutator	45/24	9922/2603	0.68/0.05
PiggyBac	0/0	0/0	0/0
Tourist/Harbinger/PIF	4/2	516/229	0.04/0
Uncharacterized TEs	10/10	5312/1634	0.36/0.03
Total		293758/13516	20.42%/0.32

[a] Number of TEs in GSSs/FL-cDNAs; [b] The combined length in bp (base pairs) of all TEs in GSSs/FL-cDNAs; [c] Combined length of all TEs in GSSs/FL-cDNAs as a proportion of the combined length of non-redundant GSS/FL-cDNA sequence data.

In contrast, the 95 cDNA-TEs comprised 54 DNA transposons (total 6.01 kb, 0.12% of the FL-cDNA data), 31 retrotransposons (total 5.87 kb, 0.12% of the FL-cDNA data) and 10 further uncharacterized TEs (total 1.63 kb, 0.03% of the FL-cDNA data).

Among the GSS DNA transposons, *Ac/Ds* elements were the most abundant (45 elements, covering 12.07 kb and 0.82% of the GSS data) followed by Mutator elements (also 45 elements, covering 9.92 kb and 0.68% of the GSS data). In contrast, among the cDNA DNA transposons, Mutator elements were the most abundant (24 elements, covering 2.60 kb and 0.05% of the FL-cDNA data) followed by the *Tc1/mariner* superfamily (11 elements, covering 1.71kb and 0.03% of the FL-cDNA data). Among GSS retrotransposons, Ty3-*gypsy* elements were most abundant (293 elements, covering 127.71 kb and 8.70% of the GSS data), followed by Ty1-*copia* elements (179 elements, covering 117.88 kb and 8.03% of the GSS data). The GSS retrotransposons were often found in clusters, e.g. in BAC GQ252887 there were 28 retrotransposons arranged in series. The Compared with 179 examples (117.9 kb, 8.03%) and 293 (127.7 kb, 8.7%) in GSS, Ty1-*copia* and Ty3-*gypsy* elements were remarkably lower in FL-cDNAs, with only 15 examples of each element covering 2.37 kb (0.05%) and 3.40 kb (0.07%) of the FL-cDNA data, respectively. These results show that GSSs have a much higher TE content than FL-cDNAs, and Ty1-copia and Ty3-gypsy are the most abundant TEs in the P. pubescens genome (Table 1).

Hence, based on the above statistical data, just over 20% of the sequence was represented by TEs in 1.47 Mb of non-redundant *P. pubescens* GSS data, which is significantly lower than the 35% reported in rice (International Rice Genome Sequencing Project, 2005), and the 85% reported in maize (Schnable et al., 2009). In contrast, the *P. pubescens* genome is relative large (approximately 2,034 Mb) and close to the maize genome in size, but more than five times larger than the genome of diploid cultivated rice (Gui et al., 2007). Given that the amplification of transposable elements (TEs) is largely responsible for the large plant genome size (Bennetzen, 2002; Feschotte et al., 2002), there should be a higher TE content in *P. pubescens* GSSs than predicted in our study. One possible reason is that *P. pubescens* GSS data currently deposited in GenBank is insufficient (covers only 0.36% of the genome) to represent the whole *P. pubescens* genome. It is likely that the proportion of TEs in *P. pubescens* genome has been underestimated by focusing on publically available GSS data.

Compared to *P. pubescens* GSS data, the FL-cDNAs data are abundant and likely to represent more than a quarter of bamboo genes and the third largest collection of FL-cDNA sequences next to those of Arabidopsis and rice. It provides the first large sequence dataset for studying the structure and function of a substantial portion of bamboo genes, and fills the gap in the grass family for comparative genomics (Peng et al., 2010). There are relatively few TEs within the FL-cDNA sequences compared to the genome average (TEs represent 0.32% of FL-cDNAs but >20% of GSSs), which indicates that most TEs are not expressed in *P. pubescens* transcriptome due to the tight regulation of TEs (Jiao and Deng, 2007). However, the presence of TEs in and around genes appears to be essential for the growth and development of the host organism (Lockton and Gaut, 2009; Pritham, 2009), because they are involved in the regulation of gene expression (Marino-Ramirez et al., 2005; Feschotte, 2008). In a seminal study, Jordan et al. (2003) reported that nearly 25% of experimentally characterized human promoters contain TE sequences, including empirically defined *cis*-regulatory elements. Furthermore, despite the strong conservation of gene expression patterns across different maize lines, *Mutator* transposition programmed by transcriptionally active

MuDR can induce a 25% change in the anther transcriptome, reflecting widespread insertion of the *Mutator* transposon into genes encoding transcription factors (Skibbe et al., 2009). Our data show that 14 of 24 *Mutator* insertion sites in *P. pubescens* FL-cDNAs are located in genes encoding transcription factors, which indicates that *Mutator* transposons may influence host growth and development by influencing the regulatory factor of gene expression (Marino-Ramirez et al., 2005).

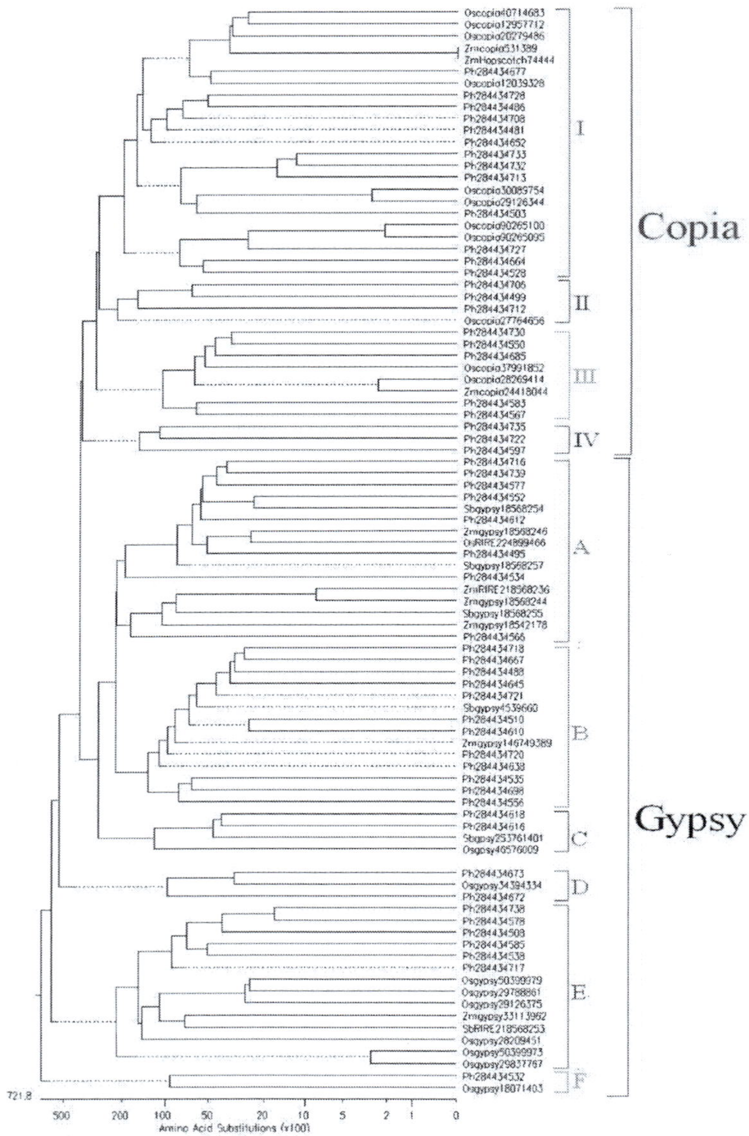

Figure 1. Phylogenetic analysis of Ty1-*copia* and Ty3-*gypsy* elements in *P. pubescens.* Groupings defining lineages and sublineages of *P. pubescens* Ty1-*copia* and Ty3-*gypsy* elements are shown in different colors and named appropriately. The phylogenetic tree was generated by aligning 55 *P. pubescens* retrotransposons and 36 related elements from other Poaceae species: 11 Ty1-*copia* elements and 10 Ty3-*gypsy* elements from rice (Os), three Ty1-*copia* elements and six Ty3-*gypsy* elements from maize (Zm), and six Ty3-*gypsy* elements from sorghum (Sb). (Zhou et al., 2011).

2.1.2. Evolution of Ty1-copia and Ty3-Gypsy Elements in P. pubescens

TE could be divided into class I and class II according to mechanism of transposition. Class I elements (retransposons, retro-elements) transpose by means of an RNA intermediate in a reaction involving several enzymes of reverse transcriptase and integrase. In contrast, class II elements (DNA transposons) move directly via DNA and the transposition reaction is catalyzed by a transposase encoded by autonomous DNA transposons (Feschotte et al., 2002). Therefore, it is not surprising that TEs of class I, of which Ty-*copia* and Ty3-*gypsy* elements accounted for major fraction of TEs in *P. pubescens* genome (Table 1) similar to rice and maize (Gui et al., 2007). The evolution of *P. pubescens* retrotransposons was investigated by aligning the complete reverse transcriptase sequences from 24 Ty1-*copia* and 31 Ty3-*gypsy* elements with 36 elements from other Poaceae species (11 Ty1-*copia* and 10 Ty3-*gypsy* elements from rice, three Ty1-*copia* and six Ty3-*gypsy* elements from maize, and six Ty3-*gypsy* elements from sorghum). In the phylogenetic tree, two retrotransposon clusters (*copia* and *gypsy*) were defined as the largest and best-supported monophyletic groups (Figure 1).

These groups were obtained regardless of the construction method of neighbor-joining (NJ), maximum parsimony (MP) or maximum likelihood (ML). The Ty1-*copia* elements could be divided into four sub-clades (I-VI) and the Ty3-*gypsy* elements into six sub-clades (A-F). Every sub-clade contained multiple retro-elements, and all but one of the sub-clades (the exception was *copia* cluster IV) contained retro-elements from more than one species, indicating that all sub-clades except *copia* IV predated the divergence of bamboo and the other grasses.

The available fossil evidence and the surviving basal lineages suggest that the Bambusoideae diverged from the rest of the Poaceae during the upper Cretaceous more than 65 Myrs ago (Guo and Li, 2002). It therefore appears that *P. pubescens* Ty1-*copia* and Ty3-*gypsy* elements are ancestral in origin, and originated more than 65 Myrs ago and have subsequently undergone extensive genetic and epigenetic diversification (Matsuoka and Tsunewaki, 1999).

2.1.3. Analysis of the Transcript Profiles of cDNA-TEs

GenBank contains 10,608 *P. pubescens* FL-cDNAs, of which 4,217 are expressed in leaf tissue, 3,072 are expressed in embryos and 3,318 are expressed in shoots (Peng et al., 2010). We investigated the distribution of the four most abundant TE families and found 24 Mutator, 11 *Tc1/mariner*, 15 Ty1-*copia* and 15 Ty3-*gypsy* among these sequences (Table 1). Among 24 *Mutator* elements, 10 were detected in leaf transcripts, seven in shoot transcripts and seven in embryo transcripts. Among 11 *Tc1/mariner*, elements, seven were detected in leaf transcripts, two in shoot transcripts and two in embryo transcripts. Among 15 Ty1-*copia* elements, eight were detected in leaf transcripts, three in shoot transcripts and four in embryo transcripts. Finally, among 15 Ty3-*gypsy* elements, seven were detected in leaf transcripts, three in shoot transcripts and five in embryo transcripts.

We also investigated the insertion sites of the 24 Mutator elements, noting that 19 elements had integrated into the 5' untranslated region (5'-UTR) and five into the coding region of the genes (Table 2). A large proportion of the inserted genes encoded putative transcription factors (14 insertions); the others were involved in the energy cycle (five insertions), post-translational regulation (two insertions) and membrane transport (one insertion).

Table 2. FL-cDNAs with integrated *Mutator* transposons. UTR = untranslated region, CDS = coding sequence (Zhou et al., 2011)

GenBank No.	Insertion	Organ	Categories of genes	Homologous genes
FP091571	5‘UTR	Leaf	Transcription factor gene	Homeobox transcription factor KNOX3 (*Hordeum vulgare*)
FP091749	5‘UTR	Shoot		Unknown
FP091954	CDS	Embryo	Transcription factor gene	Pathogenesis-related transcriptional factor (Prb1) (*O. sativa*)
FP093400	5‘UTR	Leaf	Cell energy cycle related gene	ADP-ribosylation factor (*Zea mays*)
FP093768	CDS	Leaf	Transposase gene	Mutator-like transposase (*O. sativa*)
FP094062	CDS	Shoot	Transcription factor gene	Pathogenesis-related protein 1 precursor (PR-1) (*O. sativa*)
FP095802	5‘UTR	Leaf	Transcription factor gene	Myb-like protein (*O. sativa*)
FP095913	5‘UTR	Shoot	Transcription factor gene	Transcription factor MYBS2 (*O. sativa*)
FP096707	5‘UTR	Shoot	Cell energy cycle related gene	Proton-translocating NADH-quinone oxidoreductase (*Homalodisca coagulate*)
FP096801	5‘UTR	Leaf	Post-translation related gene	Serine/Threonine protein kinases (*O. sativa*)
FP097565	5‘UTR	Embryo	Transcription factor gene	GATA transcription factor (*Ricinus communis*)
FP097737	5‘UTR	Leaf	Transcription factor gene	GlsA-related protein gene (*Chlamydomonas reinhardtii*)
FP099127	5‘UTR	Embryo	Transcription factor gene	Auxin-responsive Aux/IAA family member (IAA15) (*Z. mays*)
FP099725	5‘UTR	Shoot	Cell energy cycle related gene	Adenine phosphoribosyltransferase (*O. sativa*) catalyzes the formation of AMP from adenine and 5-phosphoribosylpyrophosphate
FP099829	5‘UTR	Embryo	Transcription factor gene	Auxin-responsive protein IAA7 (*O. sativa*)
FP100486	5‘UTR	Shoot	Transcription factor gene	Pathogenesis-related transcriptional factor and ERF (*O. sativa*)
FP100707	CDS	Leaf	Transcription factor gene	TMPIT-like protein (*Sorghum bicolor*)
FP100818	5‘UTR	Leaf	Membrane transport proteins	plastid-lipid-associated protein 2 (*Arabidopsis thaliana*)
FP100934	CDS	Embryo	Post-translation related gene	GPI-anchored protein (*O. sativa*)
FP100979	5‘UTR	Leaf	Transcription factor gene	transcription factor LcDREB3a (*Leymus chinensis*)
FP101158	5‘UTR	Embryo	Transcription factor gene	CCAAT-binding transcription factor (*O. sativa*)
FP101391	5‘UTR	Embryo	Transcription factor gene	PRLI-interacting factor G (*O. sativa*)
FP101428	5‘UTR	Shoot	Cell energy cycle related gene	GTP-binding protein sar1 (*Triticum aFL-cDNAivum*)
FP101553	5‘UTR	Leaf	Transcription factor gene	Transcription activator/transcription factor (NAM, *O. sativa*)

The 14 genes encoding putative transcription factors included two regulated by hormones (FP099127 and FP099827) and three regulated by pathogens (FP091954, FP094062 and FP100486). The results show that Mutator elements have a strong preference for the 5'-UTRs of genes encoding transcription factors.

In conclusion, we investigated the distribution, diversity and evolution of TEs in the 7.37 Mb of non-redundant *P. pubescens* sequence data available in GenBank. TEs are relatively abundant, diverse and polyphyletic, of which Ty1-*copia* and Ty3-*gypsy* are the most abundant elements in the *P. pubescens* genome. These appear to predate the divergence of the Bambusoideae from the rest of the Poaceae, and have undergone significant genetic and epigenetic differentiation. We found evidence that the distribution of some intragenic TEs is correlated with transcript profiles. Our data provide a tantalizing glimpse of the structure and evolution of *P. pubescens* genome, although large-scale sequencing of the *P. pubescens* genome would be required to fully understand the architecture of the *P. pubescens* genome.

2.2. Diversity and Evolution of *Mariner*-, *PIF*- and *Pong*- like Elements and Ty1-*Copia* Retroelements in *P. pubescens* Genome

In order to compensate for the shortage of relatively scarce sequence data, we cloned and investigated the distribution, diversity and evolution of the four super-families of TEs including *mariner*-, *PIF*- and *Pong*- like elements and Ty1-*copia* retro-elements in *P. pubescens* genome. *Mariner*-, *PIF*- and *Pong*- like elements are the founder members of a large and diverse family of class II transposons. They are widespread found in many animals and plants (Feschotte and Wessler, 2002; Zhang et al., 2001, 2004). The plant *mariner*-like elements (*MLEs*) contain the related motif DD39D, defining a distinct *mariner*-like subfamily (Feschotte and Wessler, 2002). *PIF*-like elements shares a characteristic (DD(34/37)E) catalytic motif and *Pong*-like elements contain contains an apparent DD35E motif, a signature consisting of three acidic residues found in the catalytic domains of some eukaryotic and prokaryotic transposases (Zhang et al., 2001, 2004).

The amino acid sequences surrounding the above motifs are relatively well-conserved in both animals and plants, and has therefore served to establish the evolutionary relationships among above three super-families of TEs (Feschotte and Wessler, 2002; Zhang et al., 2001, 2004). The Ty1-*copia* retro-elements belong to the ubiquitous and abundant family of class I transposons. Ty1-*copia* retro-elements are largely responsible for the huge differences in the genome size (Alix and Heslop-Harrison, 2004; Alix et al., 2005).

The Ty1-*copia* reverse transcriptase (*rt*) genes have several conserved domains in both animals and plants, and have therefore served to establish the evolutionary relationships among the *rt* (Eickbush, 1994).

Phylogenetic analysis reveals the existence of homologous *rt* in divergent species and discontinuous distributions of *rt* among species groups, and this is explained as horizontal transfer or vertical transmission followed by different evolution and stochastic loss (Alix and Heslop-Harrison, 2004; Alix et al., 2005). In this sector, we descripted 1) the identification and diversity analysis, 2) Phylogenetic position in flowering plants and 3) their evolution of these four super family TEs in *P. pubescens* genome.

2.2.1. Identification and Diversity of Transposase Genes of Mariner-, PIF- and Pong- Like Elements and Ty1-Copia Retro-Elements in the P. pubescens Genome

Based on the conserved motifs of the transposase of *mariner*-, *PIF*- and *Pong*- like elements and Ty1-*copia* retro-elements and the previously reported literatures (Hirochika et al., 1992; Hirohiko and Hirochika, 1993; Feschotte and Wessler, 2002; Zhang et al., 2001, 2004), Four pairs of degenerated PCR primers were designed for the amplification of the transposase genes of above four families of TEs (the detailed primer information was listed in Table 3). This strategy successfully isolated the fragments of transposase genes of *mariner*-like (Zhou et al., 2010b), *PIF*-like (Zhou et al., 2010a) and *Pong*-like elements (Zhong et al., 2010) and Ty1-*copia* retro-elements (Zhou et al., 2010c) from *P. pubescens* including nine intra-species cultivars or forms, which showed morphological differences in stem shape and color, or leaf color (Lin et al., 2009). Taken together, 39 *mariner*-like, 39 *PIF*-like, 39 *Pong*-like element amplification fragments and 29 Ty1-*copia* retro-element amplification fragments were identified from nine *P. pubescens* cultivars or forms were obtained and sequenced (Table 4).

The four super-families of TEs are wide-spread, abundant, and heterogeneous in *P. pubescens* genome (Zhong et al., 2010; Zhou et al., 2010a, b, c). Pairwise identities of 39 *mariner*-like transposase fragments were 54.2-100% for nucleotide sequences and 49.2-100% for amino acid sequences. Pairwise comparison of 39 *PIF*-like elements showed 38.1–100% identity at the nucleotide sequence level and 15.9–100% identity at the amino acid level. Pairwise comparison of *Pong*-like elements showed 57.8–99.6% identity at the nucleotide sequences level and 54.9-100% identity at the amino acid level. Pairwise comparisons of Ty1-*copia* retro-elements showed 37.3-100% identity at the nucleotide sequences level and 19.1-100% identity for the corresponding amino acid sequences. Southern blotting also shows that there are high abundant copies of TEs coexisting in the genome of *P. pubescens*. For example, there were four bands detected with the insertion fragments of the clone Ph-1 of *MLE* as the probes in three digestion patterns (Figure 2A).

Another Southern blotting were performed with probes was generated from PCR products amplified with the primers MLE5 and MLE3 which contained multiple amplification fragments. More than twenty (20-35) copies were detected in three digestion patterns (Figure 2B), indicating MLEs indeed are abundant in the *P. pubescens* genome. Copies of *PIF*-like elements in *P. pubescens* genomes were also investigated with PCR products as probe (Figure 3), in which more than six bold bands and faint smears were identified in all the three random selective cultivars or forms of *P. pubescens*. The above results show that TEs are abundant in *P. pubescens* genome, which are different from those by in silico analysis of genome survey sequences. The results also indicate that the proportion of TEs in *P. pubescens* genome has been underestimated by focusing on publically available GSS data and confirm that *P. pubescens* GSS data currently deposited in GenBank is too insufficient to represent the whole *P. pubescens* genome. Genome-wide analysis identified 34 different *MLE*s in rice (Feschotte et al., 2003). Given that the amplification of transposable elements is largely responsible for the huge differences in plant genome size (Bennetzen, 2002; Feschotte et al., 2002), it is reasonable to assume that even larger and more diverse families of MLEs will be found when bamboo genomes are sequenced.

Table 3. primers for amplification of *MLEs*, *PIF-* and *Pong*-like elements and Ty1-*copia* retro-elements from bamboo species

TEs	Primer names	Sequences (5`-3`)	Amplification range	References
Ty1-*copia* retro-elements	Ty1-5	CA(A/G)ATGGA(C/T)GT(A/G/C/T)AA(A/G)AC	reverse transcriptase	Hirochika et al., 1992
	Ty1-3	CAT(A/G)TC(A/G)TC(A/G/C/T)AC(A/G)TA		
PIF-like elements	PIF5	GGIGC(A/C) (A/C/T)TIGATGG(A/C/T)AC(A/T)CA	DD motif	Zhang et al., 2004
	PIF3	GGIGC(A/C) (A/C/T)TIGATGG(A/C/T)AC(A/T)CA		
Pong-like elements	Pong5	GGC(A/T)CCAT(C/T)GA(C/T)TGTATGCAC	DD35E motif	In the research
	Pong3	(C/T)TCGTC(C/T)TC(A/C/G)AC(C/T)ATCAT(A/G)TTGTG		
mariner-like *elements*	MLE5	ATT(C)GATGAA(G)AAA(G)TGGTTC	DD motif	Feschotte and Wessler, 2002
	MLE3	GCATTA(G)TCT(C)TGT(C)TGA(T)AT		

Table 4. List of species whose *mariner*-, *PIF*- and *Pong*- like elements and Ty1-*copia* retroelements sequences and their GenBank accession numbers

Taxa	Abb.	*MLE*	*PIF*-like elements		*Pong*-like elements	Ty1-*copia* retroelement
			PIF01	PIF02		
Phyllostachys pubescens	Ph	DQ528692-DQ528695	DQ861549, DQ861550	DQ861551, DQ861552	DQ528692-DQ528695	GU350894-GU350895
P. pubescens cv. Anjiensis	PhY	DQ528702-DQ528705	DQ861498, DQ861499	DQ861500-DQ861502	DQ528702-DQ528705	GU350898-GU350900
P. pubescens cv. Gracilis	PhG	DQ528706-DQ528708	DQ861503-DQ861505	DQ861506, Q861507	DQ528706-DQ528708	GU350901-GU350902
P. pubescens cv. Heterocycla	PhP	DQ528687; DQ528713-DQ528716; DQ528723-DQ528725	DQ861538-DQ861542	DQ861543-DQ861548	DQ528687; DQ528713-DQ528716; DQ528723-DQ528725	GU350903-GU350904
P. pubescens cv. Luteosulcata	PhL	DQ528699-DQ528701	DQ861508, Q861509	DQ861510-DQ861512	DQ528699-DQ528701	GU350905-GU350908
P. pubescens cv. Obliquinoda	PhO	DQ528717-DQ528720	DQ861513, Q861514	DQ861515, Q861516	DQ528717-DQ528720	GU350909-GU350911
P. pubescens cv. Tao Kiang	PhK	DQ528696-DQ528698	DQ861517-DQ861519	DQ861520-DQ861522	DQ528696-DQ528698	GU350912-GU350915
P. pubescens cv. Tubaeformis	PhT	DQ528721-DQ528722	DQ861523-DQ861525	DQ861526-DQ861528	DQ528721-DQ528722	GU350916-GU350919
P. pubescens cv. Ventricosa	PhV	DQ528688-DQ528691	DQ861529-DQ861531	DQ861532	DQ528688-DQ528691	GU350920-GU350921
P. pubescens cv. Viridisulcata	PhW	DQ528709-DQ528712	DQ861533-DQ861535	DQ861536, Q861537	DQ528709-DQ528712	GU350922-GU350923

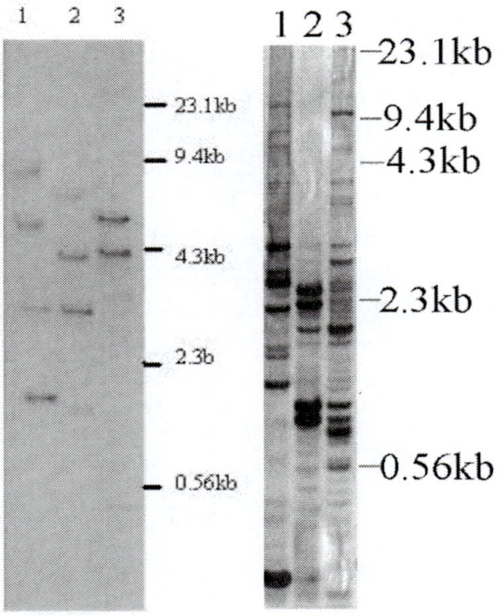

Figure 2. Southern blot analysis of MLEs present in the genomes of *P. pubescens*. 10 μg of genomic DNA digested with EcoR I (1), Hind III (2) and Xba I (3). Probes were hybridized to genomic DNA with ECL Direct Labeling and Detection System according to the manufacturer's instructions of (GE, USA). Hybridization with the multiple probes generated from PCR products amplified with the primers MLE5 and MLE3 using the *P. pubescens* genomes as the template. (Zhou et al., 2010b).

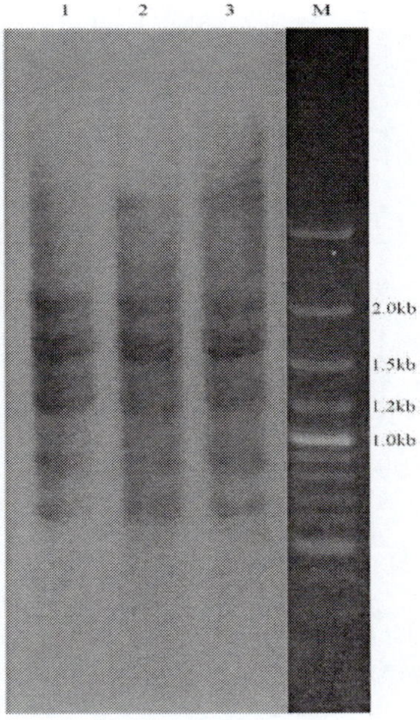

Figure 3. Southern blotting hybridized with labeled Php-1-01 of *PIF*-like elements as probes in *P. pubescens* and its two cultivars or forms. All genomic DNA were cut by *Hind* III. Line 1: *P. pubescens*, Line 2: *P. pubescens* cv. Gracilis, Line 3: *P. pubescens* cv. Viridisulcata, Line 4: molecular weight marker. (Zhou et al., 2010a).

2.2.2. Phylogenetic Positioning of Mariner-, PIF- and Pong- Like Elements and Ty1-Copia Retro-Elements from P. pubescens in Flowering Plants

TEs The have ancient origin in *P. pubescens* species by phylogenetic analysis (Zhong et al., 2010; Zhou et al., 2010a, b, c). Take *MLE*s as the example for detailed analysis. Six subfamilies of *MLE*s have thus far been distinguished in flowering plants, i.e. clades A–D, clade Y and clade Z (Feschotte and Wessler, 2002). In order to place *P. pubescens MLE*s in this phylogenetic tree, sequences representing each of the six subfamilies were aligned with the bamboo MLEs and with two animal *MLE*s (*Hsmar*2 (U49974) from *Homo sapiens* and *Sp37mar*1 (AB054644) from *Sarcophaga peregrina*) as external groups.

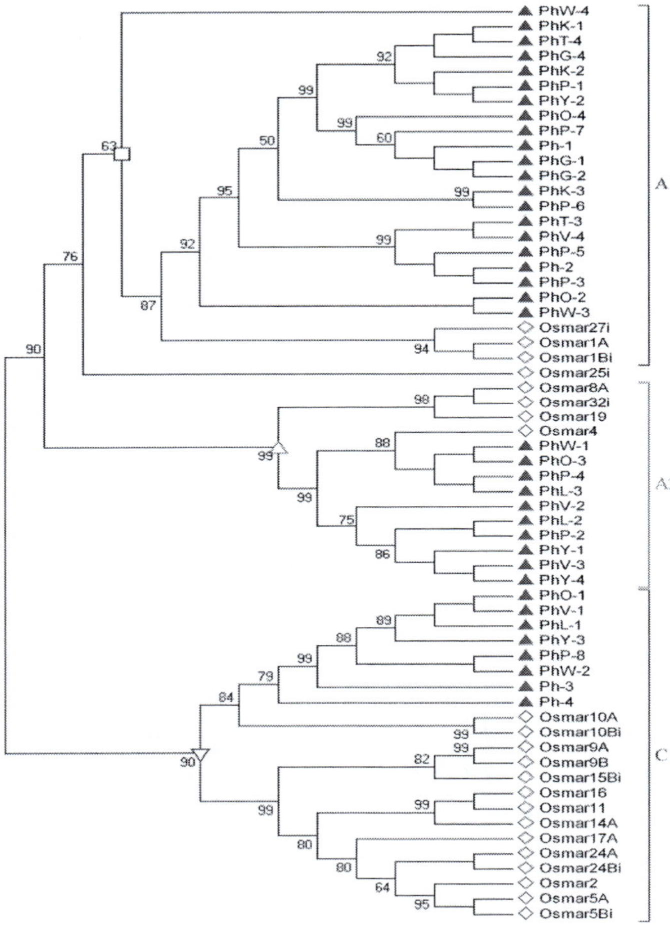

Figure 4. Phylogenetic analysis of *MLE* fragments of *P. pubescens* and 10 its cultivars or forms and 22 rice *MLE* fragments. ▲: *P. pubescens*, ◇: *O. sativa*. Main bootstrap values (1000 replicates) are shown. Groupings defining lineages and sublineages of bamboo *MLE* fragments are emphasized by capital letters and different colors.

The new *P. pubescens MLE*s clustered in two branches of the tree. Many sequences were grouped into clade A (but divided in two subclades, A1 and A2). The remainders were grouped into cade C. Since clade A and C were present in early evolving grasses more than

50 Myrs ago (Feschotte and Wessler, 2002), the *P. pubescens* MLEs are likely to represent ancient transposable elements. Genome-wide analysis identified 34 different *MLE*s in rice (Feschotte et al., 2003). The 39 *P. pubescens MLE*s were aligning with the 22 rice *MLE*s grouped into the clade A and C (Figure 4).

Every clades/sub-clades contained multiple *MLE*s, and all clades/sub-clades contained *MLE*s from two species (*O. Sativa* and *P. pubescens*) indicating that all clades/sub-clades predated the divergence of *O. Sativa* and *P. pubescens*.

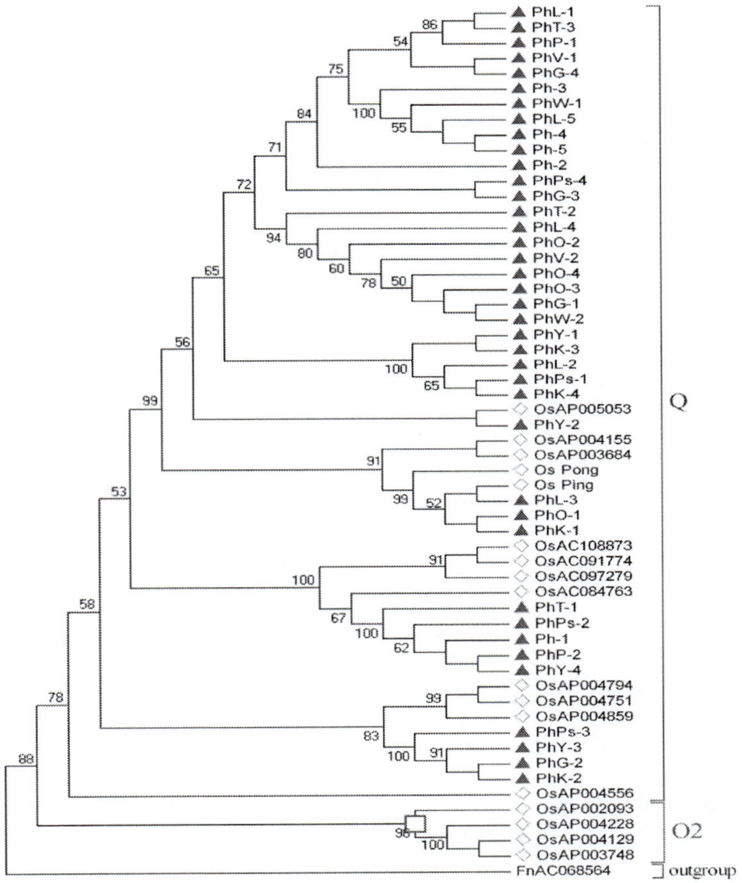

Figure 5. Phylogenetic analysis of *Pong*-like element fragments of *P. pubescens* and 10 its cultivars or forms and 17 rice *Pong*-like element fragments. ▲: *P. pubescens*, ◇: *O. sativa*. Main bootstrap values (1000 replicates) are shown. Groupings defining lineages and sublineages of bamboo *Pong*-like element fragments are emphasized by capital letters and different colors.

The available fossil evidence and the surviving basal lineages suggest that the Bambusoideae diverged from the rest of the Poaceae during the upper Cretaceous more than 65 Myrs ago (Guo and Li, 2002). It therefore appears that *P. pubescens MLE*s are ancestral in origin, and originated more than 65 Myrs ago and have subsequently undergone extensive genetic and epigenetic diversification. The similar results were obtained from the other three super-families TEs (Figure 5-7) indicating that TEs from *P. pubescens* have the multiple and ancient origin.

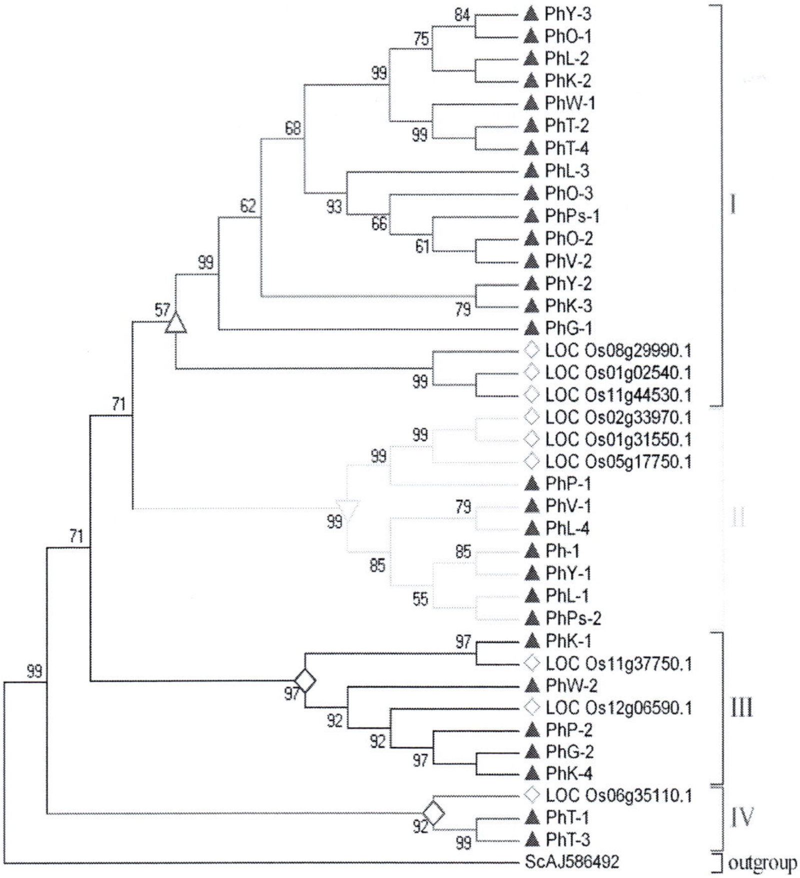

Figure 6. Phylogenetic analysis of *PIF*-like element fragments of *P. pubescens* and 10 its cultivars or forms and 14 rice *PIF*-**like element fragments.** ▲: *P. pubescens*, ◇: *O. sativa*. Main bootstrap values (1000 replicates) are shown. Groupings defining lineages and sublineages of bamboo *PIF*-like element fragments are emphasized by capital letters and different colors.

2.2.3. Evolution of Mariner-, PIF- and Pong- Like Elements and Ty1-Copia Retro-Elements from P. pubescens

The amino acids sequences surrounding the motifs of the *mariner*-, *PIF*- and *Pong*- like elements and Ty1-*copia* retro-elements is relatively well-conserved in both animals and plants, and has therefore served to establish evolution relationships among the four super family of TEs (Eickbush 1994; Robertson et al., 1998; Feschotte and Wessler, 2002; Zhang et al., 2004). To analyze the phylogenetic relationship among these four super families of TEs from *P. pubescens*, phylogenic trees were generated using the four super families of TE sequences from the nine intra-species cultivars or forms of *P. pubescens* (Figure 8-11).

Take *MLE*s as the example for detailed analysis. A phylogenetic tree was constructed from 39 amino acid sequences representing nine different cultivars or forms with two same animal *MLE*s as outgroup (Figure 8).

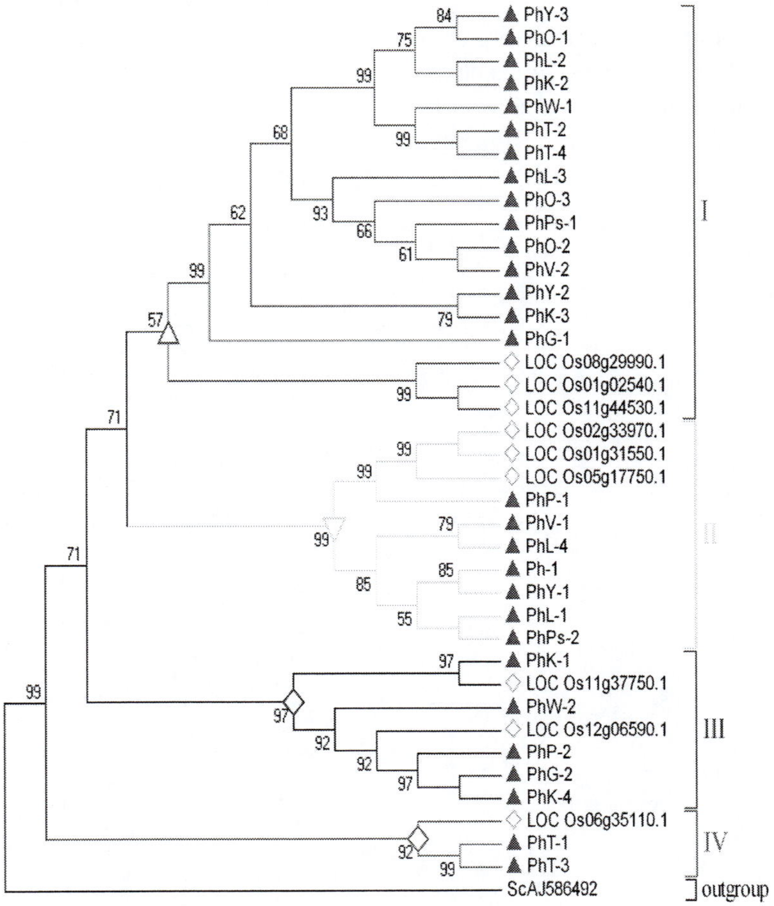

Figure 7. Phylogenetic analysis of Ty1-*copia* retro-element fragments of *P. pubescens* and 10 its cultivars or forms and 9 rice Ty1-*copia* retro-element fragments. ▲: *P. pubescens*, ◇: *O. sativa*. Main bootstrap values (1000 replicates) are shown. Groupings defining lineages and sublineages of bamboo Ty1-*copia* retro-element fragments are emphasized by number letters and different colors.

This revealed the three clades as observed in the wider analysis of the entire Bambusoideae family, i.e. clades A1, A2 and C (Zhou et al., 2010b). Most individuals display more than one kind of *MLE*s in its genome.

*MLE*s from the same individual (such as Php-7 and Php-8 from *P. pubescens* cv. Heterocycla) demonstrated a much lower level of identity than *MLE*s from different cultivars (such as PhK-1 from *P. pubescens* cv. Tao Kiang and PhT-4 from *P. pubescens* cv. Tubaeformis). It was also noted that the distribution of *MLE*s in the genomes of *P. pubescens* cultivars and forms is heterogeneous, e.g. only clade C and subclade A1 MLEs were found in the *P. pubescens* genome, whereas *P. pubescens* cultivars Anjiensis, Heterocycla, Luteosulcata, Obliquinoda, Ventricosa and Viridisulcata each contained *MLE*s from subclade A2 in addition to A1 and C. The results reflected the presence of near identical *MLE*s in distantly-related species and the presence of very diverse MLEs within the same cultivar of *P. pubescens*. The similar results were obtained from the other three super-families TEs (Figure 9-11). This probably reflects the occurrence of horizontal transfer events between phylogenetic distant species during *P. pubescens* evolution, or the existence of an ancestral TE polymorphism followed by divergent evolution and stochastic loss (Capy et al., 1994a, b; Feschotte and Wessler, 2002; Lampe et al., 2003; Silva et al., 2004).

Among the four super families of TEs, horizontal transfer of *MLE*s across wide taxonomic gaps has been well documented especially in animals (Hartl et al., 1997; Capy et al., 1998; Robertson et al., 1998).

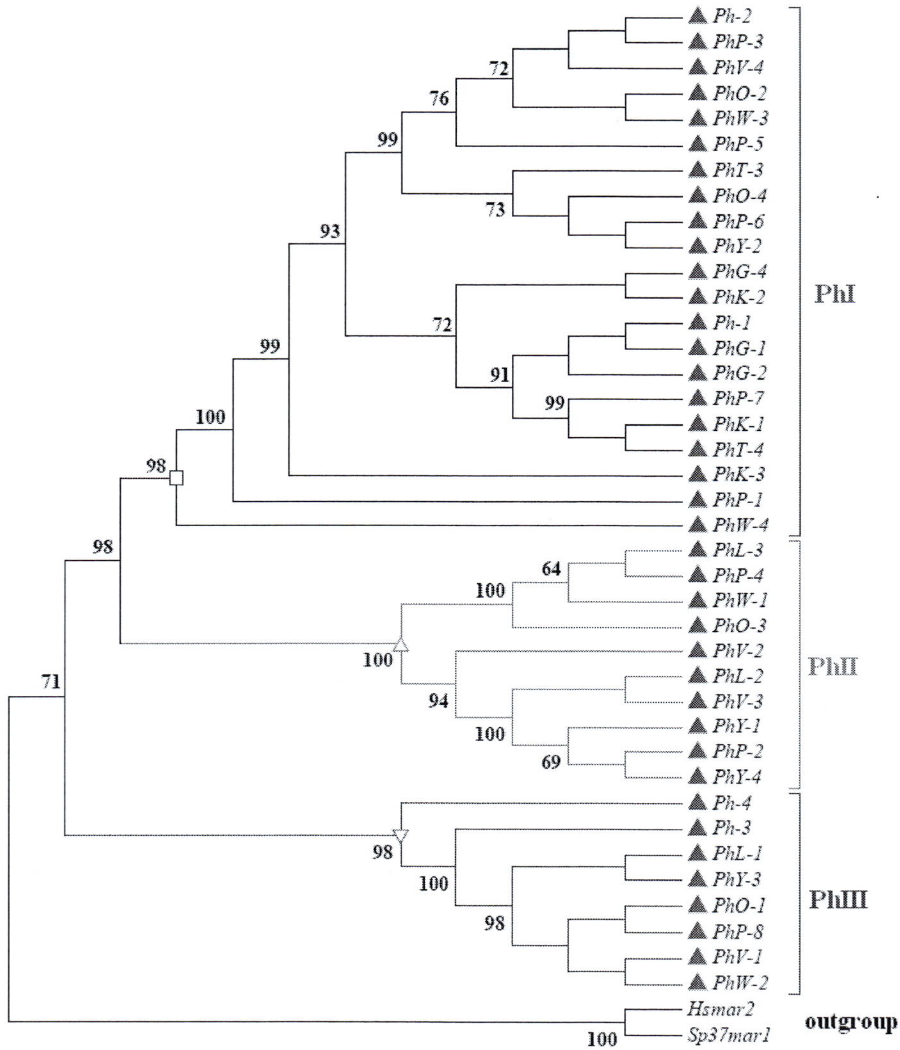

Figure 8. Phylogenetic analysis of *mariner*-like transposase fragments of *P. pubescens* and its 9 different cultivars and forms isolated by PCR. Sequences were named according to the species initials followed by the number of the clone (see Tables 2-15 for abbreviations). Main bootstrap values (1000 replicates) are shown. Groupings defining lineages and sublineages of bamboo *MLE* fragments are emphasized by capital letters and different colors. (Zhou et al., 2010b).

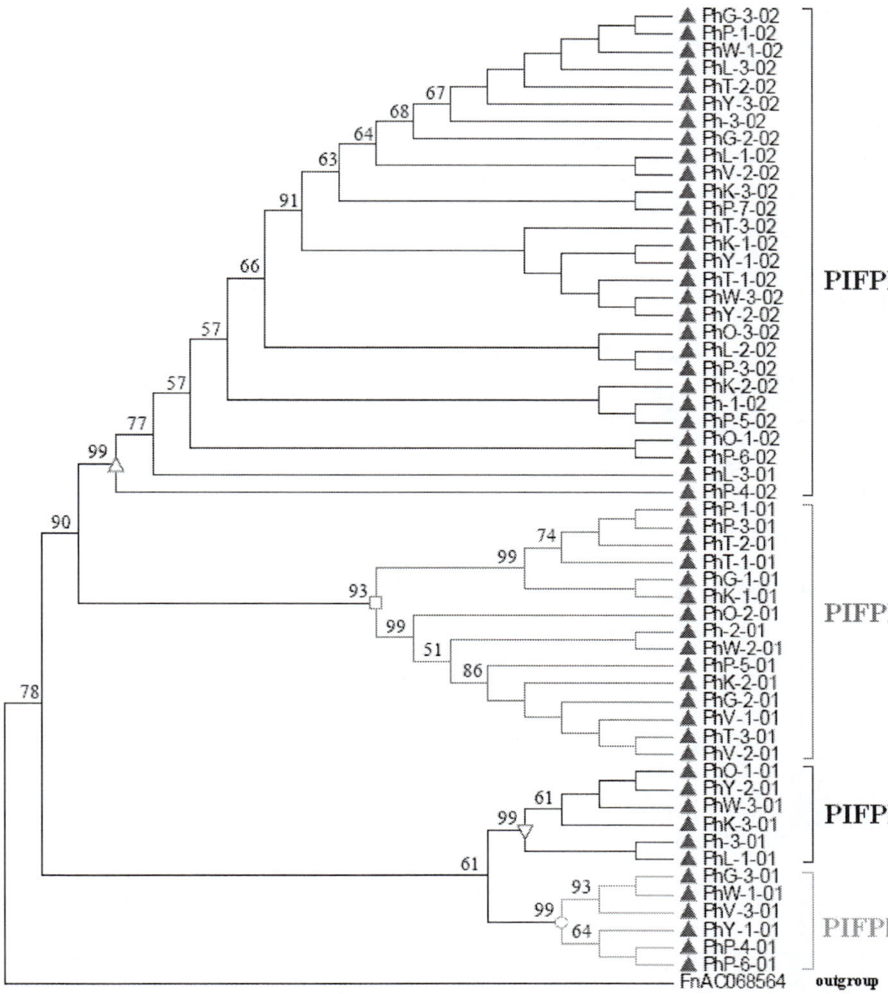

Figure 9. Phylogenetic analysis of *Pong*-like transposase fragments of *P. pubescens* and its 9 different cultivars and forms isolated by PCR. Main bootstrap values (1000 replicates) are shown. Groupings defining lineages and sublineages of bamboo *Pong*-like element fragments are emphasized by capital letters and different colors. (Zhong et al., 2010).

Feschotte and Wessler (2002) identified 37 *MLE* sequences in 54 flowering plant species; hypothesis of horizontal transfer of *MLE*s in plants has been established. Similar, in this study phylogenetic analysis of 39 *MLE* sequences from *P. pubescens* showed that the possibility of horizontal transfer existed among *P. pubescens MLE*s. Of course, except for horizontal transfer, there still are three alternative hypotheses to explain the incongruence between phylogenetic trees and TEs.

First, artifact in the phylogenetic reconstruction may introduce errors, especially when the analyzed sequences evolve at different rates (Fortune et al., 2008); second, ancestral duplication and/or polymorphism may create paralogous copies of TEs that evolve distinctly in different genomes (Capy et al., 1994a, b); and third, if transposon activity is suppressed, such elements could be eliminated from the population by stochastic loss or vertical extinction, a process that may occur in one species but perhaps not in a closely-related one (Prud'homme et al., 1995; Hartl et al., 1997).

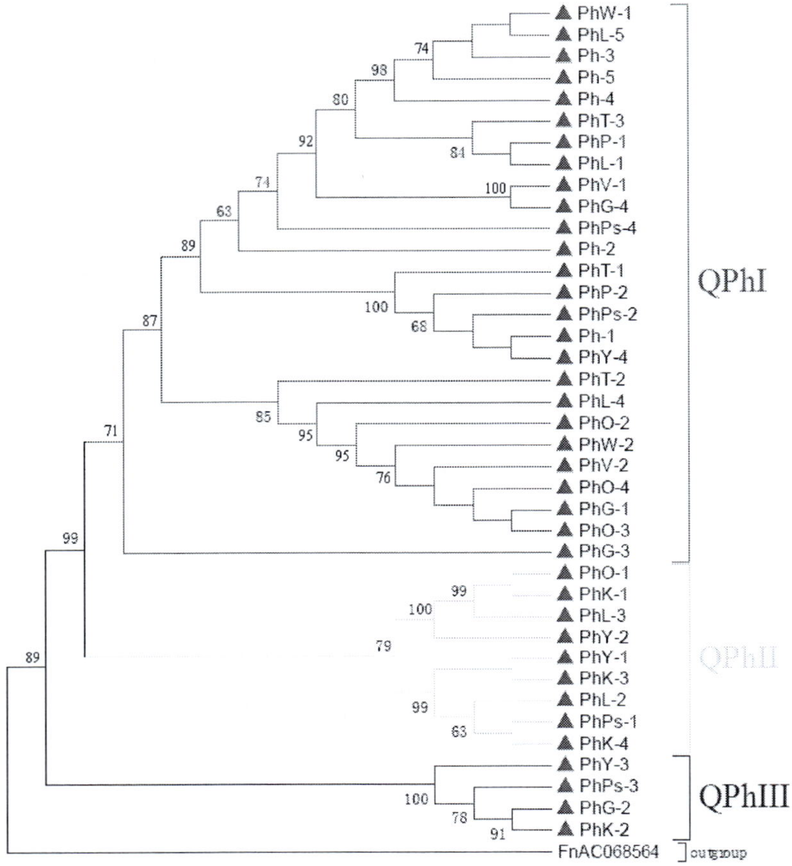

Figure 10. Phylogenetic analysis of *PIF*-like transposase fragments of *P. pubescens* and its 9 different cultivars and forms isolated by PCR. Main bootstrap values (1000 replicates) are shown. Groupings defining lineages and sublineages of bamboo *PIF*-like element fragments are emphasized by capital letters and different colors. (Zhou et al., 2010a).

All the above may contribute to the sporadic distribution of TEs. In order to provide more insight, the dn/ds ratios of host genes and TEs would need to be determined, and large-scale sequencing of *P. pubescens* would allow the divergence of host genes and TEs to be studied in more details.

In conclusion, Our research data show that the above four super-families of TEs were widespread, diverse, abundant and polyphyletic in origin in *P. pubescens*. Most identified TEs appeared and presumably occurred before separation of Bambusoideae and Oryzoideae (65 million years ago) followed by genetic and epigenetic differentiation in their respective host genomes. Additionally, distribution patterns of *P. pubescens* TEs show that the polymorphism of four super-families of TEs might be due to the divergent evolution, stochastic loss and horizontal transfer.

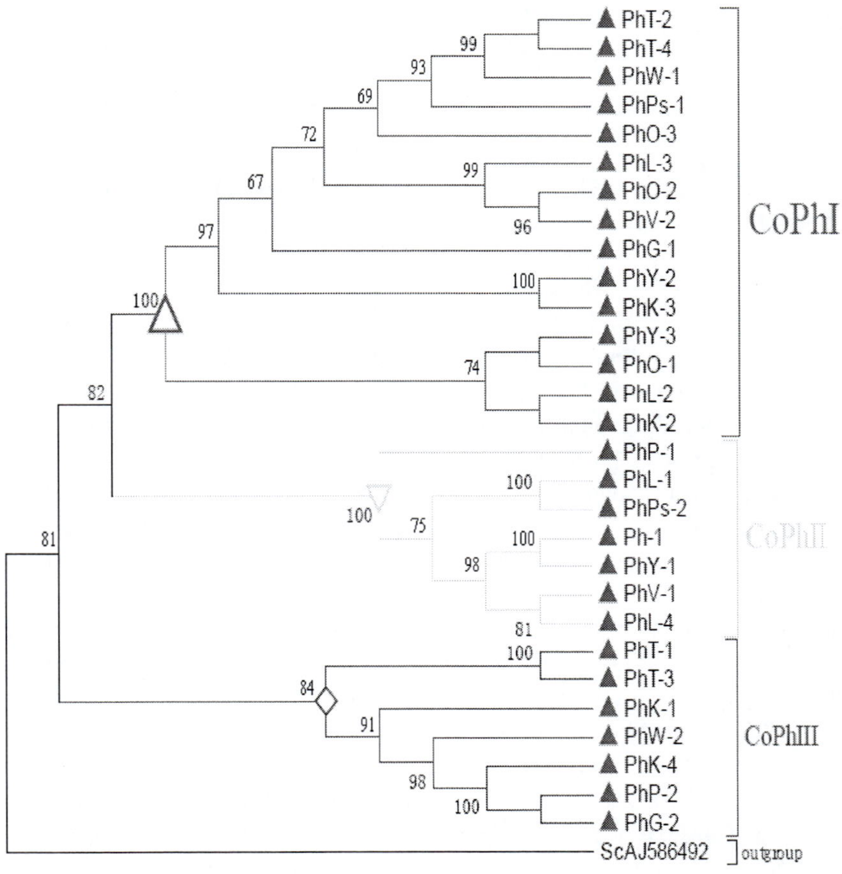

Figure 11. Phylogenetic analysis of Ty1-*copia* retro-element fragments of *P. pubescens* and its 9 different cultivars and forms. Main bootstrap values (1000 replicates) are shown. Groupings defining lineages and sublineages of bamboo Ty1-*copia* retro-element fragments are emphasized by capital letters and different colors. (Zhou et al., 2010b).

3. SSRs in *P. pubescens* Genome and Development and Utilization of SSR Markers for Marker-Assisted Selection

Genomic SSR have been detected within the genomes of every organism so far analyzed, and are often found at frequencies much higher than would be predicted purely on the grounds of base composition (Tautz and Renz, 1984). Since SSRs are abundant across the genome and show high level polymorphism, SSRs have become one of the most commonly used molecular markers in population and evolutionary biology, and are widely used in application such as parentage analysis, gene mapping, assessments of population structure and phylogenetics (Pupko and Graur,1999; Cho et al., 2000; Temnykh et al., 2000; Rajora and Rahman, 2003). In this sector, we utilized the public sequence database of *P. pubescens* available in NCBI GenBank to 1) evaluate the frequency and distribution of different classes and type of SSRs in the genome; 2) establish and validate SSR markers for the detection of polymorphism of and 3) assess cross-species transferability and identification bamboo interspecies hybrids.

3.1. Frequency and Location of FL-cDNA SSR and GSS-SSR in *P. pubescens* Genome

Total 10,608 FL-cDNA sequences (representing ~7,171 kb) and 996 GSS sequences (representing ~722 kb) from *P. pubescens* were downloaded from NCBI GenBank. EST-trimmer (http://pgrc.ipk-gatersleben.de/misa/download/est_trimmer.pl) was used to remove poly(A/T) runs, and then MISA (http://pgrc.ipk-gatersleben.de/misa/misa.html) was used to **screen for SSRs including mononucleotide repeats ≥10 bp in length, dinucleotide to hexanucleotide repeats with ≥6 repeat units, and interrupted composite SSRs with ≤100 bp of** intervening. The programs identified 2330 SSRs in 2014 of FL-cDNAs as well as 146 SSRs in 122 GSSs. After elimination of redundant sequences and splicing of overlapping sequence using CAP3, There remained non-redundant 1614 SSRs (2.60% of the total sequences in length) in 1382 FL-cDNA contigs (Tang et al., 2011) and 89 SSRs in 72 GSS contigs (8.6%) (Tang et al., 2010), which appears to be similar frequency with 1.4-4.7% previously reported in cereals (Kantety et al., 2002), the 2.65-10.6% reported in dicotyledonous plants (Kumpatla and Mukhopadhyay, 2005). This corresponded to average distances between SSRs of approximately 3.06 kb in FL-cDNAs and 7.05 kb in GSSs, which is similar to 3.4 kb in rice and 6.3 kb in barley (Thiel et al., 2003), but significantly lower than the 15.6 kb reported in wheat (Kantety et al., 2002), 13.8 kb in Arabidopsis, 11.1 kb in tomato, 20.0 kb in cotton and 14.0 kb in poplar (Cardle et al., 2000), and higher than the 1.56 kb in coffer (Aggarwal et al., 2007). In general, the observed frequency, abundance and distribution of SSRs in plants is strongly influenced by the search criteria, the size of the database and the nature of the SSR development tools (Varshney et al., 2002). For example, the frequency of putative EST-SSR markers in cereals was reduced from 50% to 1% when the search criteria were changed from a minimum of 12 bp to 30 bp (La Rota et al., 2005). With a similar database size and search criteria, the density of putative cDNA SSRs in *P. pubescens* was close to one every 3.55 kb in tea (Sharma et al., 2009), and lower than one every 2.25 kb in rubber tree (Feng et al., 2009).

More detail analysis revealed (Table 5) that trinucleotide repeats were the most abundant type of SSR in FL-cDNA sequences, accounting for 48.9% of the total, with CCG/CGG the most common repeat signature (9.7%), while mononucleotide repeats were the most in GSS sequences accounting for 48.3% of the total. Dinucleotide repeats were next in FL-cDNA and GSS sequences, accounting for 30.3% and 40.4% of the SSRs, both with AG/CT the most common repeat signature in FL-cDNAs (22.4%) and GSSs (20.2%), respectively. Tetranucleotide, pentanucleotide and hexanucleotide repeats together accounted for just 4.1% of the SSRs in FL-cDNAs and 4.5% of the SSR in GSSs. The compound type SSRs were 6.8% in FL-cDNA and 7.8% in GSS (Tang et al., 2010, 2011). BLAST searches of GSSs showed no match to almost any of the transcribed or EST regions except of ED018306 closely related to the sequence encoding a hypothetical rice protein 1074-1412 bp of Os07g0569100. The location and distribution of FL-cDNA SSR were summarized in Table 6.

Table 5. Size distribution of microsatellite motifs observed in *P. pubescens* sequences in the GenBank database

SSR motif	Number of repeat units in cDNA / GSS												
	5	6	7	8	9	10	11	12	13	14	15	>15	Total	(%)
Mono- nucleotide						92	39	36	22	13	14	55	271/43	16.8/48.3
A/T	-	-	-	-	-	79/25	19/6	19/3	2	3/1	1	39	162/35	10.0/39.3
C/G	-	-	-	-	-	13/4	20/2	17	20	10	13	16/2	109/8	6.8/9.0
Di-nucleotide		158/11	105/4	62/4	39/3	29/4	17/2	20/1	10	6	8/2	35/5	489/36	30.3/40.4
AC/GT	-	23/1	13/1	9	3/1	4	2	1	1	1	1	2/1	60/4	3.7/4.5
AG/CT	-	101/7	76/3	45/2	36/1	24/1	14/2	19	8	5	7/1	26/1	361/18	22.4/20.2
AT/AT	-	9/2	5	5/2	/1	/3	1	/1	1		/1	7/3	28/13	1.7/14.6
CG/CG	-	25/1	11	3		1							40/1	2.5/1.1
Tri-nucleotide	517/4	159	70/1	25/1	13	1	4						789/6	48.9/6.7
AAC/GTT	7												7	0.4/
AAG/CTT	24	8	3	5		1	1						42	2.9/
AAT/ATT	1												1	0.06/
ACC/GGT	33	12	8	1			1						55	3.4/
ACG/CTG	32/1	11	3	2									48/1	3.0/1.1
ACT/ATG	3	1											4	0.2/
AGC/CGT	51	9	8	2	1								71	4.4/
AGG/CCT	105/1	29	15/1	5/1	3								157/3	9.7/3.4
AGT/ATC	5	1	1	1									8	0.5/
CCG/CGG	256/2	88	32	9	9		2						396/2	24.4/2.2
Tetra-nucleotide	21/2	3/1	5	1									30/3	1.9/3.4
Penta-nucleotide	11/1	3											14/1	0.9/1.1
Hexa-nucleotide	16	4	1										21	1.3/0

Table 6. Location and size distribution of FL-cDNA SSRs in coding sequences (CDS), 5’-UTRs and 3′-UTRs

SSR class	Total number		SSR in CDS		SSR in 5'-UTR		SSR in 3'-UTR	
	No.	%	No.	Type I	No.	Type I	No.	Type I
Dinucleotide	489	36.4	105 (7.8%)	26 (1.9%)	340 (25.3%)	109 (8.1%)	44 (3.3%)	11 (0.8%)
Trinucleotide	789	58.7	532 (39.6%)	59 (4.4%)	242 (18.0%)	56 (4.2%)	15 (1.1%)	5 (0.3%)
Tetranucleotide	30	2.2	7 (0.5%)	7 (0.5%)	15 (1.1%)	13 (1.0%)	8 (0.6%)	8 (0.6)
Pentanucleotide	14	1.1	1 (<0.1%)	1(<0.1%)	12 (0.9%)	12 (0.9%)	1 (0.1%)	1 (<0.1%)
Hexanucleotide	21	1.6	9 (0.7%)	9 (0.7%)	11 (0.8%)	9 (0.6%)	2 (0.1%)	2 (0.1%)
Total	1343		654 (48.7%)	99 (7.4%)	620 (46.2%)	199 (14.8%)	70 (5.1%)	27 (2.0%)

Table 7. Candidate sequences and microsatellite marker information (Tang et al., 2010)

Marker Name	GenBank Accession No	Motif Position		Length of\ CR Fragment	5' Primer Sequence	3' Primer Sequence	Annealing Temperature
PBM001	ED018039	$(CA)_{24}(TA)_{10}$	71				
PBM002	ED018039	$(AT)_{10}$	374	298	ATGGAGCATTCATTTGAG	TTGACAAGTTTGGAGGGT	48
PBM003	ED018039	(TTTTC)4	644				
PBM004	ED018001	(AT)24	175	291	CGATGAAGTTGCCTGTGG	TCAAGTCGCATTGGAGTA	48
PBM005	ED018001	(AG)14	224				
PBM006	ED018473	(TA)10	75				
PBM007	ED018473	(AT)17	224	189	CTCGAGGTGAAGTTACATTA	TCGTAGAGCACTATCCATTT	53
PBM008	ED018674	(TA)17	1				
PBM009	ED018674	(TA)12	37				
PBM010	ED018674	(GGTTG)3	190				
PBM011	ED018475	(TA)8	295	169	TAAAGAAGTATAATGTATTCG	AGCTGTTTTAGCATCTATGTT	50
PBM012	ED018770	(AT)15	1				
PBM013	ED018016	(TC)11	191	174	TTTTACTAACGTGCCCGTATT	GGAGGATCGCTGGATTGG	53
PBM014	ED018306	(CT)11	245	294	GCCGTCCAAACGCTCCTT	CACCCATCCATCTTATGCTAT	52
PBM015	ED018426	(CT)20	245				
PBM016	ED018730	(AG)10	56	88	CCCACAGAAGGTCGCACTA	CCGCCGAGAACGTCTACAT	56
PBM017	ED018105	(CCT)8	432	237	TATGCCTCCAATAATCCG	AAGGCGTAGCCACCGAAT	55
PBM018	ED018452	(CAG)3	418	285	GACGACGAGTGTAACGCTGAG	AGGATGATGCCGGTGAGG	52
PBM019	ED018615	(CTC)8	645	202	TCCTAGCCCTACCCTGTCC	CACGGTGCTCGCTTAAATAG	59
PBM020	ED018616	(TCT)4	572	117	TCGCATCTAAACAACATC	GAGAAGAAGAAAGGAGGA	46
PBM021	ED018721	(GCC)4	827	260	AGGGTGTTATTTGCTATTGT	AAATCCGACGCTGGAGGC	55
PBM022	ED018528	(TTTG)5	255	146	AAAGATGTTGCCTTGTTC	GTGCATGTCTATTCCCTC	47
PBM023	ED017945	(TTCCG)3	199	208	TTTCATGGAGGGCGACAG	GCCGACCAGGAAACAAACGAA	58
PBM024	ED018034	(ATGG)4-6	379				
PBM025	ED018083	(ATTTC)5	359	281	GTTGTGCCCAGCACTTTA	TGACGCCACTCATTTCTT	47
PBM026	ED018344	(GCTCG)4	395	144	CTATCATCAAATGTGCCTCAA	CCGAGCCTCACTTTCTGC	50
PBM027	ED018387	(TTCTCC)3	725	287	AAAGCAACCACGCCATTA	TCGGAGGAGAAGGAGAAG	49
PBM028	ED018541	(CGAGC)5	684	157	CAGGTAGTCGTCTTATCGAGT	ATAAGTCAATTTCGAATGAA	53
PBM029	ED018589	(CTCC)5	795				
PBM030	ED018773	(ACAT)4	161	224	GTTGCCCGTGGTTACTGT	CGTGAGCCATCTGAGTTT	50

SSRs were preferentially found in coding sequences (48.7%) and 5′-UTRs (46.2%), with only 5.1% found in 3′-UTRs. Trinucleotide SSRs were predominantly located in coding sequences (39.6%) followed by 5′-UTRs (18.0%), whereas most dinucleotide SSRs were located in 5′-UTRs (25.3%) followed by coding sequences (7.8%). Trinucleotide repeats are generally the most common SSRs in plants (Morgante et al., 2002; Varshney et al., 2005), although in the present case it should be noted that 48.7% of all the cDNA-SSRs were located in the coding region, where trinucleotide would be better tolerated than other repeat units because expansion and contraction would not alter the reading frame. When considering specific motifs, the trinucleotide $(CCG/GGC)_n$ was the most common in *P. pubescens* (24.4%) followed by the dinucleotide $(AG/TC)_n$. The trinucleotide $(CCG/GGC)_n$ is strongly favored in cereals (Kantety et al., 2002; Morgante et al., 2002), e.g. it accounts for half of all trinucleotide repeats in rice, which may reflect the relatively high GC content of monocot genomes (Morgante et al., 2002). The dinucleotide $(AG/TC)_n$ is the most common SSR in many plants, including coffee (Aggarwal et al., 2007), tea (Sharma et al., 2009) and barley (Thiel et al., 2003) and rice (secondary most common, International Rice Genome Sequencing Project, 2005).

3.2. Development of SSR Markers from FL-cDNA and GSS Sequences

SSRs can be categorized as class I (SSRs ≥20 nucleotides) or class II (SSRs ≥12 but <20 nucleotides) (Temnykh et al., 2001). Class I SSRs tend to be highly polymorphic, whereas class II SSRs show less variability (Weber, 1990; Cho et al., 2000; Temnykh et al., 2000). SSRs <12 bp tend to mutate at the same rate as unique sequences and therefore demonstrate stochastic variation (Pupko and Graur, 1999). Excluding mononucleotide repeats, the average number and length of tandem repeats per SSR was 6.7 and 17.5 bp in FL-cDNA, 9.0 and 19.8 bp in GSSs, respectively. The results indicate SSR in FL-cDNA are rather shorter than that in GSS, which could be reflected by that 77.4% of all SSR were smaller than 20 bp (Type I SSR) in FL-cDNAs but 56.5% in GSSs. To develop polymorphic microsatellite markers for bamboo, we followed the procedure already adopted for rice (Temnykh et al., 2001). We selected 100 most promising FL-cDNA sequences (107 SSR loci) and 30 GSS sequences for PCR validation, including dinucleotide repeats with ≥12 repeat units, trinucleotide repeats with ≥8 repeat units, tetranucleotide repeats with ≥6 repeat units, pentanucleotide and hexanucleotide repeats with ≥5 repeat units and some compound SSRs with >24 repeats. As the results, 68 (63.6%) FL-cDNA SSR and 19 (63.3%) GSS SSR markers were developed (Table 7) (Tang et al., 2010, 2011). This was towards the lower end of the 60–90% success rate previously reported in sugarcane (Cordeiro et al., 2001), barley (Thiel et al., 2003), wheat (Gupta et al., 2003; Yu et al., 2004) and peanut (Liang et al., 2009).

3.3. Polymorphic Assessment of SSR Markers and Its Application for Hybrid Identification

To assess the polymorphism of the developed SSR markers, one *P. pubescens* seedling, nine cultivars or forms and 17 provenances were used and little size or sequence variation were detected from 19 GSS SSR markers. The results indicate that the 19 GSS SSR loci display little polymorphism and the genetic diversity of cultivars and provenances of *P. pubescens* is limited (Tang et al., 2010), which typed with AFLP and ISSR markers (Ruan et al., 2008; Lin et al., 2009). We also surveyed the allelic variability of the FL-cDNA SSR markers by genotyping 30 open-pollinated seedlings germinated from year 2008 seed lots. Among the 68 FL-cDNA SSR markers, only 22 (32.4%) showed polymorphism. The polymorphism information content (PIC) values of the 68 markers ranged from 0 to 0.51 with a mean value of 0.12. For the 22 polymorphic loci, the PIC values ranged from 0.19 to 0.51 with a mean value of 0.36 (Tang et al., 2011). The value is a higher than that obtained using GSS-SSRs without regard to the materials tested, but much lower than observed in other plants, including cereals (Thiel et al., 2003; Gao et al., 2004; Yu et al., 2004), coffee (Aggarwal et al., 2007) and the rubber tree (Feng et al., 2009).

Rich polymorphism is the norm for microsatellites (Tautz, 1989), e.g. in rice, the average polymorphism in microsatellites developed from the GenBank database is 54%, whereas that observed in genomic libraries is 83.8% at the intra-species level (Cho et al., 2000). Therefore, the limited polymorphism observed in this study appears to be another distinguishing feature for *P. pubescent*. Replication slippage (Schlotterer and Tautz, 1992; Richards and Sutherland, 1994) and recombination (Jakupiak and Wells, 1999) during DNA replication are the principal methods for the diversification of microsatellite alleles. However, the long flowering interval (Janzen, 1976; Watanabe et al., 1982) means that *Phyllostachys* species are propagated clonally more often than sexually, reducing the likelihood of allele diversification as described for *Bambusa arundinacea* (Nayak and Rout, 2005).

SSR primers developed for one species can be used to detect polymorphism at homologous sites in relative species. Ten of 19 GSS SSRs (52.6%), corresponding to loci PBM004, PBM014, PBM016, PBM018, PBM020, PBM022, PBM023, PBM025, PBM027 and PBM028 successfully transferred to all six other *Phyllostachys* species, of which PBM004, PBM014 and PBM016 showed rich polymorphism among the these 6 *Phyllostachys* species (Tang et al., 2010).

Primer pairs for 68 FL-cDNA SSRs produced 83.1% transferability within the genus *Phyllostachys*, 69.9% to other genera within subtribe Shibataeeae, and 49.3% to more distant species and showed 79.4% polymorphism within the genus *Phyllostachys*, 91.3% among different genera within the subtribe Shibataeeae, and 92.8% among different Bambusoideae subtribes (Tang et al., 2011).

The different SSR copy numbers among these bamboo species provide a series of SSLP (simple sequences length polymorphism) markers that may be useful for the genetic analysis of interspecies hybrids. Figure 12 showed the polymorphism of PBM014 in all the tested *Phyllostachys* species (*P. kwangsiensis*, *P. bambusoides* and *P. pubescens*) species-specific alleles at this locus (Figure 12A).

As indicated in Figure 13A, the heterozygosity of clones I and II revealed *P. kwangsiensis* and *P. bambusoides* parents, whereas homozygous clone III was derived from the female parent *P. kwangsiensis*. Sequencing the corresponding bands demonstrated that all of the bands contained the $(CT)_n$ and $(CAT)_n$ SSR motifs. These data show that clones I and II are interspecies hybrids of *P. kwangsiensis* and *P. bambusoides*, but clone III is not. Similarly, clone IV was confirmed as an interspecies hybrid of *P. bambusoides* and *P. pubescens* (Tang et al., 2010).

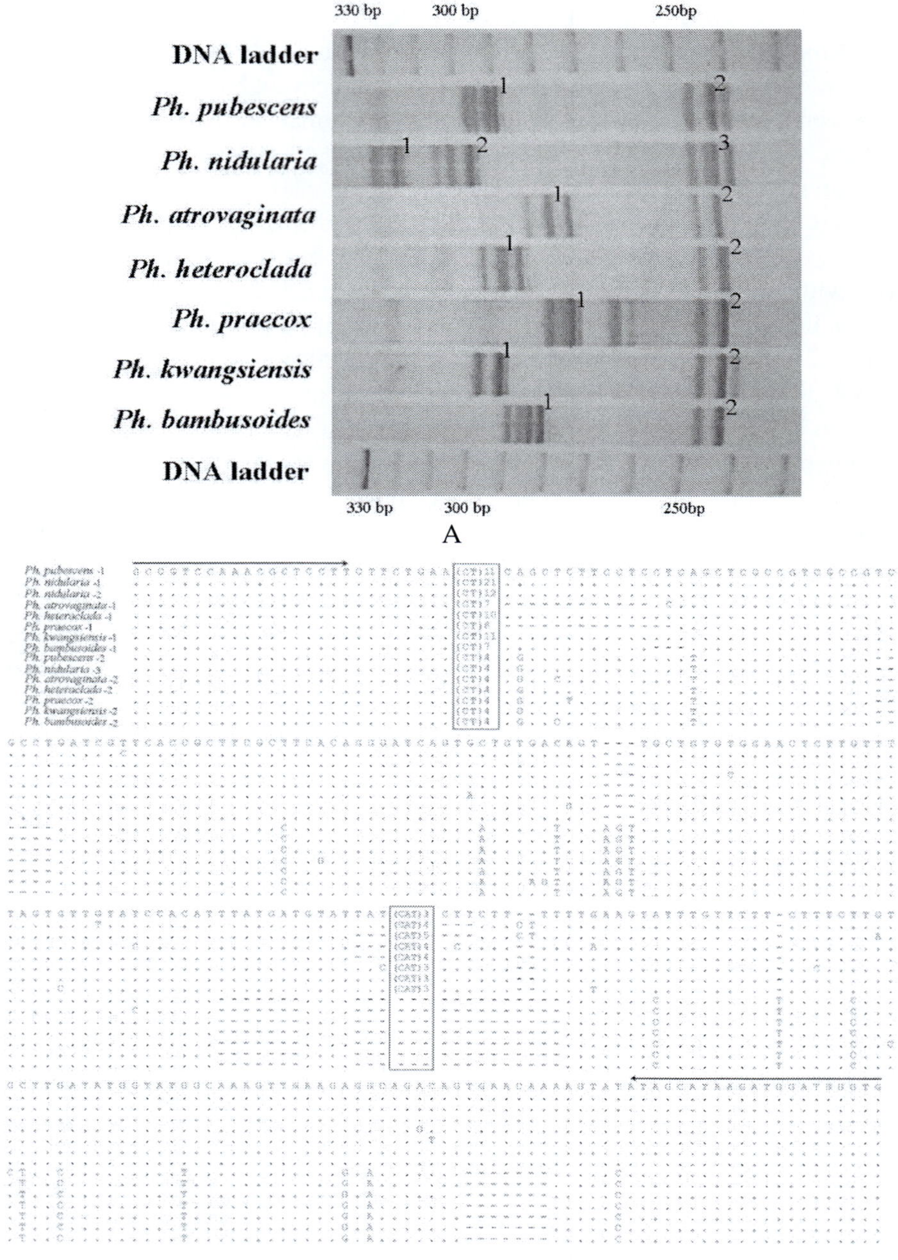

Figure 12. A. Polyacrylamide gel electrophoresis patterns of microsatellite alleles derived from locus PBM014. B: Alignment of the nucleotide sequences of the microsatellite alleles at locus PBM014 amplified from seven representative bamboo species of genus *Phyllostachys*. Nucleotides conserved among these sequences (relative to *P. pubescens*) are shown by dots. The dashes indicate deletions. The arrows indicate the primer sequences used to amplify this microsatellite locus. The box highlights the microsatellite motif. The suffix numbers after bamboo species correspond to the DNA bands marked in Fig. 12A. (Tang et al., 2010).

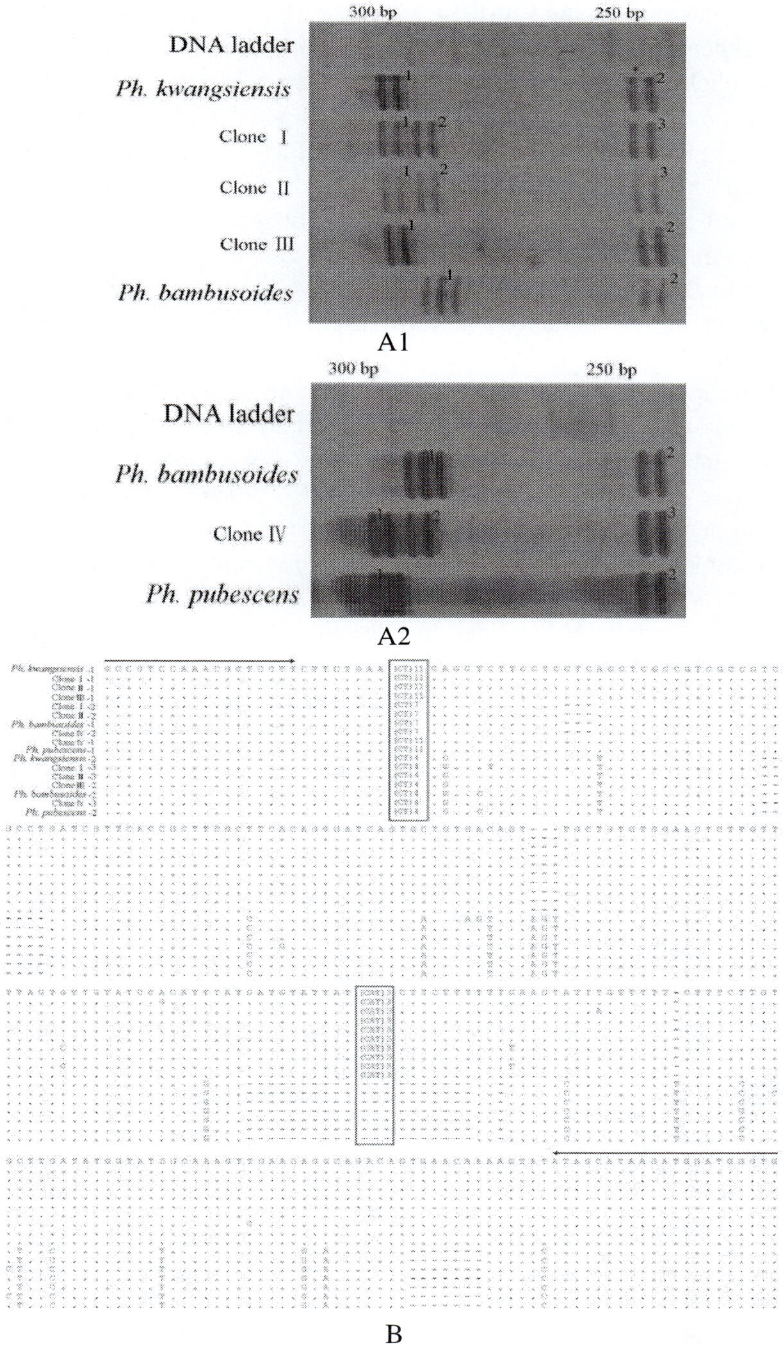

Figure 13. A. Microsatellite DNA fingerprints of *P. pubescens*, *P. kwangsiensis* and *P. bambusoides* and their presumable hybrids at PBM014 locus. B: Alignment of the nucleotide sequences of the microsatellite alleles at locus PBM014 amplified from *P. pubescens, P. kwangsiensis* and *P. bambusoides* and their presumable hybrids. Nucleotides conserved among these sequences (relative to *P. kwangsiensis*) are shown by dots. The dashes indicate deletions. The arrows indicate the primer sequences used to amplify this microsatellite locus. The box highlights the microsatellite. The suffix numbers after bamboo species correspond to the DNA bands marked in Figure 13A. (Tang et al., 2010).

In another of our studies, the PBM014, PBM022, PBM025 and PBM028 loci transferred successfully to other bamboo species beyond *Phyllostachys*. PBM014 and PBM025 served as species-specific alleles for the identification of interspecies hybrids of *Sinobambusa tootsik* × *Pleioblastus distichus, Sasa tokugawana* × *S. borealis* and *Pleioblastus simoni* × *P. praecos* (Lu et al., 2009). SSR marker derived from FL-cDNA PMB063 identified the parental species in one hybrid as *P. kwangsiensis* and *P. bambusoides*, because all sequenced bands contained the $(TCCA)_n$ motif although with a variable number of repeats. Similarly, marker PMB064 identified the parental species *B. pervariabilis* and *Dendrocalamus latiflorus*, which are distantly related to *P. pubescens* (Tang et al., 2011). We have also developed several putative EST-SSR markers in *B. oldhamii* and have used these to identify caespitose bamboo interspecies hybrids (Wu et al., 2009; Dong et al., 2011). These results demonstrate that SSR markers are ideal tools for bamboo hybrid identification as reported in poplar (Rajora and Rahman, 2003) and wheat-barley (Malysheva et al., 2003). Until recently it has been difficult to characterize bamboo populations by identifying the association between markers and traits of interest because the uncertain flowering intervals. When a large set of polymorphic markers becomes available, we can use genome wide association mapping in bamboo, in the absence of structured populations, to identify markers for traits of interest that can be used to select plus culms for large-scale propagation.

In conclusion, Our data has provided insight into the frequency, type and distribution of SSRs in *P. pubescens* GSSs and FL-cDNAs. It allowed us to develop 19 GSS SSR and 68 FL-cDNA SSR markers that can be used in *P. pubescens* and many other bamboo species. These SSR markers enriched the available molecular marker resources for bamboo and could be used for marker-assisted selection, and genetic diversity studies in the Bambusoideae subfamily.

4. Intimate Association between TEs with SSRs in *P. pubescens* Genome

SSR genesis appears to be related to mobile elements such as retro-element and often tails of retro-elements (Kaukinen and Varvio, 1992). In many animals and plants, SSRs are distributed throughout the genome, but many number are located within TEs (Ramsay et al., 1999; Richard et al., 2008), e.g. 50% of human SSRs distributed within TEs (Scherer, 2008). We therefore investigated the distribution of SSRs among the 769 *P. pubescens* TEs and compared their abundance in TEs, GSSs and FL-cDNAs by software MISA (Table 8). We found 69 SSRs in 63 TEs, covering 14,011 bp of TE sequences (4.56% of the total TE sequences). In contrast, there were 204 SSRs in the non-redundant GSS data (1.47 Mb, 0.10% of the total GSS data) and 1,614 SSRs in the non-redundant FL-cDNA data (4.90 Mb, 2.60% of the total FL-cDNA data). These results clearly show that *P. pubescens* SSRs are more abundant in TEs than in FL-cDNAs and in GSS.

Among the DNA transposons, SSRs were most likely to occur in *En/Spm* elements (5.00% of the total *En/Spm* DNA sequence), with 11 elements containing one SSR, and two containing two SSRs (Table 9). *Mutator* elements were the next most likely to contain SSRs, with 12 of 69 elements (3.96% of the total *Mutator* DNA sequences) containing at least one SSR, and one element containing two SSRs (Table 9).

Table 8. Association between simple sequences repeats (SSRs) and transposable elements (TEs) in the *P. pubescens* genome (Tang et al., 2011)

Names of TE families	No [a]	Length (bp) [b]	No. of SSR loci [c]	No. of TE-SSR sequences [d]	SSR proportion (%) [e]	No. of SSR motifs with different repeat units					
						Mono-	Di-	Tri-	Tetra-	Penta-	Hexa-
TEs	769	307274	69	63	4.56	12	49	6	1	1	0
En/Spm	40	11226	13	11	5.00	0	11	1	0	1	0
Mutator	69	12525	13	12	3.96	1	10	2	0	0	0
Ty1/*copia*	194	121272	2	2	0	0	1	1	0	0	0
Ty3/*gyspy*	308	130086	8	8	0.52	5	2	1	0	0	0
Other TEs	158	32165	33	30	2.41	6	25	1	1	0	0
GSSs/FL-cDNAs	907/7089	1467132/4942281	204/1614	111/1614[f]	0.098/2.60[g]	101/271	76/489	21/789	4/30	2/14	0/21

[a] Number of TE sequences identified in GSSs/FL-cDNAs; [b] The combined length of TEs identified in GSSs/FL-cDNAs; [c] Number of SSR loci identified in TEs; [d] Number of TEs that contain SSR loci; [e] SSR sequence length as a proportion (%) of TE sequence length; [f] Number of GSSs/FL-cDNAs that contain SSR loci; [g] SSR sequence length as a proportion (%) of GSS/FL-cDNA sequence length.

The situation was very different among the retro-elements, with 0% of Ty1-*copia* DNA and only 0.52% of Ty3-*gypsy* DNA sequences made up of SSRs. Ten of the 12 SSR in *En/Spm* elements were TA/AT repeats, and six of the 13 SSR in *Mutator* elements were CT/AG repeats, revealing a strong preference for dinucleotide repeat sequences. All 13 of the *Mutator* SSRs and 10 of the *En/Spm* SSRs were located in the 5'-UTR, revealing the strong preference for this location, at least in these transposon families. Some studies have shown that some types of SSRs are intimately associated with some families of TEs, e.g. $(TA)_n$ dinucleotide repeats are frequently found in the 5'-UTR of the Micron (a microsatellite-targeting transposable element) in rice (Akagi et al., 2001; Temnykh et al., 2001). Similarly, TA/AT repeats and CT/AG repeats were found to be intimately associated with *En/Spm* and *Mutator* elements of *P. pubescens*, respectively (Table 9).

Table 9. Distribution of SSRs in *En/Spm* and *Mutator* transposons (Zhou et al., 2011)

ID	SSR motifs	Length (bp)	Starting	Ending	Location
SSR distribution in *En/Spm* transposons					
gi112350848	$(AT)_{17}$	34	9	42	5'UTR
gi284434591	$(AT)_9$	18	22	39	5'UTR
gi284434649	$(AT)_6$	12	19	30	5'UTR
gi284434671	$(AT)_{11}$-$(AT)_{15}$	87	28	114	5'UTR+CDS
gi284434701	$(TA)_{19}$	38	1	38	5'UTR
FP091991	$(GAGGA)_6$	30	109	138	CDS
FP091422	$(TA)_{22}(CA)_9$	62	12	73	5'UTR
FP097776	$(TA)_{23}$	46	1	46	5'UTR
FP100462	$(TA)_{31}$	62	14	75	5'UTR
FP100841	$(CGG)_6$	18	38	55	5'UTR
FP100858	$(AT)_{29}$	58	5	62	5'UTR
SSR distribution in *Mutator* transposons					
FP100733	$(TC)_8$-$(GGC)_5$	89	20	108	5'UTR
FP100664	$(AG)_{17}$	34	32	65	5'UTR
FP094905	$(CT)_{17}$	34	1	34	5'UTR
FP099988	$(CT)_{19}$	38	1	38	5'UTR
FP094782	$(CT)_{12}$	24	7	30	5'UTR
FP099842	$(CT)_{15}$	30	4	33	5'UTR
FP091749	$(CT)_{23}$	46	2	47	5'UTR
FP093400	$(GA)_{16}$	32	23	54	5'UTR
FP096707	$(GAA)_8$	24	36	59	5'UTR
FP096801	$(TC)_8$	16	2	17	5'UTR
FP099127	$(AG)_{18}$	36	40	75	5'UTR
FP099725	$(C)_{13}$	13	1	13	5'UTR

Squirrell et al. (2003) have defined the successive loss of sequenced fragments and designed primers, until arriving at a final number of "working SSRs" producing discrete bands of the expected size, as the "attrition rate". Similarly, our original collection of 10,680 *P. pubescens* FL-cDNA sequences produced 1,382 cDNA contigs containing SSRs, and primer pairs could be designed for 583 of these sequences. Finally, 100 promising sequences were selected, yielding 68 FL-cDNA SSR markers. The attrition rate therefore appears to be high for FL-cDNA SSR markers in *P. pubescens*. Kofler et al. (2008) reported a high attrition rate when developing SSR markers from enriched libraries, BAC-end sequences and ESTs in rye, possibly by reflecting the large number of TEs in the rye genome. Tero et al. (2006) found that the number of useful SSR markers was reduced when the markers were predominantly located within TEs. Squirrell et al. (2003) suggested that SSR marker development would be challenging in polypoid species and species such as wheat and rye with large numbers of TEs. Only a small fraction of the *P. pubescens* genome has been sequenced thus far, so the number of known TEs is likely to increase in the future. Previous studies suggest that the size of the *P. pubescens* genome >2,000 Mb, more than five times that of diploid cultivated rice (Gui et al., 2007), and it contains a large number of TEs, including *mariner*-like, *Pong*/*PIF*-like elements and Ty1-*copia* elements (Zhong et al., 2010; Zhou et al., 2010 a, b, c). *P. pubescens* has 2n = 48 chromosomes and is thought to be tetraploid (Li et al., 1999). Detailed analysis of the association between SSRs and TEs shown that specific dinucleotide repeats tend to associate with particular TE families, $(TA)_n$ with *En/Spm* and $(CT)_n$ with *Mutator*, as reported in rice (Akagi et al., 2001; Temnykh et al., 2001). The slightly higher attrition rate we encountered therefore seems reasonable when considering the size, polyploidy and TE content of the *P. pubescens* genome. We also found a similarly high attrition rate in another bamboo species (*B. oldhamii*), in which we developed 15 EST-SSR markers from 3406 non-redundant ESTs (Dong et al., 2011).

Conclusion

Bamboo occupies an important phylogenetic node in the grass family, of which *P. pubescens* plays a significant role in the forest industry and rural economy. Repetitive DNA accounts for significant proportions of *P. pubescens* genome. Analysis of these repetitive DNA sequences provides a tantalizing glimpse of the structure and evolution of genome for *P. pubescens*, and enriches the molecular markers for bamboo marker-assisted selection. It compensates the nodal data for comparative genomics for grass family and becomes the necessary precondition for large-scale sequencing of the *P. pubescens* or other bamboo species genome in the future.

Acknowledgments

This work was supported by special grants from the National Science Foundation of China (nos 31170623 and 31070590), Priority Topics on Agricultural Projects of Zhejiang Province (2010C12011) and "973" Program (2012CB723008).

References

Aggarwal RK, Hendre PS, Varshney RK, Bhat PR, Krishnakumar V, Singh L (2007) Identification, characterization and utilization of EST-derived genic microsatellite markers for genome analyses of coffee and related species. *Theor. Appl. Genet.* 114(2):359–372.

Akagi H, Yokozeki Y, Inagaki A, Mori K, Fujimura T (2001) Micron, a microsatellite-targeting transposable element in the rice genome. *Mol. Genet. Genomics* 266: 471–480.

Alix K, Heslop-Harrison JS (2004) The diversity of retroelements in diploid and allotetraploid *Brassica* species. *Plant Mol. Biol.* 54: 895–909.

Alix K, Ryder CR, Moore JM, King G, Heslop-Harrison JS (2005) The genomic organization of retrotransposons in *Brassica oleracea*. *Plant Mol. Biol.* 59: 839–851.

Bennetzen JL (2002) Mechanisms and rates of genome expansion and contraction in flowering plants. *Genetica* 11: 5 (2002) 29–36.

Bystriakova N, Kapos V, Lysenko I, Stapleton C (2003a) Distribution and conservation status of forest bamboo biodiversity in the Asia–Pacific region. *Biodiversity and Conservation* 12:1833–1841.

Bystriakova N, Kapos V, Stapleton C, Lysenko I (2003b) Bamboo Biodiversity: Information for Planning Conservation and Management in the Asia–Pacific Region. UNEP–World Conservation Monitoring Center and International Network for Bamboo and Rattan, Cambridge, UK. pp. 1–71.

Capy P, Anxolabéhère D, Langin T (1994a) The strange phylogenies of transposable elements: are horizontal transfers the only explanation? *Trends Genet.* 10:7–12.

Capy P, Bazin C, Higuet D, Langin T (1998) Dy*namics and evolution of transposable elements.* Springer-Verlag, Austin, Texas.

Capy P, Langin T, Bigot Y, Brunet F, Daboussi MJ, Périquet G, David JR, Hartl DL (1994b) Horizontal transmission versus ancient origin: *mariner* in the witness box. *Genetica* 93:161–170.

Cardle L, Ramsay L, Milbourne D, Macaulay M, Marshall D, Waugh R (2000) Computational and experimental characterization of physically clustered simple sequence repeats in plants. *Genetics* 156(2):847–854.

Cho YG, Ishii T, Temnykh S, Chen X, Lipovich L, McCouch SR, Park WD, Ayres N, Cartinhour S (2000) Diversity of microsatellites derived from genomic libraries and GenBank sequences in rice (*Oryza sativa* L.). *Theor. Appl. Genet.* 100:713–722.

Clark LG, Zhang W, Wendel JF (1995) A phylogeny of the grass family (Poaceae) based on *ndh*F sequence data. *Syst. Bot.* 20:436–460.

Cordeiro GM, Casu R, McIntyre CL, Manners JM, Henry RJ (2001) Microsatellite markers from sugarcane (*Saccharum* spp.) ESTs cross transferable to erianthus and sorghum. *Plant Sci.* 160(6):1115–1123.

Dong WJ, Wu MD, Lin Y, Tang DQ (2011) Evaluation of 15 caespitose bamboo EST-SSR markers for cross-species/genera transferability and ability to identify interspecies hybrids. *Plant Breed.* 130:596–600.

Eickbush T (1994) Origin and evolutionary relationships of retroelements. In: Morse S (Ed.) *The Evolutionary Biology of Viruses*. Raven Press, New York, pp 121–157.

Feng S, Li W, Huang H, Wang J, Wu Y (2009) Development, characterization and cross-species/genera transferability of EST-SSR markers for rubber tree (*Hevea brasiliensis*). *Mol. Breed.* 23(1):85–97.

Feschotte C (2008) TEs and the evolution of regulatory networks. *Nat. Rev. Genet.* 9:397–405.

Feschotte C, Jiang N, Wessler SR (2002) Plant transposable elements: where genetics meets genomics. *Nat. Rev. Genet.* 3:329–341.

Feschotte C, Swamy L, Wessler SR (2003) Genome-wide analysis of *mariner*-like transposable elements in rice reveals complex relationships with Stowaway MITEs. *Genetics* 163:747–758.

Feschotte C, Wessler SR (2002) *Mariner*-like transposases are widespread and diverse in flowering plants. *Proc. Natl. Acad. Sci. USA* 99:280–285.

Fortune P, Roulin A, Panaud O (2008) Horizontal transfer of transposable elements in plants. *Commun. Integr. Biol.* 1(1):74–77.

Fu G, Perona-Wright G, Barker DC (1998) *Leishmania braziliensis*: characterization of a complex specific subtelomeric repeat sequence and its use in the detection of parasites. *Exp. Parasitol.* 90(3):236–243.

Fu J (2001) Chinese Moso Bamboo: Its importance. *Bamboo* 22(5):5–7.

Gao LF, Jing RL, Huo NX, Li Y, Li XP, Zhou RH, Chang XP, Tang JF, Ma ZY, Jia JZ (2004) One hundred and one new microsatellite loci derived from ESTs (EST-SSRs) in bread wheat. *Theor. Appl. Genet.* 108(7):1392–1400.

Gerber HP, Seipel K, Georgiev O, Hofferer M, Hug M, Rusconi S, Schaffner W (1994) Transcriptional activation modulated by homopolymeric glutamine and proline stretches. *Science* 263:808–811.

Ghorai A, Sharma A (1980) Cyto taxonomy of Indian Bambuseae II: *Dendrocalameae* and *Melocanneae*. *Acta Bot. Indica* 8(2):134–138.

Gielis J, Valente P, Bridts C, Verbelen JP (1997) Estimation of DNA content of bamboos using flow cytometry and confocal laser scanning microscopy (CLSM). In Chapman G (eds): *The Bamboos, Linnean Society Symposium Series 19*, Academic press, London. pp 215–223.

Gui YJ, Wang S, Quan LY, Zhou CP, Long SB, Zheng HJ, Jin L, Zhang XY, Ma NX, Fan LJ (2007) Genome size and sequence composition of moso bamboo: A comparative study. *Sci. China Ser. C: Life Sci.* 50(5):1–6.

Gui Y, Wang Y, Zhou Y, Wang S, Wang SY, Hu Y, Bo SP, Chen H, Zhou CP, Ma NX, Zhang TZ , Fan L (2010) Insights into the bamboo genome: Syntenic relationships to rice and sorghum. *J. Integr. Plant Biol.* 52 (11):1008–1015.

Guo ZH, Li DZ (2002). Advances in the systematics and biogeography of the Bambusoideae (Gramineae) with remarks on some remaining problems. *Acta Bot. Yunnan.* 24(4): 431–438. (in Chinese with English abstract).

Gupta PK, Rustgi S, Sharma S, Singh R, Kumar N, Balyan HS (2003) Transferable EST-SSR markers for the study of polymorphism and genetic diversity in bread wheat. *Mol. Genet. Genomics* 270(4):315–323.

Hartl DL, Lohe AR, Lozovskaya ER (1997) Modern thoughts on an ancient *mariner*: function, evolution, regulation. *Ann. Rev. Genet.* 31:337–358.

Hirochika H, Fukuchi A, Kikuchi F (1992) Retrotransposon families in rice plants. *Mol. Gen. Genet.* 233:209–216.

Hirochika H, Hirochika R (1993) Ty1-group retrotransposons as ubiquitous components of plant genomes. *Jpn. J. Genet.* 68:35–46.

Huang X, Madan A (1999) CAP3: A DNA sequence assembly program. *Genome Res.* 9:868–877.

Hunziker JH, Wulff AF, Soderstrom TR (1982) Chromosome studies on the Bambusoideae (Gramineae). *Brittonia* 34(1):30–35.

International Rice Genome Sequencing Project (2005) The map-based sequence of the rice genome. *Nature* 436:793–800.

Jakupiak JP, Wells RD (1999) Genetic instabilities in (CTG.CAG) repeats occur by recombination. *J. Biol. Chem.* 274:23468–23479.

Janzen DH (1976) Why bamboos wait so long to flower. Ann. Rev. Eco. Syst. 7:347–391.

Jiao YL, Deng XW (2007) A genome-wide transcriptional activity survey of rice TE-related genes. *Genome Biol.* 8: R28.

Jordan IK, Rogozin IB, Glazko GV, Koonin EV (2003) Origin of a substantial fraction of human regulatory sequences from TEs. *Trends Genet.* 19:68–72.

Kaukinen J, Varvio SL (1992) Artiodactyl retroposons: association with microsatellites and use in SINEmorph detection by PCR. *Nucleic Acids Res.* 20:2955 - 8.

Kantety RV, La Rota M, Matthews DE, Sorrells ME (2002) Data mining for simple sequence repeats in expressed sequence tags from barley, maize, rice, sorghum and wheat. *Plant Mol. Biol.* 48(5–6):501–510.

Kashi Y, King D, Soller M (1997) Simple sequence repeats as a source of quantitative genetic variation. *Trends Genet.* 13:74–78.

Kellogg EA, Watson L (1993) Phylogenetic studies of a large data set. Bambusoideae, Andropogonodae, and Pooideae (Gramineae). *Bot. Rev.* 59(4):273–343.

Khatta V, Katoch BS (1983) Nutrient composition of some fodder tree leaves available in sub-mountainous region of Himachal Pradesh. *Indian Forester* 109:17–24.

Kofler R, Bartos J, Gong L, Stift G, Sucha´nkova P, Suchankova P, Simkova H, Berenyi M, Burg K, Dolezel J, Lelley T (2008) Development of microsatellite markers specific for the short arm of rye (*Secale cereale* L.) chromosome 1. *Theor. Appl. Genet.* 117(6): 915–926.

Kumpatla SP, Mukhopadhyay S (2005) Mining and survey of simple sequence repeats in expressed sequence tags of dicotyledonous species. *Genome* 48:985–998.

La Rota M, Kantety RV, Yu JK, Sorrells ME (2005) Nonrandom distribution and frequencies of genomic and EST-derived microsatellite markers in rice, wheat, and barley. *BMC Genomics* 6(1):23.

Lampe DJ, Witherspoon DJ, Soto-Adames FN, Robertson HM (2003) Recent horizontal transfer of mellifera subfamily mariner transposons into insect clades representing four different orders shows that selection acts only during horizontal transfer. *Mol. Biol. Evol.* 20:554–562.

Lander ES, Linton LM, Birren B, Nusbaum C, et al. (2001) Initial sequencing and analysis of the human genome. *Nature* 409:860–92.

Li R, Zhong ZC, Werger MJA (1997) Studies on the dynamics of the bamboo shoots in *Phyllostachys pubescens*. *Acta Phytoecol. Sinica* 21:53–59.

Li XL, Liu S, Shong WQ, Chen RY, Wang WZ (1999) Chromosome number of forty species of scattered bamboos. *Acta Phytotax Sinica* 37(6):541–544. (in Chinese with English abstract)

Li YC (2001) Microsatellite: genesis, genomic distribution, function and evolutionary dynamics. *J. Sichuan Agri. Univ.* 19(4): 303–316. (in Chinese with English abstract)

Liang X, Chen X, Hong Y, Liu H, Zhou G, Li S, Guo B (2009) Utility of EST-derived SSR in cultivated peanut (*Arachis hypogaea* L.) and *Arachis* wild species. *BMC Plant Biol.* 9:35.

Lin XC, Ruan XS, Lou YF, Guo XQ, Fang W (2009) Genetic similarity among cultivars of *Phyllostachys pubescens*. *Plant Syst. Evol.* 277:67–73.

Lockton S, Gaut BS (2009) The Contribution of TEs to expressed coding sequence in *Arabidopsis thaliana*. *J. Mol. Evol.* 68:80–89.

Lu JJ, Yoshinaga K, Fang W, Tang DQ (2009) Identification of the hybrid bamboo F1 by SSR markers. *Sci. Silv. Sin.* 45(3):29–34. (in Chinese with English abstract)

Malysheva L, Sjakste T, Matzk F, Roder M, Ganal M (2003) Molecular cytogenetic analysis of wheat-barley hybrids using genomic in situ hybridization and barley microsatellite markers. *Genome* 46:314–322.

Marino-Ramirez L, Lewis KC, Landsman D, Jordan IK (2005) TEs donate lineage-specific regulatory sequences to host genomes. *Cytogenet. Genome Res.* 110:333–341.

Matsuoka Y, Tsunewaki K (1999) Evolutionary dynamics of Ty1-*copia* group retrotransposons in grass shown by reverse transcriptase domain analysis. *Mol. Biol. Evol.* 16:208–217.

Morgante M, Hanafey M, Powell W (2002) Microsatellites are preferentially associated with non-repetitive DNA in plant genomes. *Nature Genet.* 30:194–200.

Naito K, Zhang F, Tsukiyama T, Saito H, et al. (2009) Unexpected consequences of a sudden and massive transposon amplification on rice gene expression. *Nature* 461(7267):1130–1134.

Nayak S, Rout GR (2005) Isolation and characterization of microsatellites in *Bambusa arundinacea* and cross species amplification in other bamboos. *Plant Breed* 124:559–602.

Park SB, Kwon SD (1998) Development of new uses of bamboos (II): Development of carbonization kiln and schedule investigation for bamboo charcoal making. *FRI J. Forest Sci.* 59:17–24.

Peng Z, Lu T, Li L, Liu X, Hu T, Yang X, Feng Q, Guan J, Weng Q, Fan D, Zhu C, Lu Y, Han B, Jiang Z (2010) Genome wide characterization of the biggest grass, bamboo, based on 10,608 putative full-length cDNA sequences. *BMC Plant Biol.* 10:116.

Perutz MF, Johnson T, Suzuki M, Finch JT (1994) Glutamine repeats as polar zippers: Their possible role in inherited neurodegenerative diseases. *Proc. Natl. Acad. Sci.*91:5355–5358.

Pohl RW, Clark LG (1992) New chromosome counts for *Chusquea* and *Aulonemia* (Poaceae: Bambusoideae). *Amer. J. Bot.* 79(4):478–480.

Pritham EJ (2009) TEs and factors influencing their success in eukaryotes. *J. Heredity* 100(5):648–655.

Prud'homme N, Gans M, Masson M, Terzian C, Bucheton A (1995) *flamenco*, a gene controlling the gypsy retrovirus of *Drosophila melanogaster*. Genetics 139(2):697–711.

Pupko T, Graur D (1999) Evolution of microsatellites in the yeast *Saccharomyces cerevisiae*: Role of length and number of repeated units. *J. Mol. Evol.* 48:313–316.

Rajora P, Rahman H (2003) Microsatellite DNA and RAPD fingerprinting, identification and genetic relationships of hybrid poplar (*Populus* x *canadensis*) cultivars. *Theor. Appl. Genet.* 106:470–477.

Ramsay L, Macaulay M, Cardle L, Morgante M, Ivanissevich S, Maestri E, Powelll W, Waugh R (1999) Intimate association of microsatellite repeats with retrotransposons and other dispersed repetitive elements in barley. *Plant J.* 17(4):415–425.

Richard GF, Kerrest A, Dujon B (2008) Comparative genomics and molecular dynamics of DNA repeats in eukaryotes. *Micr. Mol. Bio. Rev.* 72:686–727.

Richards RI, Sutherland GR (1994) Simple repeat DNA is not replicated simply. *Nature Genet.* 6:114–116.

Robertson HM, Soto-Adames FN, Walden KO, Avancini RM, Lampe DJ (1998) The *mariner* transposons of animals: horizontally jumping genes. In: Syvanen M, Kido CI (eds) *Horizontal Gene Transfer*. Chapman and Hall, London, pp 268–284.

Ruan X, Lin X, Lou Y, Guo X, Fan W, Chen C (2008) Genetic diversity of *Phyllostachys pubescens* provenances by AFLP and ISSR. *J. Zhejiang For. Sci. Tech.* 28:29–33. (in Chinese with English abstract)

Scherer S (2008). *A short guide to the human genome*. Cold Spring Harbor University Press, Cold Spring, NY.

Schlotterer C, Tautz D (1992) Slippage synthesis of simple sequence DNA. *Nucleic Acids Res.* 20:2211–215.

Schnable PS, Ware D, Fulton RS, Stein JC, et al. (2009) The B73 maize genome: complexity, diversity and dynamics. *Science* 326:1112–1115.

Scurlock JMO, Dayton DC, Hames B (2000) Bamboo: An overlooked biomass resource? *Biomass and Bioenergy* 19:229–244.

Sharma R, Bhardwaj P, Negi R, Mohapatra T, Ahuja P (2009) Identification, characterization and utilization of unigene derived microsatellite markers in tea (*Camellia sinensis* L.). *BMC Plant Biol.* 9(1):53.

Silva JC, Loreto EL, Clark JB (2004) Factors that affect the horizontal transfer of transposable elements. *Curr. Issues Mol. Biol.* 6:57–72.

Sinzelle L, Izsvaka Z and Ivicsa Z (2009) Molecular domestication of TEs: From detrimental parasites to useful host genes. *Cell Mol. Life Sci.* 66:1073–1093.

Skibbe DS, Fernandes JF, Medzihradszky KF, Burlingame AL, et al. (2009) *Mutator* transposon activity reprograms the transcriptomes and proteomes of developing maize anthers. *Plant J.* 59(4):622–633.

Soderstrom TR (1981) Some evolutionary trends in the Bambusoideae (Poaceae). Annals *Missouri Bot. Garden* 68:15–47.

Squirrell J, Hollingsworth PM, Woodhead M, Russell J, Lowe AJ, Gibby M, Powell W (2003) How much effort is required to isolate nuclear microsatellites from plants? *Mol. Ecol.* 12(6):1339–1348.

Tang DQ (2009) Genomic sequencing and its application for biological and evolutional research in bamboo. *Bamboo J.* 26:1–10.

Tang DQ, Lu JJ, Fang W, Zhou MB (2010) Development, characterization and utilization of GenBank microsatellite markers in *Phyllostachys pubescens* and related species. *Mol. Breed.* 25: 299–311.

Tang DQ, Wu MD, Dong WJ, Zhou MB, Lin Y, Lu JJ (2011) Identification of full-length cDNA SSR markers in *Phyllostachys pubescens* for marker-assisted selection and cross-taxon transferability analysis. *BMC Genomics* (in press)

Tautz D (1989) Hypervariability of simple sequences as a general source for polymorphic DNA markers. *Nucleic Acids Res.* 17:6463–6471.

Tautz D, Renz M (1984) Simple sequences are ubiquitous repetitive components of eukaryotic genomes. *Nucleic Acids Res.* 12:4127–4138.

Tautz D, Trick M, Dover G (1986) Cryptic simplicity in DNA is a major source of genetic variation. *Nature* 322:652–656.

Temnykh S, DeClerck G, Lukashova A, Lipovich L, Cartinhour S, McCouch SR (2001) Computational and experimental analysis of microsatellites in rice (*Oryza sativa* L.): frequency, length variation, transposon associations, and genetic marker potential. *Genome Res.* 11:1441–1452.

Temnykh S, Park WD, Ayers N, Cartinhour S, Hauck N, Lipovich L, Cho YG, Ishii T, McCouch SR (2000) Mapping and genome organization of microsatellite sequences in rice (*Oryza sativa* L.) *Theor. Appl. Genet.* 100:697–712.

Tero N, Neumeier H, Gudavalli R, Schlo¨tterer C (2006) *Silene tatarica* microsatellites are frequently located in repetitive DNA. *J. Evol. Biol.* 19(5):1612–1619.

Thiel T, Michalek W, Varshney RK, Graner A (2003) Exploiting EST databases for the development and characterization of gene-derived SSR-markers in barley (*Hordeum vulgare* L.). *Theor. Appl. Genet.* 106:411–422.

Ussery DW, Wassenaar T, Borini S (2008) Word frequencies, repeats, and repeat-related structures in bacterial genomes. In: *Computing for Comparative Microbial Genomics: Bioinformatics for Microbiologists. Computational Biology* 8 (1 eds), Springer. pp 133–144.

Varshney RK, Graner A, Sorrells ME (2005) Genic microsatellite markers in plants: features and applications. *Trends Biotechnol.* 23(1): 48–55.

Varshney RK, Thiel T, Stein N, Langridge P, Graner, A (2002) In silico analysis on frequency and distribution of microsatellites in ESTs of some cereal species. *Cell Mol. Biol. Lett.* 7(2A):537–546.

Watanabe M, Ueda K, Manabe I, Akai T (1982) Flowering, seeding, germination and flowering periodicity of *Phyllostachys pubescens. J. Jpn. For. Soc.* 64:107–111.

Weber JL (1990) Informativeness of human (dC-dA)n (dG-dT)n polymorphisms. *Genomics* 7:524–530.

Wickstead B, Ersfeld K, Gull K (2003) Repetitive elements in genomes of parasitic protozoa. *Microbiol Mol. Biol. Rev.* 67(3):360–375.

Wu M, Dong W, Tang DQ (2009) Identification of four caespitose hybrid bamboos by using SSR markers. *Mol. Plant Breed.* 7(5): 959–965. (in Chinese with English abstract)

Yu JK, Dake TM, Singh S, Benscher D, Li W, Gill B, Sorrells ME (2004) Development and mapping of EST-derived simple sequence repeat markers for hexaploid wheat. *Genome* 47(5):805–818.

Zhang X, Feschotte C, Zhang Q, Jiang N, Eggleston WB, Wessler SR (2001) *P* Instability factor: an active maize transposon system associated with the amplification of Tourist-like MITEs and a new superfamily of transposases. *Proc. Natl. Acad. Sci. USA* 98:12572–12577.

Zhang X, Jiang N, Feschotte C, Wessler SR (2004) *PIF*- and *Pong*-like transposable elements: distribution, evolution and relationship with *Tourist*-like miniature inverted-repeat transposable elements. *Genetics* 166:971–986.

Zhong H, Zhou MB, Xu CM, Tang DQ (2010). Diversity and evolution of *Pong*-like elements in Bambusoideae subfamily. *Biochem. Syst. Ecol.* 38:750–758.

Zhou MB, Liu XM, Tang DQ (2011) Transposable elements in *Phyllostachys pubescens* (Poaceae) genome survey sequences and the full-length cDNA sequences, and their association with simple sequence repeats. *Genet. Mol. Res.* 10:3026–3037.

Zhou MB, Lu JJ, Zhong H, Liu XM, Tang DQ (2010a). Distribution and diversity of *PIF*-like transposable elements in the Bambusoideae subfamily. *Plant Sci.* 179:257–266.

Zhou MB, Lu JJ, Zhong H, Tang KX, Tang DQ (2010b). Distribution and polymorphism of *mariner*-like elements in the Bambusoideae subfamily. *Plant Syst. Evol.* 289:1–11.

Zhou MB, Zhong H, Zhang QH, Tang KX, Tang DQ (2010c) Diversity and evolution of Ty1-*copia* retroelements in representative tribes of Bambusoideae subfamily. *Genetica* 138:861–868.

In: Bioinformatics
Editor: Renu Tuteja

ISBN: 978-1-62100-913-9

Chapter II

Transcriptional Networks: Dynamic Information Fluxes

***Absalom Zamorano-Carrillo*[1,2]*, Beatriz Zamora-López*[3]*, Claudia-Guadalupe Benítez-Cardoza*[1,2]*, Rosas-Trigueros Jorge Luis*[2],[4]*, César López-Camarillo*[5] *and Laurence A. Marchat*[1,2,*]**

[1]Instituto Politécnico Nacional, Escuela Nacional de Medicina y Homeopatía. Programa Institucional de Biomedicina Molecular, México D.F., México
[2]Instituto Politécnico Nacional, Doctorado en Biotecnología en Red, México D.F., México
[3]Universidad Nacional Autónoma de México, Facultad de Medicina, Departamento de Psiquiatría y Salud Mental, México D.F., México
[4]Instituto Politécnico Nacional, Escuela Superior de Cómputo, México D.F., México
[5]Universidad Autónoma de la Ciudad de México, Posgrado en Ciencias Genómicas, México D.F., México

Abstract

Cells are alive, but which essential features do they have that make them so? Which properties do they possess that allow them to persist in an environment, replicating themselves and producing mRNA and proteins? Cells are self-organized: processes and structures emerge from local interactions of many modules. Particularly, abstract models permit us to study connections and regulation of cell components clustered to two goals, i.e., to accomplish cell survival and to achieve a successful differentiation. Contexts where a cell exists are differentiated by defined groups of expressed genes. Furthermore, a transcriptional network could emerge to complement another or substitute it. In recent years, high-throughput technology has accelerated the sequencing of a large variety of genomes including human. Microarrays and serial analysis of gene expression (SAGE) have allowed a parallel analysis of the expression levels of thousands of genes, creating a

* Correspondence to: Laurence A. Marchat, Programa Institucional de Biomedicina Molecular, ENMyH-IPN. Guillermo Massieu Helguera No. 239, Fracc. La Escalera Ticoman. D.F. 07320, Mexico City, Mexico. Tel.: +52 55 57296000 ext. 55543; E-mail address: lmarchat@ipn.mx, lmarchat@gmail.com.

strong demand for suitable theoretical and computational tools to analyze such data. This chapter reviews the current knowledge on trends of reconstruction of transcriptional regulatory networks. This information is of interest for understanding reverse engineering in transcriptional networks, how groups of molecules within functional modules carry out cellular functions, and how they might interact within a network.

1. Introduction

The process employed by living organisms to transfer genetic information from DNA to messenger RNA (mRNA) is known as transcription. A DNA sequence is read by RNA polymerase (RNAP), which produces a complementary RNA strand. Transcription is the initial step leading to gene expression where regulatory DNA sequences, including promoters or enhancers, and proteins called transcription factors (TF) can modulate mRNA synthesis [1]. Suitable computational tools to analyze the high-throughput gene-expression experiments have become a strong demand. Initially, regulatory elements participating in control of particular gene-expression profiles have been studied to find patterns sequences in 3' and 5' UTR regions, while level expression data have been analyzed by inferential statistics. Reconstruction of transcriptional regulatory networks is one of the most common analyses, where nodes are associated with genes or transcription factors, while edges represent some kind of communication that regulate gene function. Mathematical models of gene-proteins regulatory networks play a key role in archiving and advancing our interpretations of the molecular basis of cellular processes. A set of molecular wiring diagrams provide connections between physiological events. Furthermore, a mathematical representation of the temporal behavior of a regulatory network can be codified by a set of rate equations with associated parameters. Methods to reconstruct transcriptional regulatory networks from gene-expression data comprise Boolean networks [2], Probabilistic networks [3] and Differential equations [4]. Here, we present the frequent methodology to integrate transcriptional regulatory information for our understanding of the gene expression data.

2. Finding Regulatory Pattern Sequences

The first level of gene expression regulation is the transcription step, which involves the interaction of specific *cis*-regulatory elements in the DNA with *trans*-acting nuclear proteins called TF in response to diverse external and internal stimuli. Computational methods are useful tools for the identification of these regulatory elements through the processing of large scale data obtained from genome sequencing projects, transcriptomes and proteomes results.

The main part of the human genome corresponds to regulatory signals that are generally clustered in enhancers, silencers and promoter regions and act in a cooperative way to control gene expression [5]. Only 1.5% of the human genome is thought to encode proteins and DNA and RNA-binding proteins represent approximately 13.5% of these predicted genes. The characterization of DNA and RNA-binding proteins, that means the identification of their specific nucleic acid target and the characterization of their expression profiling, is considered as one of the most important objective in functional genomics [6].

The characterization of *cis*-regulatory sequences first involves the identification of promoters located upstream the coding region. The existence of almost 2000 TF in human allows a large number of combinatorial transcription factor binding sites (TFBS) that can be analyzed by computational algorithms to delineate promoters regions. For example, the signal-based approach uses a neural network and relies on the recognition of relatively conserved signals (TATA box, CAAT box and other TFBS) and conserved spacing among them [7][8].

The content-based approach distinguishes promoter sequences from non-promoter ones by comparing nucleotide contents such as triplet base-pair preferences around transcription start sites or hexamer frequencies in consecutive 100 bp upstream regions, using linear discriminant function and quadratic discriminant analysis.

Other approaches consider the presence of CpG islands that are characteristics of many vertebrate promoters [9]. Additionally, it is possible to compare genomic sequence with their corresponding full-length mRNA or expressed sequence tags (EST) to successfully predict promoter regions.

The next step of promoter characterization is the identification of TFBS. Several programs such as SIGNAL SCAN, MATRIX SEARCH, MatInspector, ConsInspector and TF Search make use of a position weight matrix (PWM) to quantitatively represent the specificity of *cis*-regulatory elements [10]. Most of these programs use the TRANSFAC database that includes data on known TF, TFBS and promoter information elements [11][12]. Other programs, such as Consensus [13], MEME [14], Gibbssampler [15], RSA tool analysis [16], Align ACE [17], projection [18], MDSCAN [19], YMF [20][18], TRED [21] and SwissRegulon [22] look for overrepresented motifs to cluster co-regulated genes. Additionally, the Bussemaker *et al.* regression model predicts the statistical significance of a motif based on gene expression levels [23]).

The identification of new *cis*-regulatory elements can be based on the assumption that mRNA levels of TF and their target genes are co-regulated [24]. Another strategy consists in looking for conserved motifs in co-regulated genes [25]. The TRANSCompel database also considers the presence of multiple TFBS in co-regulated genes [26].

The relevance of a functional regulatory sequence can be confirmed through comparative genomics by BLAST [27], PIPMaker [28] or CLUSTAL W, based on its conservation in genomes from distinct species. These phylogenetic footprinting studies can also evidence evolutionary changes in *cis*-regulatory sequences. Finally, the knowledge of regulatory pattern sequences can be used to identify genes regulated by similar mechanisms.

3. Gene Expression: Microarray Data Analysis

Microarray technology is a powerful tool for large-scale analysis of gene expression analysis. Briefly, Cy3 and/or Cy5-labeled RNA of an experiment are hybridized with DNA fragments corresponding to thousands of genes immobilized on a chip so that gene expression level can be deduced from fluorescence intensities. The analysis of the large volume of data obtained is critical, since any error introduced in this step will result in biologically insignificant results [29] (Figure 1).

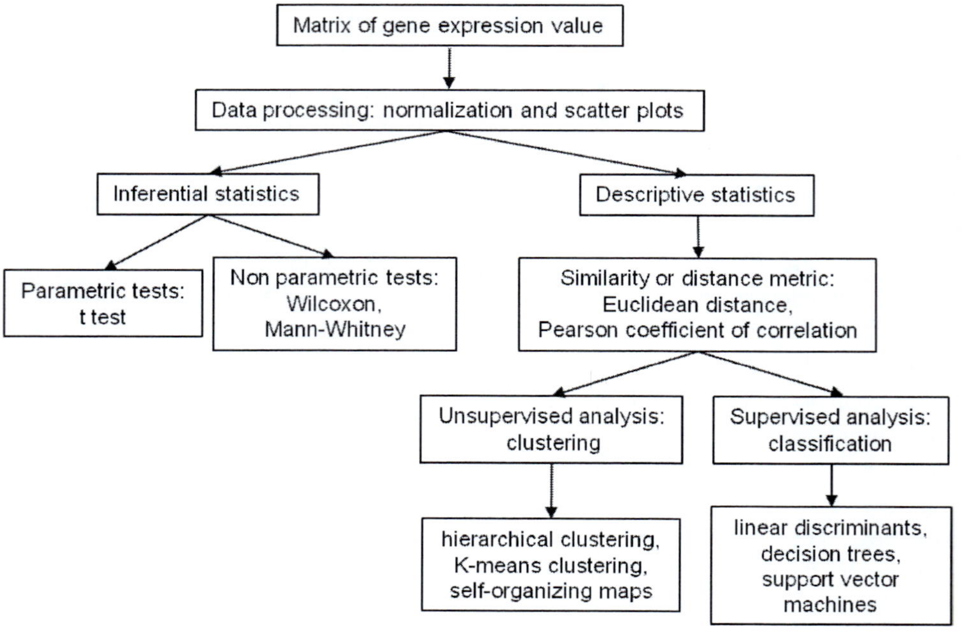

Figure 1. Overview of microarray data analysis. Adapted from [29].

3.1. Microarray Data Processing and Analysis

Microarray data analysis initiates with rough data processing to remove experimental artifacts without affecting the biologically relevant changes in gene expression. The first step of global normalization consists in subtracting the background intensity signal corresponding to empty spots on the array, from the signal for each gene [30]. Then it is possible to raw array element intensities so that the average ratio for gene expression is 1 assuming that the average gene does not change in its expression level in the biological samples being tested. Another strategy is to divide each gene expression value from a single array experiment by the mean expression value of a set of housekeeping genes present on the array, assuming that the expression of such genes does not change between two conditions. Good candidates for human housekeeping genes can be found in the Human Gene Expression (HuGE) Index database [31]. Then, microarray data can be visualized through a scatter plot which allows the comparison of gene expression values in two experimental conditions according to the Cy3 and Cy5 signals of each gene on the X and Y axes, respectively. Most data points lie along a 45° line that bisects the data and differentially regulated genes can be easily identified as isolated dots. Additionally, axes conversion from linear to logarithmic scale will help to analyze the data set and the fold regulation of genes. Artifactual variation present in gene expression can be addressed by local normalization processes which consist in plotting the ratio of the two intensities for each array element (Y axis) against the geometric mean intensity (X axis), that is, the arithmetic mean of the $\log_{10}$ intensities of the two experiments being compared, which is a measure of the average intensity of a particular element across the control and experimental conditions. In this format, up regulated genes are to the right (higher value on the Y axis) while down-regulated genes are to the left (smaller values on the Y axis).

The skewing of the data can be corrected by fitting a local regression curve to the data, providing a "best-fit" line. The new Y-axis consists of a corrected log ratio, which is created by taking each log ratio value and subtracting the local mean intensity value to obtain the residual.

To confirm the differential regulation of genes in a microarray assays, it is necessary to perform inferential statistics analysis [32]. T-tests are useful to compare two experimental groups and gene expression will be considered as significantly different at $p < 0.05$. The analysis-of-variance (ANOVA) method should be used for the comparison of more than two conditions. Both tests may be parametric (when data sets show a normal or Gaussian distribution) or nonparametric (when the data distribution is variable). Significance analysis of microarrays (SAM) is a Microsoft Excel plug-in for determining whether changes in gene expression are statistically significant from data about the expression of thousands of genes in a single hybridization experiment [33].

The input is a matrix of normalized gene expression values of control and experimental conditions according to a cut off significance which corresponds to a specific fold-change measurement. SAM identifies statistically significant genes by carrying out gene specific t-tests and computes a statistic d_j for each gene j, which measures the strength of the relationship between gene expression and a response variable. This analysis uses non-parametric statistics, since the data may not follow a normal distribution, and permutations to determine if the expression of any gene is significant related to the response. Moreover, SAM relies on a *"q* value," which is the lowest false discovery rate at which a gene is described as significantly regulated. The SAM output includes a plot of expected versus observed expression values for a set of significantly up- and down-regulated genes.

3.2. Gene Clustering and Annotation

The main objective of microarray data analysis is to cluster genes with similar expression profiles together in order to make meaningful biological inference. Descriptive statistics allow the visualization of co-expressed gene through the determination of the distance between gene expression data using the Euclidean distance and the Pearson coefficient of correlation. The most common form for finding relatively homogeneous genes groups is hierarchical clustering, which results in a dendrogram also called tree that depicts the relationships between the genes. In divisive clustering, the algorithm begins at step 1 with all the data in one cluster ($k = 1$); in each subsequent step a cluster is split off, until there are n clusters. In agglomerative clustering, all the objects start apart, thus there are n clusters at step 0 and each object forms a separate cluster; in each subsequent step, two clusters are merged, until only one cluster is left. Many programs perform cluster analysis including Cluster and Treeview [34]. K-means clustering is an alternative algorithm for classifying genes based on pattern into K groups. Grouping is done by minimizing the sum of the squares of distances between the data and the corresponding cluster centroid. Thus the purpose of K-mean clustering is to classify data based on similar expression [35]. This analysis requires a prior knowledge of the number of clusters, i.e. k value, represented in the data. Each element in the array is randomly assigned to a cluster and an average expression vector is calculated for each cluster and used to compute the average Euclidean distances between clusters. Elements are then moved between clusters and allowed to remain in a cluster if the new computed distance is lower.

This process is carried out iteratively in order to minimize the distances within and between clusters, so that each cluster contains genes with similar expression profiles. The self-organizing map (SOM) algorithm is another type of mathematical cluster analysis that imposes a partial structure on the data, with neighboring nodes tending to define related clusters, being an excellent tool for the analysis and visualization of gene expression profiles [36]. Principal Components Analysis (PCA) is an exploratory multivariate statistical technique that reduces the dimensionality of microarray data to visualize the relationship between genes or samples [37][38]. The final product of PCA is usually a two- or three-dimensional plot consisting of points in space that correspond to either genes or samples. These unsupervised clustering algorithms classify all genes according to the same criteria, regardless of the functions of the genes. The supervised method in contrast uses the expression profiles of genes with known functions as training samples. Several supervised clustering techniques can be applied to microarray data, i.e. the support vector machines [39] and multilayer perceptrons [40].

Another challenge of microarray data analysis is to understand the function of the large number of genes that are differentially expressed in the conditions tested. The biological information of these genes can be retrieved from specialized scientific literature. Moreover, specific software tools provide annotation of genes, such as Gene Ontology (http://www.geneontology.org), DRAGON [41] and A-MADMAN [42].

3.3. Other Applications of Microarray Data

Finally, it is clear that microarray data can be linked to external information to gain insight into biological processes and to make new discoveries. Thus, it is reasonable to hypothesize that genes with similar expression profiles are regulated by the same set of TF and should have therefore similar TFBS in their promoter [43][23][44]. Several works have demonstrated that proteins that are evolutionarily conserved in different organisms have similar gene expression profiles and are part of the same complex or biological pathway [45][46]. Additionally, genes that have similar expression profiles in various organisms can be considered as members of a functional conserved module [47][48]. Expression data can also be used to identify modules of coregulated genes, their transcriptional regulators and conditions that influence regulation. This 'reverse-engineering' of gene regulatory networks strategy allows generating further hypotheses, which are experimentally testable [49].

4. Pathway Analysis of Genome-Wide Data

Genome-wide association studies (GWAS) have improved the reliability to localize and identify the precursor for common human diseases and has delivered an unprecedented number of DNA loci associated with a human diseases as type 1 diabetes (T1D) [50], coronary artery disease [51], HIV-1 infection [52], and type 2 diabetes (T2D) [53]. Currently, gene expression studies based on microarrays and GWAS are common to test the relationship of each individual (for microarray or proteomic studies) or single nucleotide polymorphism (SNP) (for GWAS) with diseases. However, interest has grown in testing association of set of genes or SNP with disease simultaneously, where these genes are related each other within a biological pathway. Two types of studies are formulated: 1) competitive tests that compare the association between a gene-set and disease with that of all other gene sets being studied; 2) self-contained tests that evaluate whether there is significant association between the gene-set and disease [54].

Pathway analysis involves several challenges, such as: 1) overrepresentation of genes; 2) selection of significant genes or SNP; 3) association between any gene in the pathway and the phenotype or disease; 4) finding a model that relates a trait phenotype and a set of genes or SNP, e.g., a combination of SNP associated with a disease; 5) representation by networks of SNP whose genotypes are significantly correlated with gene expression [55]. Some of the more commonly databases providing information to analyze pathways are Gene ontology database (GO) [56], Kyoto encyclopedia of genes and genomes database (KEGG) [57], Protein Analysis Through Evolutionary Relationships (PANTHER) [58], Molecular Signatures Database (MSIGDB) [59], Database for annotation, visualization, and integrated Discovery (DAVID) [60], and Mouse genome informatics database (MGI) [61]. Regarding software analysis, the list is as follow: ALIGATOR [62], GenGen [63], ProxyGeneLD [64], GSEA-SNP [65], SRT [66], EVA [67], and GRAIL [68].

5. Network Thermodynamics

Living systems have evolved developing complex (and often overlapping) transcriptional regulatory networks. The question of how living systems orchestrate these processes has been recently assessed by thermodynamics [69]. Thermodynamics offers a rigorous quantitation of gene transcription probability under regulation, especially complex combinatorial regulation governed by multiple regulators. The quantitation of gene expression is focused in four characteristics: a) the level of expression relative to some reference value; b) how fast a given gene is expressed after induction; c) the precise relative timing of expression of different genes and d) the spatial location of expression activators [70].

Cellular signaling networks comprise biological switches that operate under nonequilibrium steady state. These switches operate in open (heat dissipation and entropy production) and nonlinear (bifurcation) regimes that exhibit ultrasensitivity, bistability, and hysteresis [69]. In 2003, Qian and Reluga derived a set of thermodynamical relations for an open system of nonlinear biochemical reactions responsible for cell signal processing. They showed that the availability of free energy ΔG from the adenosine triphosphate (ATP) hydrolysis reaction is crucial to the integrity of the transcriptional network in living systems [71]

$$\Delta G = k_B T \ln \frac{[ATP]}{K_{eq}[ADP]} \tag{1}$$

where k_B is the Boltzman constant, T is a given temperature and K_{eq} is the equilibrium constant.

Bintu *et al.* reviewed some thermodynamics models and their applications founded on the idea that the instantaneous disposition of the gene of interest can be established from the probability that various molecules such as RNAP, activators, repressors and inducers are bound to their appropriate DNA targets [70][72].Since thermodynamics is based in the assumption that equilibrium is always established, then using equilibrium models to describe processes that operate under nonequilibrium steady state is an approximation. However, deviations from equilibrium can be calculated on the basis of the rate constants associated to each step (TF binding, RNAP binding, open complex formation, transcript formation and translation). And sometimes equilibrium approximations are good estimation. For example, if the rate for open complex formation is much smaller than the rates for RNAP binding and unbinding from the promoter, then the probability of finding the polymerase on the promoter will be given by its equilibrium value [70]. Another assumption is the idea that the probability of promoter occupancy by RNAP is purely proportional to the level of expression of a given gene. This is intricate because several mechanisms can mediate concomitantly gene transcription.

With this in mind, the mathematical description for various regulatory motifs can be derived for a transcription regulation scenario where activators as wells as repressors and RNAP can have non-specific and specific binding to DNA. When RNAP binds specifically at the promoter, the transcription of target gene will start. The RNAP binding probability (p_{bound}) to DNA, can be calculated by [70]:

$$p_{bound} = \frac{1}{1 + \frac{N}{PF_{reg}} e^{\Delta\varepsilon_{pd}/k_BT}} \quad (2)$$

where N is the genome background size, and De_{pd} = $\varepsilon^{s}_{pd} - \varepsilon^{ns}_{pd}$ is the difference of RNAP binding energy at specific site (promoter) and non-specific sites; P is the effective number of RNAP molecules available for binding to the promoter. The regulation factor (F_{reg}) describes the variation of the number of RNAP molecules that are available to bind the promoter, and is given by [73]:

$$F_{reg} = \frac{\sum_{\{k\}}^{S} \prod_{i}^{n} \left(\frac{R_i}{N}\right)^{k_i} e^{-\sum_j^{k_i} \Delta\varepsilon_{r_ij}d/k_BT} e^{-k_i \Delta\varepsilon_{r_ip}/k_BT} \prod_i e^{-\varepsilon_{g_i}/k_BT} e^{-\varepsilon_{g_ip}/k_BT}}{\sum_{\{k\}}^{S} \prod_{i}^{n} \left(\frac{R_i}{N}\right)^{k_i} e^{-\sum_j^{k_i} \Delta\varepsilon_{r_ij}d/k_BT} \prod_i e^{-\varepsilon_{g_i}/k_BT}} \quad (3)$$

where, n is the number of regulator species and R_i is the number of molecules for i-th regulator. S is the complete set of binding site combinations (excluding forbidden ones). For each combination in S, a vector $\{k\}$ of length n is used to denote the number of times the binding has occurred for each regulator. is the binding energy of regulator i at binding site j, compared with this regulator's background binding energy. is the interaction energy between regulator *i* with RNAP. The summation $S\varepsilon_g$ and $S(\varepsilon_g+\varepsilon_{gp})$ are performed for any groups of interactive regulators in the current binding state, ε_g is the group interaction energy of the regulators and ε_{gp} is their interaction energy with RNAP as a whole.

It is worth to point out that the thermodynamic models for any regulatory motifs of gene transcription have the same conceptual basis, regardless of regulatory motif and they involve summing all of the relevant states. In 2008, Zhou and Su implemented a transcriptional probability calculator in a Python module named 'tCal' This module can be used to build transcription units, and either compute the transcription probabilities, or construct models of regulatory networks (SBML) [73] (http://bioinformatics.cau.edu.cn/tCal/).

6. Deterministic Models

6.1. Boolean Networks

Boolean networks provide a framework to build simple yet powerful models of transcriptional networks where genes can only be in one of two possible states of activity (ON or OFF), and can change their states at discrete intervals of time by means of rules for acting on each other.

Boolean networks for modeling genetic regulatory networks were developed by Kauffman [74]. This model is adequate for modeling large genetic networks due to the reduced number of parameters needed and the simplicity of the computations involved, yet it is powerful enough to represent the structure of network motifs [75]. This approach has recently been used to predict novel genes associated with cancer [76].

A Boolean network consists of a set of variables, a directed graph and a set of rules. The state of each gene is associated with one of the variables, thus the terms gene and variable are used interchangeably. In the graph, the vertices represent genes, or rather their state, while the directed edges represent the actions of genes on each other. For instance, a directed edge from gene A into gene B indicates that the product of gene A acts on gene B. In this way, the input-output relationships of the rules are represented by the graph. The rules define the interaction between the variables. For example, consider three genes A, B and C, and the function C=A AND B. The directed graph for this example is depicted in Figure 2a. Boolean functions can be defined by means of a so called truth table, which lists all possible combinations for the inputs and the corresponding outputs. The truth table for the basic Boolean functions is shown in Figure 2b. For this model of a transcriptional Boolean network, for a given point in time (t), gene C will be in state 1 (ON) in the next point in time (t+1) if and only if both A and B are in state 1 at time t. In any other case, the state for gene C in t+1 will be 0 (OFF). A formal definition follows.

A Boolean network is defined by a set of genes $\{v_1, v_2,\ldots, v_n\}$ where $v_i \in \{0,1\}$ for all i, and a set of functions $\{f_1, f_2, \ldots, f_n\}$. For each gene v_i, there are ki regulatory genes ($v_{j1}, v_{j2},\ldots, v_{jki}$) assigned to it. These regulatory genes determine the state of gene v_i at time t+1 by means of a Boolean function $f_i:\{0,1\}^{ki} \rightarrow \{0,1\}$. Thus, we have:

$$v_i(t+1) = f_i(v_{j1}(t),v_{j2}(t),\ldots,v_{jki}(t)) \quad (4)$$

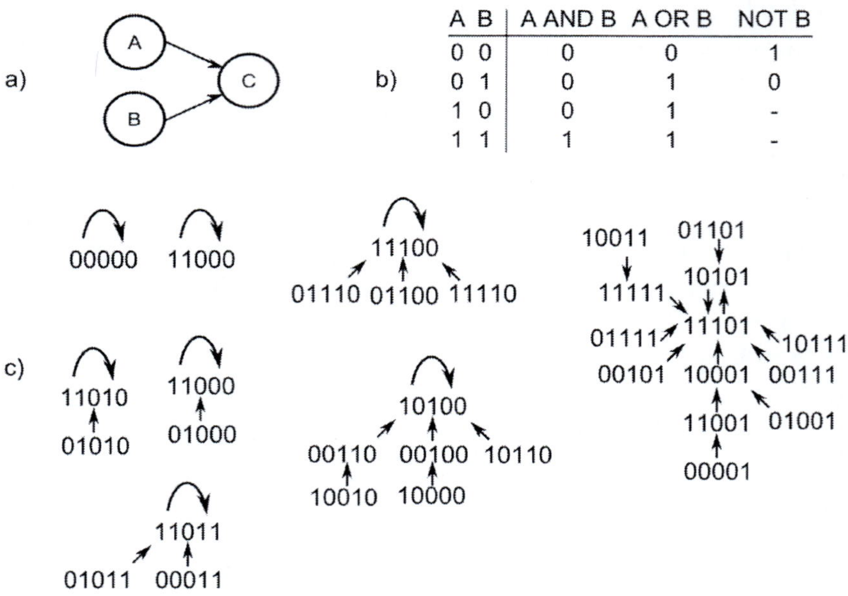

Figure 2. Example of Boolean network. a) Directed graph of the network. Genes A and B regulate gene C. b) Basic Boolean functions. If C=A AND B, then C reaches an active state only when both A and B are active. c) The state-transition diagram for the Boolean network defined in Table 1.

Table 1. Truth tables of the functions in a Boolean network with five genes

v_{j1}	v_{j2}	v_{j3}	f_1	f_2	f_3	f_4	f_5
0	0	0	0	0	0	0	0
0	0	1	1	1	1	1	1
0	1	0	1	0	0	0	-
0	1	1	1	1	1	0	-
1	0	0	1	1	1	-	-
1	0	1	1	0	1	-	-
1	1	0	1	1	0	-	-
1	1	1	1	1	1	-	-
		j_1	2	2	1	-	-
		j_2	3	4	2	3	-
		j_3	5	5	3	4	5

All genes are synchronously updated and the iteration of this process gives rise to the dynamics of the network. The synchronicity ensures that the system is entirely deterministic (given an initial state, the system will always reach the same state after a fixed number of steps). For example, consider a Boolean network with five genes $\{v_1, v_2, v_3, v_4, v_5\}$ and Boolean functions given by the truth tables shown in Table 1. Notice that not all functions necessarily have the same number of inputs. For example, f_4 is a function of only two variables and the dependence of v_5 only on itself is an example of autoregulation.

Depending on the values of the five genes, this Boolean network can only be in one of 32 possible states (e.g. v_1=0, v_2=0, v_3=1, v_4=0 and v_5=1 would be state 00101). The state transition diagrams that show the dynamics of this network are depicted in Figure 2c. Some paths in the diagrams are infinitely repeated. For instance, if the network visits state 11100, this state will only transition into itself. States that behave in this manner are called stationary states. A similar behavior is observed if the network visits state 11101, for it will transition to state 10101 and then back to 11101. This is a cycle of length 2. The cycle length is the number of transitions that return the system to a given state. Stationary states can be regarded as cycles of length 1. Cycles in this context are called attractors, and together with the states that lead into them, they form their basins of attraction. Attractors are important because they capture the stable, long-term behavior of the system and can be related to homeostasis. Since the number of states is finite, the system is bound to converge into an attractor eventually regardless of the initial state. Rules to represent real systems can be systematically inferred from gene expression data [77]. Boolean models for genetic regulatory networks can be easily implemented in software designed for this purpose such as BooleanNet [78].

6.2. Differential Equation Models

In order to avoid abrupt transitions in a genetic network model, the functions associated with the states of the genes can be defined as differential equations, which can take into account the concentrations of elements in the system and are able to model their rates of change. The general form of a nonlinear time-dependent differential equation is:

$$x' = f(x,u,t), \tag{5}$$

where x is a vector formed with the values of the physical variables in the system, u is a vector of external inputs, t is the time and x' = dx/dt is the vector of derivatives of the elements of x with respect to t. A common approximation is to consider the system to be linear and time-invariant, and then the functions can be written as:

$$x'=Ax+Bu, \tag{6}$$

where A and B are constant matrices.

The system is thought to reach a steady state when x'=0, which is analogous to reaching an attractor in a Boolean network. Non-linearity can be added to the model by means of sigmoid functions or step functions. The latter yields piecewise linear differential equations and models the switch-like character of many of the interactions in gene expression [79]. One model for such equations takes boolean relationships as a starting point and introduces a continuous variable to model the concentration of the gene products at every point in time, so that every gene A would have an associated concentration C_A. Two continuous parameters are also introduced, one to determine the minimum concentration (threshold) needed for the activity of the gene (t_A) and another one (d_A) associated to free (uncatalyzed) degradation (decay). Thus, every gene can include these parameters in its definition with a triplet of the form (A, t_A, d_A) and the boolean rule A= B AND C would be replaced by the differential equation:

$$\frac{dC_A}{dt} = ((C_B > t_B) AND (C_C > t_C)) - d_A C_A \quad (7)$$

where the inequalities and boolean functions numerically evaluate to 1 if they are True and 0 otherwise. When the boolean part of the previous equation is True, C_A increases asymptotically to $1/d_A$, unless the initial concentration was already this value, which would then be maintained. If the boolean term is False, C_A exponentially decreases to zero, or maintains an initial value of zero. For example, consider a Boolean network with four genes {(A, 1, 0.5), (B, 1, 0.5), (C, 1, 0.1), (D, 1, 0.9)} and the following Boolean functions:

B = (NOT A) AND (B AND C)
C = A OR (NOT D)
D = B AND C

Depending on the initial state, this system can show oscillatory behavior or converge to steady states **(since the values do not change, x'=0) as illustrated in Figure 3. When the initial** state of A is 1, the system always reaches the stationary state 1010, which can be identified as an attractor. This sort of system, as well as more sophisticated ones, can be implemented in BooleanNet [78][80][81].

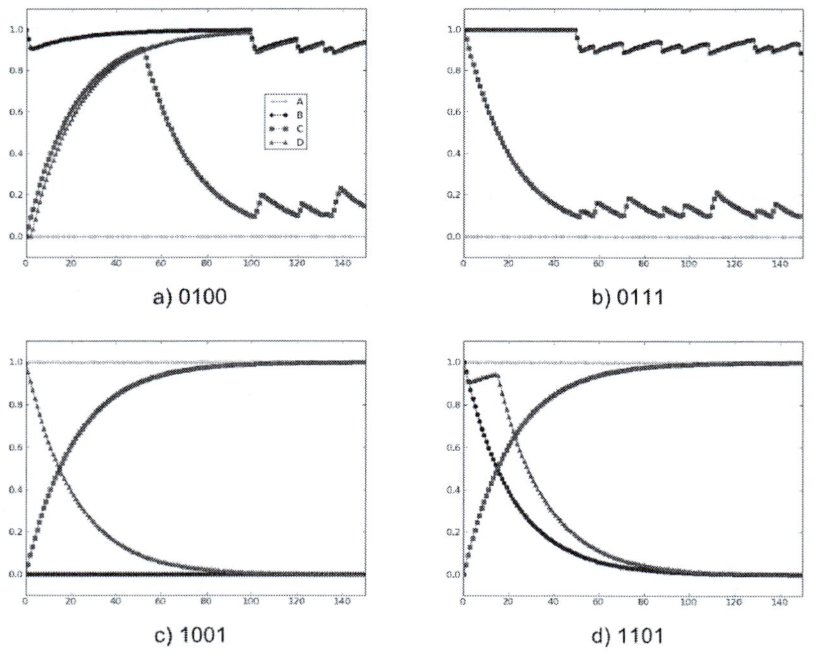

Figure 3. Behavior of a boolean network with partial linear differential equations as interaction rules for representative initial states (see text for the definition of the system). The legend in a) is the same for all graphics. Initial states 0100 (a) and 0111 (b) lead to oscillatory behavior of the system. Both initial states 1001 (c) and 1101 (d) lead to stationary state 1010, which is an attractor.

7. Probabilistic Models

This class of models affords flexibility since data contain an uncertainty reflected in the dynamical behavior of biological networks. Since, a high noise level in the data, due to large number of genes and a small number of samples, is recognized and must be dealt.

In this probabilistic framework, Babu *et al.* pioneered an evolutionary approach to the study of regulatory networks in *E. coli* and *S. cerevisiae* [48], which adds or removes edges to the network according to how well such model explains the dynamic evolution of two well-studied regulatory networks. Other approaches consider a strategy where orthologous regulatory networks have been (separately) deduced for a number of related organisms whose phylogenetic relationships are known [82].

The transitions from one state to another (possibly itself) are specified by state transition probabilities. Inspired by Markov chains, stochastic networks have the property that next state only depends on the present state, and not on the past states. As a special case of Boolean networks, where state transition probabilities are either 1 or 0, matrices represent probabilistic networks; each element of the i-th row is the probability of connecting the i-th node with a node associated with a number of entry of the row-vector (Figure 4).

Once the state transition probabilities for Markov chain are determined, it is possible to investigate the steady state (long-run) behavior of the system, which can be produced as an attractor. Also, this exponentiation of the initial stochastic matrix could evidence the ergodicity of the system, i.e., it is possible to reach any state from any other state after a number of steps. Thus, the stochastic dynamics of Markov chain are commonly used to model gene expression dynamics [83][84][85]. From this perspective, a set of genes can predict the activation of a gene.

Formally, a probabilistic boolean network can be considered as a sequence: $V=\{x_i\}$ with $i=1$ to n, of n nodes, where $x_i \in \{0,1\}$, and a sequence $\{f_l\}$ with $l=1$ to m, of vector-valued functions. Thus, each element x_i represents the expression value of a gene, and $f_l=(f_l(1), f_l(2),...,f_l(n))$ determines the network. The function $f_l(i) : \{0,1\}\ n \dashrightarrow \{0,1\}$ is a predictor of gene i. and l is the network selected [86].

Since most regulatory molecules are synthesized at very low intracellular concentrations there is a large variability in the reaction rates. This molecular noise has been reported to be important for many biological functions and processes [87][88].

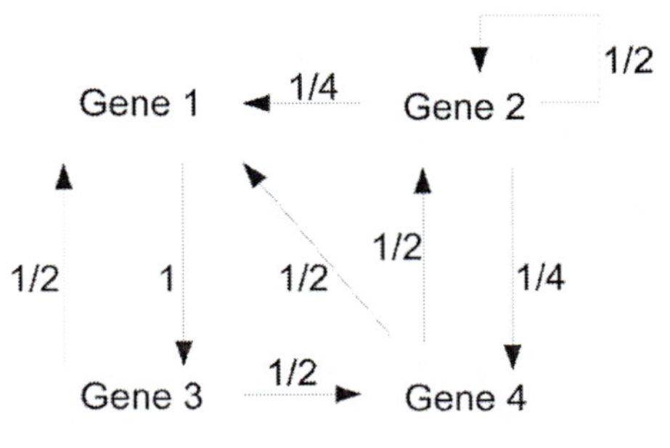

	Gene 1	Gene 2	Gene 3	Gene 4
Gene 1	0	0	1	0
Gene 2	1/4	1/2	0	1/4
Gene 3	1/2	0	0	1/2
Gene 4	1/2	1/2	0	0

Figure 4. Probabilistic Network and its Markov Matrix. Transitions states are associated with a probability. Self-multiplication of matrix permits the evolution of the transition states.

Additionally, variability factors in gene and protein expression due to spatial heterogeneity or fluctuations in cellular components have been also described [89].

Langevin equation is an elemental model with multiplicative noise, so that for a single species x_i [90], we have:

$$\dot{x} = f_i(x,u,t) + g(x_i)\xi_i Z(t) \tag{8}$$

where $f_i(x, u, t)$ is the deterministic model and $\xi_i(t)$ is zero mean unit variance Gaussian white noise. The function $g(x_i)$ is the contribution of the fluctuations and proportional to the square root of the concentration, i.e., $g(x_i) \sim (x_i)^{1/2}$. This kind of stochastic differential equations can be solved by numerical integration.

Conclusion

In integrating genome-wide data on transcript abundance into a dynamic view of gene networks, studies have focused on finding the basis of the architecture and interplay of these networks. Mathematical models of gene-proteins regulatory networks play a key role in archiving and advancing our interpretations of the molecular basis of cellular processes. A set of molecular wiring diagrams provides connections between physiological events. Furthermore, a mathematical representation of the temporal behavior of a regulatory network can be codified by a set of rate equations with associated parameters. On the other hand, an apparent modularity might be ruled by environment-dependent connections, where each piece of the diagram implements a biological function. Moreover, clustering and feedback abundance probably optimizes the circuits at the transcriptional level. For instance, a diagram of interaction might be selected through evolution if it decreases the biosynthetic cost, reduces the response delay of an intracellular signal, and stabilizes the fluctuation of expression of the involved gene [91].

Acknowledgments

AZC, CGBC, JLRT and LAM received a fellowship from COFAA-IPN, Mexico.

References

[1] Berg J. *Biochemistry*. 6th ed. New York N. Y: W. H. Freeman and Co Ltd; 2007.

[2] Akutsu T, Miyano S, Kuhara S. Identification of genetic networks from a small number of gene expression patterns under the Boolean network model. *Pac. Symp. Biocomput.* 1999;:17-28.

[3] Friedman N, Linial M, Nachman I, Pe'er D. Using Bayesian networks to analyze expression data. *J. Comput. Biol.* 2000;7(3-4):601-620.

[4] Chen T, He HL, Church GM. Modeling gene expression with differential equations. *Pac. Symp Biocomput*. 1999;:29-40.

[5] Bortoluzzi S, Coppe A, Bisognin A, Pizzi C, Danieli GA. A multistep bioinformatic approach detects putative regulatory elements in gene promoters. *BMC Bioinformatics*. 2005;6:121.

[6] Ramanathan Y, Zhang H, Aris V, Soteropoulos P, Aaronson SA, Tolias PP. Functional cloning, sorting, and expression profiling of nucleic acid-binding proteins. *Genome Res*. 2002 Aug;12(8):1175-1184.

[7] Zheng J, Wu J, Sun Z. An approach to identify over-represented cis-elements in related sequences. *Nucleic Acids Res*. 2003 Apr 1;31(7):1995-2005.

[8] Solovyev VV, Shahmuradov IA. PromH: Promoters identification using orthologous genomic sequences. *Nucleic Acids Res*. 2003 Jul 1;31(13):3540-3545.

[9] Ioshikhes IP, Zhang MQ. Large-scale human promoter mapping using CpG islands. *Nat. Genet*. 2000 Sep;26(1):61-63.

[10] Tomovic A, Oakeley EJ. Position dependencies in transcription factor binding sites. *Bioinformatics*. 2007 Apr 15;23(8):933-941.

[11] Frith MC, Hansen U, Weng Z. Detection of cis-element clusters in higher eukaryotic DNA. *Bioinformatics*. 2001 Oct;17(10):878-889.

[12] Frith MC, Spouge JL, Hansen U, Weng Z. Statistical significance of clusters of motifs represented by position specific scoring matrices in nucleotide sequences. *Nucleic Acids Res*. 2002 Jul 15;30(14):3214-3224.

[13] Hertz GZ, Stormo GD. Identifying DNA and protein patterns with statistically significant alignments of multiple sequences. *Bioinformatics*. 1999 Aug;15(7-8):563-577.

[14] Bailey TL, Elkan C. Fitting a mixture model by expectation maximization to discover motifs in biopolymers. *Proc. Int Conf.Intell Syst. Mol. Biol*. 1994;2:28-36.

[15] Lawrence CE, Altschul SF, Boguski MS, Liu JS, Neuwald AF, Wootton JC. Detecting subtle sequence signals: a Gibbs sampling strategy for multiple alignment. *Science*. 1993 Oct 8;262(5131):208-214.

[16] van Helden J, André B, Collado-Vides J. Extracting regulatory sites from the upstream region of yeast genes by computational analysis of oligonucleotide frequencies. *J. Mol. Biol*. 1998 Sep 4;281(5):827-842.

[17] Roth FP, Hughes JD, Estep PW, Church GM. Finding DNA regulatory motifs within unaligned noncoding sequences clustered by whole-genome mRNA quantitation. *Nat. Biotechnol*. 1998 Oct;16(10):939-945.

[18] Sinha S, Tompa M. YMF: A program for discovery of novel transcription factor binding sites by statistical overrepresentation. *Nucleic Acids Res*. 2003 Jul 1;31(13):3586-3588.

[19] Liu R, States DJ. Consensus promoter identification in the human genome utilizing expressed gene markers and gene modeling. *Genome Res*. 2002 Mar;12(3):462-469.

[20] Buhler J, Tompa M. Finding motifs using random projections. *J. Comput. Biol*. 2002;9(2):225-242.

[21] Jiang C, Xuan Z, Zhao F, Zhang MQ. TRED: a transcriptional regulatory element database, new entries and other development. *Nucleic Acids Res*. 2007 Jan;35(Database issue):D137-140.

[22] Pachkov M, Erb I, Molina N, van Nimwegen E. SwissRegulon: a database of genome-wide annotations of regulatory sites. *Nucleic Acids Res.* 2007 Jan;35(Database issue):D127-131.

[23] Bussemaker HJ, Li H, Siggia ED. Regulatory element detection using correlation with expression. *Nat. Genet.* 2001 Feb;27(2):167-171.

[24] Birnbaum K, Benfey PN, Shasha DE. cis element/transcription factor analysis (cis/TF): a method for discovering transcription factor/cis element relationships. *Genome Res.* 2001 Sep;11(9):1567-1573.

[25] Zhu Z, Pilpel Y, Church GM. Computational identification of transcription factor binding sites via a transcription-factor-centric clustering (TFCC) algorithm. *J. Mol. Biol.* 2002 Apr 19;318(1):71-81.

[26] Kel-Margoulis OV, Kel AE, Reuter I, Deineko IV, Wingender E. TRANSCompel: a database on composite regulatory elements in eukaryotic genes. *Nucleic Acids Res.* 2002 Jan 1;30(1):332-334.

[27] Altschul SF, Gish W, Miller W, Myers EW, Lipman DJ. Basic local alignment search tool. J. Mol. Biol. 1990 Oct 5;215(3):403-410.

[28] Schwartz S, Zhang Z, Frazer KA, Smit A, Riemer C, Bouck J, et al. PipMaker--a web server for aligning two genomic DNA sequences. *Genome Res.* 2000 Apr;10(4):577-586.

[29] Brazma A, Vilo J. Gene expression data analysis. *FEBS Lett.* 2000 Aug 25;480(1):17-24.

[30] Beissbarth T, Fellenberg K, Brors B, Arribas-Prat R, Boer J, Hauser NC, et al. Processing and quality control of DNA array hybridization data. *Bioinformatics.* 2000 Nov;16(11):1014-1022.

[31] Haverty PM, Weng Z, Best NL, Auerbach KR, Hsiao L-L, Jensen RV, et al. HugeIndex: a database with visualization tools for high-density oligonucleotide array data from normal human tissues. *Nucleic Acids Res.* 2002 Jan 1;30(1):214-217.

[32] Motulsky AG. Jewish diseases and origins. Nat. Genet. 1995 Feb;9(2):99-101.

[33] Tusher VG, Tibshirani R, Chu G. Significance analysis of microarrays applied to the ionizing radiation response. *Proc. Natl. Acad. Sci. USA.* 2001 Apr 24;98(9):5116-5121.

[34] Eisen MB, Spellman PT, Brown PO, Botstein D. Cluster analysis and display of genome-wide expression patterns. *Proc. Natl. Acad. Sci. USA.* 1998 Dec 8;95(25):14863-14868.

[35] Tavazoie S, Hughes JD, Campbell MJ, Cho RJ, Church GM. Systematic determination of genetic network architecture. *Nat. Genet.* 1999 Jul;22(3):281-285.

[36] Kohonen T. The Self-Organizing Map. *Proceedings of the IEEE.* 1990;78:1464-1468.

[37] Basilevsky A. Statistical factor analysis and related methods: theory and applications. New York: Wiley; 1994.

[38] Everitt B. *Applied multivariate data analysis.* New York: Oxford University Press; 1992.

[39] Brown MP, Grundy WN, Lin D, Cristianini N, Sugnet CW, Furey TS, et al. Knowledge-based analysis of microarray gene expression data by using support vector machines. *Proc. Natl. Acad. Sci. USA.* 2000 Jan 4;97(1):262-267.

[40] Mateos A, Dopazo J, Jansen R, Tu Y, Gerstein M, Stolovitzky G. Systematic learning of gene functional classes from DNA array expression data by using multilayer perceptrons. *Genome Res.* 2002 Nov;12(11):1703-1715.

[41] Bouton CMLS, Pevsner J. DRAGON View: information visualization for annotated microarray data. Bioinformatics. 2002 Feb;18(2):323-324.

[42] Bisognin A, Coppe A, Ferrari F, Risso D, Romualdi C, Bicciato S, et al. A-MADMAN: annotation-based microarray data meta-analysis tool. *BMC Bioinformatics*. 2009;10:201.

[43] Brazma A, Jonassen I, Vilo J, Ukkonen E. Predicting gene regulatory elements in silico on a genomic scale. *Genome Res*. 1998 Nov;8(11):1202-1215.

[44] Conlon EM, Liu XS, Lieb JD, Liu JS. Integrating regulatory motif discovery and genome-wide expression analysis. *Proc. Natl. Acad. Sci. USA*. 2003 Mar 18;100(6):3339-3344.

[45] Teichmann SA, Babu MM. Conservation of gene co-regulation in prokaryotes and eukaryotes. *Trends Biotechnol.* 2002 Oct;20(10):407-410; discussion 410.

[46] van Noort V, Snel B, Huynen MA. Predicting gene function by conserved co-expression. *Trends Genet*. 2003 May;19(5):238-242.

[47] Stuart JM, Segal E, Koller D, Kim SK. A gene-coexpression network for global discovery of conserved genetic modules. *Science.* 2003 Oct 10;302(5643):249-255.

[48] Babu MM, Luscombe NM, Aravind L, Gerstein M, Teichmann SA. Structure and evolution of transcriptional regulatory networks. *Curr. Opin. Struct. Biol.* 2004 Jun;14(3):283-291.

[49] Segal MR, Dahlquist KD, Conklin BR. Regression approaches for microarray data analysis. *J. Comput. Biol*. 2003;10(6):961-980.

[50] Todd JA, Walker NM, Cooper JD, Smyth DJ, Downes K, Plagnol V, et al. Robust associations of four new chromosome regions from genome-wide analyses of type 1 diabetes. *Nat. Genet.* 2007 Jul;39(7):857-864.

[51] Samani NJ, Erdmann J, Hall AS, Hengstenberg C, Mangino M, Mayer B, et al. Genomewide association analysis of coronary artery disease. *N. Engl. J. Med*. 2007 Aug 2;357(5):443-453.

[52] Fellay J, Shianna KV, Ge D, Colombo S, Ledergerber B, Weale M, et al. A whole-genome association study of major determinants for host control of HIV-1. *Science.* 2007 Aug 17;317(5840):944-947.

[53] Saxena R, Voight BF, Lyssenko V, Burtt NP, de Bakker PIW, Chen H, et al. Genome-wide association analysis identifies loci for type 2 diabetes and triglyceride levels. *Science*. 2007 Jun 1;316(5829):1331-1336.

[54] Goeman JJ, Bühlmann P. Analyzing gene expression data in terms of gene sets: methodological issues. *Bioinformatics*. 2007 Apr 15;23(8):980-987.

[55] Holmans P. Statistical methods for pathway analysis of genome-wide data for association with complex genetic traits. *Adv. Genet*. 2010;72:141-179.

[56] Ashburner M, Ball CA, Blake JA, Botstein D, Butler H, Cherry JM, et al. Gene ontology: tool for the unification of biology. The Gene Ontology Consortium. *Nat. Genet.* 2000 May;25(1):25-29.

[57] Kanehisa M. Representation and analysis of molecular networks involving diseases and drugs. Genome Inform. 2009 Oct;23(1):212-213.

[58] Thomas PD, Campbell MJ, Kejariwal A, Mi H, Karlak B, Daverman R, et al. PANTHER: a library of protein families and subfamilies indexed by function. *Genome Res*. 2003 Sep;13(9):2129-2141.

[59] Subramanian A, Tamayo P, Mootha VK, Mukherjee S, Ebert BL, Gillette MA, et al. Gene set enrichment analysis: a knowledge-based approach for interpreting genome-wide expression profiles. *Proc. Natl. Acad. Sci. USA*. 2005 Oct 25;102(43):15545-15550.

[60] Sherman BT, Huang DW, Tan Q, Guo Y, Bour S, Liu D, et al. DAVID Knowledgebase: a gene-centered database integrating heterogeneous gene annotation resources to facilitate high-throughput gene functional analysis. *BMC Bioinformatics*. 2007;8:426.

[61] Bult CJ, Eppig JT, Kadin JA, Richardson JE, Blake JA. The Mouse Genome Database (MGD): mouse biology and model systems. *Nucleic Acids Res*. 2008 Jan;36(Database issue):D724-728.

[62] Holmans P, Green EK, Pahwa JS, Ferreira MAR, Purcell SM, Sklar P, et al. Gene ontology analysis of GWA study data sets provides insights into the biology of bipolar disorder. *Am. J. Hum. Genet*. 2009 Jul;85(1):13-24.

[63] Wang K, Li M, Bucan M. Pathway-Based Approaches for Analysis of Genomewide Association Studies. *Am. J. Hum. Genet* [Internet]. 2007 Oct 26 [cited 2011 May 11];81(6). Available from: http://www.ncbi.nlm.nih.gov/pubmed/17966091

[64] Hong M-G, Pawitan Y, Magnusson PKE, Prince JA. Strategies and issues in the detection of pathway enrichment in genome-wide association studies. *Hum. Genet*. 2009 Aug;126(2):289-301.

[65] Holden M, Deng S, Wojnowski L, Kulle B. GSEA-SNP: applying gene set enrichment analysis to SNP data from genome-wide association studies. *Bioinformatics*. 2008 Dec 1;24(23):2784-2785.

[66] O'Dushlaine C, Kenny E, Heron E, Donohoe G, Gill M, Morris D, et al. Molecular pathways involved in neuronal cell adhesion and membrane scaffolding contribute to schizophrenia and bipolar disorder susceptibility. *Mol. Psychiatry*. 2011 Mar;16(3):286-292.

[67] Askland K, Read C, Moore J. Pathways-based analyses of whole-genome association study data in bipolar disorder reveal genes mediating ion channel activity and synaptic neurotransmission. *Hum. Genet*. 2009 Feb;125(1):63-79.

[68] Raychaudhuri S, Plenge RM, Rossin EJ, Ng ACY, Purcell SM, Sklar P, et al. Identifying relationships among genomic disease regions: predicting genes at pathogenic SNP associations and rare deletions. *PLoS Genet*. 2009 Jun;5(6):e1000534.

[69] Soh KC, Hatzimanikatis V. Network thermodynamics in the post-genomic era. Curr. *Opin. Microbiol*. 2010 Jun;13(3):350-357.

[70] Bintu L, Buchler NE, Garcia HG, Gerland U, Hwa T, Kondev J, et al. Transcriptional regulation by the numbers: models. *Curr. Opin. Genet. Dev*. 2005 Apr;15(2):116-124.

[71] Qian H, Reluga TC. Nonequilibrium thermodynamics and nonlinear kinetics in a cellular signaling switch. *Phys. Rev. Lett*. 2005 Jan 21;94(2):028101.

[72] Bintu L, Buchler NE, Garcia HG, Gerland U, Hwa T, Kondev J, et al. Transcriptional regulation by the numbers: applications. *Curr. Opin. Genet. Dev*. 2005 Apr;15(2):125-135.

[73] Zhou X, Su Z. tCal: transcriptional probability calculator using thermodynamic model. *Bioinformatics*. 2008 Nov 15;24(22):2639-2640.

[74] Kauffman SA. Metabolic stability and epigenesis in randomly constructed genetic nets. *J. Theor. Biol*. 1969 Mar;22(3):437-467.

[75] Bornholdt S. Systems biology. Less is more in modeling large genetic networks. *Science*. 2005 Oct 21;310(5747):449-451.

[76] Nagaraj SH, Reverter A. A Boolean-based systems biology approach to predict novel genes associated with cancer: Application to colorectal cancer. *BMC Syst Biol*. 2011;5:35.

[77] Maucher M, Kracher B, Kühl M, Kestler HA. Inferring Boolean network structure via correlation. *Bioinformatics* [Internet]. 2011 Apr 5 [cited 2011 May 11];Available from: http://www.ncbi.nlm.nih.gov/pubmed/21471013

[78] Albert I, Thakar J, Li S, Zhang R, Albert R. Boolean network simulations for life scientists. *Source Code Biol Med*. 2008;3:16.

[79] Glass L, Kauffman SA. The logical analysis of continuous, non-linear biochemical control networks. *J. Theor. Biol.* 1973 Apr;39(1):103-129.

[80] Klebanoff CA, Gattinoni L, Restifo NP. CD8+ T-cell memory in tumor immunology and immunotherapy. *Immunol. Rev*. 2006 Jun;211:214-224.

[81] Lamy T, Liu JH, Landowski TH, Dalton WS, Loughran TP Jr. Dysregulation of CD95/CD95 ligand-apoptotic pathway in CD3(+) large granular lymphocyte leukemia. *Blood.* 1998 Dec 15;92(12):4771-4777.

[82] Zhang X, Moret BME. Refining transcriptional regulatory networks using network evolutionary models and gene histories. *Algorithms Mol Biol*. 2010;5:1.

[83] Kim SY, Imoto S, Miyano S. Inferring gene networks from time series microarray data using dynamic Bayesian networks. *Brief. Bioinformatics*. 2003 Sep;4(3):228-235.

[84] Yu J, Smith VA, Wang PP, Hartemink AJ, Jarvis ED. Advances to Bayesian network inference for generating causal networks from observational biological data. *Bioinformatics.* 2004 Dec 12;20(18):3594-3603.

[85] Zou M, Conzen SD. A new dynamic Bayesian network (DBN) approach for identifying gene regulatory networks from time course microarray data. *Bioinformatics*. 2005 Jan 1;21(1):71-79.

[86] Shmulevich I, Aitchison JD. Deterministic and stochastic models of genetic regulatory networks. *Meth. Enzymol*. 2009;467:335-356.

[87] Ozbudak EM, Thattai M, Kurtser I, Grossman AD, van Oudenaarden A. Regulation of noise in the expression of a single gene. *Nat. Genet*. 2002 May;31(1):69-73.

[88] Raser JM, O'Shea EK. Noise in gene expression: origins, consequences, and control. *Science*. 2005 Sep 23;309(5743):2010-2013.

[89] Swain PS, Elowitz MB, Siggia ED. Intrinsic and extrinsic contributions to stochasticity in gene expression. *Proc. Natl. Acad. Sci. USA*. 2002 Oct 1;99(20):12795-12800.

[90] Rao CV, Wolf DM, Arkin AP. Control, exploitation and tolerance of intracellular noise. *Nature*. 2002 Nov 14;420(6912):231-237.

[91] Becskei A, Serrano L. Engineering stability in gene networks by autoregulation. *Nature.* 2000 Jun 1;405(6786):590-593.

In: Bioinformatics
Editor: Renu Tuteja

ISBN: 978-1-62100-913-9

Chapter III

A New Method for the Identification of Leucine-Rich Repeats by Incorporating Protein Secondary Structure Prediction

Norio Matsushima*, Hiroki Miyashita, Tomoko Mikami and Keiko Yamada
Division of Biophysics, Center for Medical Education,
Sapporo Medical University, Chuo-ku, Sapporo, Japan

Abstract

LRRs (leucine rich repeats) are present in over 14,000 proteins in data bases. Most LRR repeating units are 20-30 residues long. LRRs are tandemly repeated 2 - 74 times. All known structures of repeated LRRs adopt an arc shape. All LRRs contain LxxLxLxxNxL, in which "L" is Leu, Ile, Val, or Phe and "N" is Asn, Thr, Ser, or Cys and "x" is any amino acid. Eight classes of LRRs have been identified. There are some prediction programs such as PFAM, SMART and LRRfinder to identify LRRs. However, the repeat number of LRRs and their "phasing" (that is, what segment or residue corresponds to the beginning of a repeating unit) in individual LRR proteins frequently differ among data bases (or researchers) and species.

We developed a new method to recognize and align LRR motifs (called LRRpred). In addition to PFAM, SMART and LRRfinder, LRRpred utilizes protein secondary structure predictions and then uses multiple sequence alignment when an LRR protein has been identified from various species. This article describes LRRpred and its application. LRRpred identified LRRs that were first recognized in the crystal structure or are not recognized by data bases, and also recognized a nested LRR (called "IRREKO" LRR) as a new LRR class. Furthermore, LRRpred was used to identify non-LRR, island regions (IRs) interrupting LRRs.

Keywords: leucine rich repeat, LRRpred, PFAM, SMART, LRRfinder, PRAME, oogenesin, flagellar autotomy protein, non-LRR, isaland region, IRREKO LRR,

* Tel.: +81-11-611-2111; Fax: +81-11-612-3617; E-mail: matusima@sapmed.ac.jp

LRR (leucine rich repeat) regions are present in over 14,000 proteins. LRR-containing proteins (LRR proteins) have been identified in viruses, bacteria, archae, and eukaryotes. Human and mouse contain 400-550 LRR proteins. LRR proteins are listed in data bases of PFAM, SMART, PRINTS, InterPro, PANTHER, UniProKB and LRRfinder [1-3]. Most LRR proteins are involved in protein, ligand and in protein, protein interactions; these include plant immune response and the mammalian innate immune response (for reviews, see [4-8]).

Most LRR repeating units are 20-30 residues in length. All LRR units can be divided into a HCS (highly conserved segment) and a VS (variable segment). The HCS part consists of an eleven residue stretch, LxxLxLxxNxL, or a twelve residue stretch, LxxLxLxxCxxL, in which "L" is Leu, Ile, Val, or Phe, "N" is Asn, Thr, Ser, or Cys, and "C" is Cys, Ser or Asn.

Eight classes of LRRs have been recognized; they are characterized by different lengths and consensus sequences of the VS part of the repeats [5,9-11]. They are "RI-like", "CC", "bacterial", "SDS22-like", "plant specific", "typical", "TpLRR" [5,9,10] and "IRREKO" [11].

Descriptions for some of the eight classes of LRRs were initially incorporated in several of the sequence databases (PFAM, SMART and InterPro) but they are slowly becoming abandoned in favor of more generic descriptors that can be applied automatically to as many LRR sequences as possible [8].

All of the LRR domains in one protein form a single continuous structure and adopt an arc or horseshoe shape. On the inner, concave face there is a stack of parallel β-strands and on the outer, convex face there are a variety of secondary structures such as α–helix, 3_{10}-helix, polyproline II helix, or a tandem arrangement of β-turns, which are connected by two loops. The short β-strands are usually formed by three residues at positions 3 through 5 in the HCS part. In the known structures the minimum repeat number of the LRR domains is four and the lengths of LRRs range from 20 to 43 residues.

The known LRR structures also revealed that the LRR motif is surprisingly variable and thus indicated that there are many LRRs that can not be easily assigned to one of the eight classes. These indicate inadequate, insufficient predictions in which the indicated repeat number of LRRs and "phasing" (that is, what segment or residue corresponds to the beginning of a repeating unit) in individual LRR proteins are frequently different among the databases (or researchers) and species.

As noted, some LRRs were first recognized in the crystal structure. These observations emphasize the need to develop a method to identify LRRs with higher reliability.

We have recently developed a method to recognize and align LRR motifs that predicts the repeat number and "phasing" of LRRs with greater reliability [12]. We call this method LRRpred. This article describes LRRpred and how to identify LRR in proteins. We apply LRRpred to tropomodulin, plant TIR1, mononocyte differentiation antigen CD14, glycoprotein hormone receptors (FSHR and TSHR), human PRAMEs, mouse oogenesins, flagellar anatomy protein (Falpp), leimodin, and plant F-box/LRR-repeat protein 21 (FBL21). Finally we demonstrate its validity and reliability. LRRpred should become a useful tool of sequence analyses of LRR proteins.

Four Methods

PFAM and SMART

PFAM and SMART use profile-HMMs to identify LRRs [1,2]. They run HMM searches against the public databases (UniProt etc) and report all the hits with a score above a specific threshold. SMART mainly estimates the sequence of the HCS. PFAM is available for E-value ≦ 10 (the default is 1.0) (http://pfam.sanger.ac.uk/). SMART predicts not only LRRs with confidence but also the possibilities LRRs whose scores are less significant than the required threshold; these scores sometimes exceed E-value ≥ 1,000 (http://smart.embl-heidelberg.de/smart/
set_mode.cgi?GENOMIC=1).

The data base of PFAM classifies LRRs in three categories (LRR_1, LRR_2, and LRR_3). The data base of SMART classifies LRRs in categories such as LRR, LRR_TYP, LRR_RI, LRR_CC, LRR_BAC, and LRR_SD22. InterPro classifies LRRs in categories such as Leu-rich_rpt, Leu-rich_rpt_Cys-con_subtyp, Leu-rich_rpt_RNase_inh_sub-typ, Leu-rich_rpt_typical-subtyp and Leu-rich_rpt_2 (http://www.ebi.ac.uk/interpro/).

LRRfinder

Offord *et al.* [3] recently developed LRRfinder (http://www.lrrfinder.com)_and generated an integrated Toll-like receptor database. LRRfinder uses a position-specific scoring matrix (PSSM) to find the most probable location of the 11-residue HCS within a given sequence. An E-value < 0.05 is preferred.

LRRpred

The known LRR structures reveal that the most conserved segment in the HCSs is LxxLxLxx; the remaining segment of NxL is replaced by (L/I/M)xL, (R/K/E)xL, (N/S/T/A)xx, and xxx. Also the HCSs have rareK examples, xVxxLxLxxNxL and PxxLxLxxNxL, and LxxLxGxxS/PxI, as well as an irregular LRR with (L/x)xx(L/A)xCxx(L/R)xLxxVPxxIPxx. The programs of PFAM and SMART do not recognize amino acid sequence having these features as an LRR motif, because the E-value scores of these irregular LRRs are very large.

From the analyses of the known LRR structures, the secondary structure predictions and the multiple sequence alignments of Toll-like receptor 2 (TLR2), we obtained strong evidence that allows us to identify LRRs over an extended range of sequences and inferred structures [12,13]. Therefore, we developed LRRpred for the identification of LRRs in each LRR protein [12]. LRRpred consists of four steps (Figure 1)

In *Step 1*, all motifs with E-value ≦ 10 (detected by PFAM) and less significant motifs (detected by SMART and LRRfinder) are tentatively regarded as an LRR.

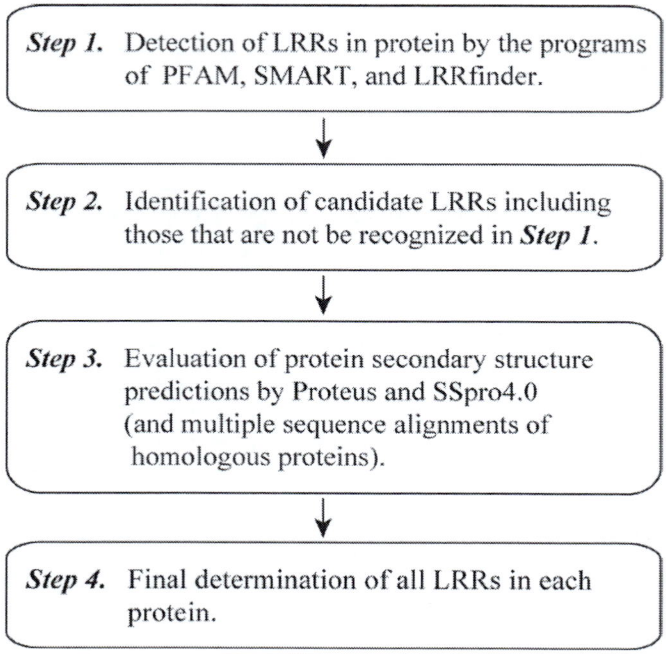

Figure 1. Procedures for identifying LRRs in proteins by LRRpred.

Tandem periodicity of LRRs is honored. In *Step 2*, based on tentative LRRs detected by *Step 1*, the LRR candidates are selected using the criterion that they are longer than 18 residues and that the HCS part consisting of LxxLxLxxNxL contains at least hydrophobic residues at positions 4 and 6. In order to investigate the conservation of the VSs in the LRR candidates, the sequence alignments of the LRRs within LRR proteins are performed. As a result, the HCSs sometimes consist of LxxLxLxxxxx or xxxLxLxxxxx.

In *Step 3*, the preference of a β–strand in the HCS part of the LRR candidate selected by *Step 1* and *Step 2* is investigated by the secondary structure predictions; the β–strand preference at positions 3 through 5 and/or its neighboring positions are investigated [14,15]. In case there are homologous LRR proteins from many species, multiple sequence alignments are also considered for identification.

In *Step 4*, when positions 3 through 5 and/or their neighboring positions in the HCS part of the LRR candidate prefers β-strand by either Proteus or SSpro4.0 (at least in one species) [14,15], it is identified as an LRR. In this step, the sequence alignments of the LRRs within proteins are completed. Finally, predicted LRRs include irregular, non-canonical motifs that are similar to LRRs first recognized by the known structures. Positions 4 and 6 in their HCSs are rarely occupied by Ala, Tyr and Gly.

We first applied LRRpred to Toll-like receptors (TLRs) that play a central role in innate immunity. LRRpred indicated that TLRs contain 20-28 LRRs [12]. The predicted repeat number of LRRs and their "phasing" in TLR1, TLR2, TLR3 and TLR4 are consistent with those revealed by their crystal structures which were determined later [13,16-18].

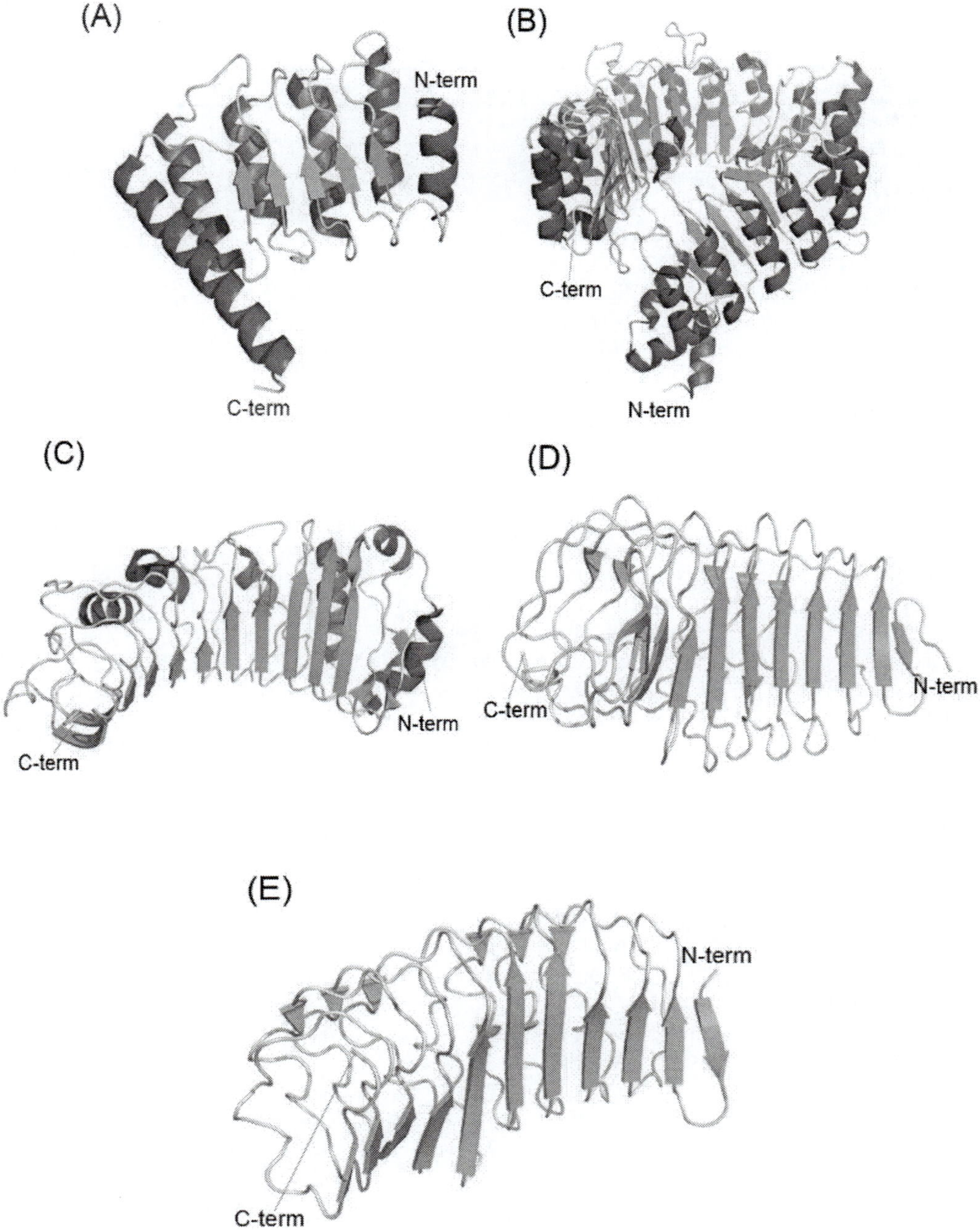

Figure 2. Tertiary structures of leucine-rich repeat (LRR) domains viewed from the sides of the respective solenoids. All drawings are shown at scale. (A) The *Caenorhabditis elegans* tropomodulin structure (PDB ID, 1PGV); (B) The Ar*abidopsis thaliana* TIR1 structure (PDB ID, 2P1O); (C) The mouse CD14 structure (PDB ID, 1WWL); (D) The human FSHR structure (PDB ID, 1XWD). The β-strands are colored green and α-helices red.

Table 1. The identification of LRRs in proteins by PFAM, SMART, LRRfinder, and LRRpred

Protein	ID	Length[a]	PFAM	SMART[b]	LRR_{finder}[c]	LRR_{pred}	Others
The Tropmodulin Family							
Tropomodulin	TMOD_CAEEL	392	0	0 (2)	1	5	5[d]
"	Q9DEA6_CHICK	345 (F)	0	0 (1)	1	5	5[d]
"	C1BTR8_9MAXI	343	0	0 (2)	1	5	
"	B8ZWJ7_ECHMU	361	1	0 (1)	2 (3)	5	
"	C4PXY0_SCHMA	357	0	0 (2)	2 (4)	5	
TRIADDRAFT_49736	B3RI12_TRIAD	309	0	0 (2)	2 (3)	5	
Tropomodulin-2	TMOD2_HUMAN	351	0	0	1	5	
"	Q17823_CAEEL	308	0	0 (1)	0	5	
"	A8XDD4_CAEBR	641	0	0	0 (1)	5	
The Plant TIR1 Family							
TIR1	TIR1_ARATH	594	1	6	2	18	18[d]
"	B9S3W2_RICCO	585	1	7	1 (3)	18	
"	Q9AUH6_9ROSI	635	0	4	0 (3)	18	
"	A4KA30_GOSHI	586	2	8	2 (4)	18	
"	B0LXW5_ORYSI	585	0	3 (5)	2 (4)	18	
TIR1A	TIR1A_ORYSJ	597	0	3	2 (4)	18	
TIR1B	TIR1B_ORYSJ	575	0	6	2 (4)	18	
"	A9SYG2_PHYPA	570	0	7 (8)	2 (4)	18	
Hypothetical	A5JVC9_BRACM	590	1	0	1 (2)	18	
AFB2	AFB2_ARATH	575	0	7	1 (3)	18	
Hypothetical	B4FJG6_MAIZE	573	0	5 (7)	2 (3)	18	
Hypothetical	B8LQ74_PICSI	603	1	7	2 (5)	18	
POPTRDRAFT_572746	B9I9W7_POPTR	584	0	6 (7)	1 (3)	18	
Hypothetical	C0PDU1_MAIZE	594	0	5 (8)	2 (3)	18	
Sb09g003870	C5Z046_SORBI	591	1	4 (6)	3 (4)	18	
GRH1	GRH1_ARATH	585	0	6	1 (2)	18	
COI1	B2BDT6_ARALY	592	0	0 (4)	0 (1)	18	
"	Q45FY8_SOYBN	590	0	4	2 (3)	18	
"	COI1_ARATH	592	0	4	0 (1)	18	16[d]
"	B6C7B5_PEA	563	0	6	2 (3)	18	
Monocyte differentiation antigen CD14							
CD14	CD14_MOUSE	360	2	0 (4)	4 (6)	11	11[d]
"	Q9TTT3_HORSE	369	1	0 (4)	5 (7)	11	
"	Q06AW0_OVICA	373	3	0 (4)	6	11	
"	Q06AV9_SHEEP	373	3	0 (4)	6	11	
"	Q1PBC9_CAPHI	373	3	0 (3)	3 (5)	11	
"	CD14_RAT	372	2	0	4	11	
"	CD14_RABIT	372	1	0 (3)	5 (7)	11	
"	CD14_HUMAN	315	2	0	5 (6)	11	
"	CD14_BOVIN	373	3	0	5 (6)	11	
"	B3Y6B8_MACM	375	2	0	5	11	
"	B3Y6B6_PONPY	375	2	0 (4)	5 (6)	11	
"	B3Y6B5_9PRIM	375	2	0 (4)	5 (6)	11	
"	B3Y6B3_PANTR	375	2	0 (3)	5 (6)	11	
"	B2X875_MACEU	380	2	0 (5)	4 (7)	11	
"	A7BG66_PIG	373	1	0	5 (6)	11	

Protein	ID	Length[a]	PFAM	SMART[b]	LRR_{finder}[c]	LRR_{pred}	Others
"	B0BL87_CHICK	465	3	6 (7)	6 (7)	13	
"	XP_002190482	450	3	6	6 (7)	13	
Human FSHR and TSHR							
FSHR	FSHR_HUMAN	695	2	0 (4)	2 (3)	11	10[d]
TSHR	TSHR_HUMAN	764	0	0 (2)	0 (1)	11	10[d]
Human PRAMEs							
MAPE	MAPE_HUMAN	509	0	0	4	12	7[e]
PRAME1	PRAM1_HUMAN	474	0	0	4 (5)	12	3[e]
PRAME2	PRAM2_HUMAN	474	0	0	4	12	"
PRAME3	PRAM3_HUMAN	479	0	0	2 (3)	12	"
PRAME4	PRAM4_HUMAN	478	1	0	3 (6)	12	"
PRAME5	PRAM5_HUMAN	476	1	0	0 (1)	12	"
PRAME6	PRAM6_HUMAN	476	1	0	2 (6)	12	"
PRAME7	PRAM7_HUMAN	474	1	0	3	12	"
PRAME8	PRAM8_HUMAN	474	1	0	3	12	"
PRAME9	PRAM9_HUMAN	478	1	0	2 (7)	12	"
PRAME10	PRA10_HUMAN	474	0	0 (1)	3	12	"
PRAME11	PRA11_HUMAN	435	1	0	2 (6)	12	"
PRAME12	PRA12_HUMAN	483	0	0	3	12	"
PRAME18	PRA18_HUMAN	479	0	0	2 (3)	12	"
PRAME19	PRA19_HUMAN	474	0	0	2 (3)	12	"
Mouse Oogenesins							
Oogenesin-1	Q810D9_MOUSE	326	0	0 (1)	0 (3)	12	
"	Q3UTB0_MOUSE	325	0	0	0 (2)	12	
"	Q5RJV9_MOUSE	496	0	0	1 (3)	12	
"	A2A958-MOUSE	426	0	0 (3)	2 (3)	12	
Oogenesin-2	Q7TPX8_MOUSE	476	0	0	2	12	
"	A2A8N1_MOUSE	432	0	0 (1)	2	12	
"	A2AGW7_MOUS	455	0	0	1 (2)	12	
Oogenesin-3	Q3UWY1_MOUS	500	0	0	0	12	9[e]
"	Q7TPT9_MOUSE	448	0	0	1 (3)	12	
"	A2AGW6_MOUS	460	0	0	0	12	
Oogenesin-4	Q4G0C7_MOUSE	502	0	0	1 (3)	12	
"	Q8BHX9_MOUS	430	0	0	1 (2)	12	
Flagellar autotomy protein (Fa1p)							
Fa1p	Q9M4X9_CHLRE	1787	0	11	8 (9)	12	
"	C1FHW4_9CHLO	1414	0	8~10	7	12	
"	C1N4E3_9CHLO	1615	0	9	9 (11)	12	
Leimodin							
Leimodin-1	LMOD1_HUMAN	600	0	0 (1)	2	5	
"	LMOD1_MOUSE	595	0	0 (1)	2	5	
Leimodin-2	C5NMH2_RAT	549	0	0 (2)	0	5	
Leimodin-3	LMOD3_HUMAN	560	0	0 (1)	1	5	
The Plant FBL21 Family							
FBL21	FBL21_ARATH	304	1	0 (2)	1 (2)	7	6[e]
OsJ_0645	A3A637_ORYSJ	317	0	3	2 (3)	7	
Hypothetical	B4G0W0_MAIZE	327	2	4	3 (4)	7	

Table 1. (Continued)

Protein	ID	Length[a]	PFAM	SMART[b]	LRR_{finder}[c]	LRR_{pred}	Others
Sb09g002930	C5YZK2_SORBI	332	2	3	3 (4)	7	
Sb07g005600	C5YIE0_SORBI	470	1	0 (2)	2	7	
Hypothetical	B7FIU5_MEDTR	234	2	4	2 (3)	7	
Cyclin-like F-box	Q2HSK3_MEDTR	262	1	0	2 (3)	7	
RCOM_1091170	B9STU0_RICCO	313	1	4	3	7	
SKIP1	SKIP1_ARATH	300	2	3	2 (3)	7	
FBL19	FBL19_ARATH	301	2	4	1 (2)	7	
N7 protein	Q6Z846_ORYSJ	328	0	3	2 (3)	7	
"	O82023_MEDTR	314 (F)	0	0	1	7	
OsI_06943	A2X411_ORYSI	317	0	3	2 (3)	7	
FB221	FB221_ARATH	322	0	0	1 (2)	7	
FBL23	FBL23_ARATH	449	3	5	4 (5)	9	5[e]
SKI19	SKI19_ARATH	302	1	3	3	7	4[e]
FBL22	FBL22_ARATH	307	0	0	1 (2)	7	
VITISV_041807	A5ALT6_VITVI	321	0	3	1 (2)	7	
RCOM_1091170	B9SKT5_RICCO	305	1	4	3	7	
OSJNBa0064H22.13	Q7XV23_ORYSJ	413	0	0	2 (3)	7	
OSIGBa0152K17.13	Q01J74_ORYSA	274	1	0	2	7	

[a] "F" indicates fragment. [b] The number in parentheses indicates the repeat number of LRRs that was predicted confidently and/or plausibly. [c] The number in parentheses indicates the repeat number of LRRs that was predicted significantly and/or insignificantly. [d] The number is the repeat number of LRRs determined by the structure analysis. [e] The number is the repeat number of LRRs described in InterProKB or researchers.

Comparison among the Four Methods

Here in order to demonstrate the validity of LRRpred we apply LRRpred to other LRR proteins and compare the prediction results by this method with those by PFAM, SMART and LRRfinder (Table 1). These include some proteins in which LRRs were first recognized in the crystal structure or are not recognized by PFAM, SMART and InterPro. One group is LRR proteins (tropomodulin, TIR1, CD14, TSHR and FSHR) whose structures are already determined (Figure 2). The other group is LRR proteins (PRAME, oogenesin, leiomodin, Fa1p, and FBL21) whose structures are unknown.

Tropomodulin

Tropomodulin (TMOD) binds to the pointed, capping end of the actin-tropmyosin filament and regulates the architecture and the dynamics of the filament (for review, see [19,20]). TMOD is 300~360 residues long. Data bases - InterPro, PANTHER, PFAM, and PROSITE - do not identify LRRs; they do identify "tropomodulin" and "WH2" domains).
The amino acid sequences of TMOD have been identified from ten species. PFAM predicts that *Echinococcus multilocularis* TMOD contains only one LRR and the others do not contain LRR. SMART predicts that *Caenorhabditis elegans* TMOD [TMOD_CAEEL] with 392 residues contains two plausible LRRs at residues 264-291 ("RI" class, E-value = 49.7) and 292-316/309 ("CC", "RI", "bacterial" or "Typical", E-value = 7.25 ~ 270); the two HCSs are IEKFSLANTAI and LRVLNVESNFL (Table 1).

TMODs from other species contain plausibly 0 ~ 2 LRRs. LRRfinder predicts 0 ~ 4, significant or insignificant LRRs (Table 1). Finally, *Step 1* suggests strongly that TMODs is an LRR protein. Taking account of multiple alignments of the plausible VSs, *Step 2* indicate that *C. elegans* TMOD contains five LRRs (Figure 3A).

In *Step 3*, the secondary structure predictions by Proteus and SSpro4.0 indicate that all HCSs in the five LRRs prefer β-strand is preferred at positions 3 through 5 and/or its neighboring positions (Figure 3A). Finally, it can be concluded that LRRpred identifies five LRRs. This prediction by LRRpred is completely consistent with the results revealed by the crystal structure of chicken and *C. elegans* TMOD (Figure 3A) [21,22].

Plant TIR1

Plant TIR1 is an F-box/LRR-repeat (FBL) protein and is a receptor for auxin (or indole-3-acetic acid, IAA); TIR1 is a component of a cellular protein complex known as SCFTIR1 [23,24]. The amino acid sequences of twenty-one TIR homologs from fifteen species are avilable. The predicted repeat number and "phasing" of LRRs are different between PFAM and SMART (Table 1). For example, PFAM predicts that *Arabidopsis thaliana* TIR [TIR_ARATH] with 594 residues contains one LRR with 456-484 residues (E-value = 8.4), while SMART predict six LRRs with 129-154 residues (E-value =103), 155-192 (438), 313-336 (401), 371-395 (0.809), 431-454 (171) and 480-504 (29.1). Finally, 0 ~ 2 LRRs in 21 TIR's are predicted by PFAM, 0 ~ 8 repeats by SMART, and 1 ~ 5 repeats by LRRfinder (Table 1).

In *Step 2*, conservation of the VSs in the LRR candidates was considered. LRRpred predicts 16 LRRs in *A. thaliana* TIR and 18 LRRs in AFB2 [AFB2_ARATH] (Figures 3B and 3C). Together with multiple sequence alignments of the TIR homologs, it can be concluded that LRRpred identifies 18 LRRs in the TIR homologs; LRRpred identifies LEELGTGGYTA EVRPDVYSGLSVALSGCKE (*LRR8*) and LESVLYFCRQM TNAALITIARNRPN (*LRR13*) in *A. thaliana* TIR as LRRs. The crystal structure of *A. thaliana* TIR reveals 18 LRRs (Figure 3B) [25,26]. Thus, LRRpred is clearly the best method.

```
(A) Caenorhabditis elegans Tropomodulin
                     LxxLxLxxNxL
LRR1   seq.          LKEVNINNMKRV   SKERIRSLIEAACNSKH
       proteus       ceeeeecccccc   chhhhhhhhhhhhccc
       sspro         hhhhcccccccc   ccccccceeeeeeecccc
LRR2   seq.          IEKFSLANTAI    SDSEARGLIELIETSPS
       sroteus       eeeeeeecccc    cchhhhhhhhhhhccc
       sspro         cchhhhhhhhe    ehhhhcccccceeeeeee
LRR3   seq.          LRVLNVESNFL    TPELLARLLRSTLVTQS
       proteus       ceeeeeecccc    chhhhhhhhhhhhcccc
       sspro         eccccccceeee   eeeeecccccccccccc
LRR4   seq.          IVEFKADNQRQ    SVLGNQVEMDMMMAIEENES
       proteus       eeeeeeecccc    ccchhhhhhhhhhhhhhccc
       sspro         cccccccccch    cccccccccccccccccccc
LRR5   seq.          LLRVGISFASM    EARHRVSEALERNYERVRLRRLGKDPNV
       proteus       eeeeeecccch    hhhhhhhhhhhhhhhhhhhhhhcccccc
       sspro         chhhhhhhhch    hhhhhhhhhhhhhhhhhhhhhccccccc
                                                             3A
```

Figure 3 (continued on next page)

(B) *Arabidopsis thaliana* TIR1

		LxxLxLxxNxL	
LRR1	seq.	RRKVFIGNCYA	VSPATVIRRFPK
	proteus	ccccccccccc	ccchhhhhhccc
	sspro	cceeeeehhec	ccchhhhhhccc
LRR2	seq	VRSVELKGKPHF	ADFNLVPDGWGGYVYPWIEAMSSSYTW
	proteus	eeeeeecccccc	cccccchhccccccchhhhhhhhhhccc
	sspro	ceeeeecccccc	cccccccccccccccchhhhhhhcchhh
LRR3	seq.	LEEIRLKRMVV	TDDCLELIAKSFKN
	proteus	ceeeecccee	cchhhhhhhhhccc
	sspro	hhhhhhccec	cchhhhhhhhhccc
LRR4	seq.	FKVLVLSSCEGF	STDGLAAIAATCRN
	proteus	ceeeeeeccccc	chhhhhhhhhhccc
	sspro	cceeeeeccccc	chhhhhhhhhhccc
LRR5	seq.	LKELDLRESDV	DDVSGHWLSHFPDTYTS
	proteus	ceeeeeecccc	cccchhhhhhhhcccc
	ssro	ceeeeccccc	ccccchhhhhhccccccce
LRR6	seq.	LVSLNISCLAS	EVSFSALERLVTRCPN
	proteus	ccceeccccccc	cccccchhhhcccccc
	sspro	eeeeeeccccc	ccchhhhhhhhhcccc
LRR7	seq.	LKSLKLNRAVPL	EKLATLLQRAPQ
	proteus	eeeeeecccccc	ccccchhhcccc
	sspro	ceeeeeccccch	hhhhhhhhhchh
LRR8	seq.	LEELGTGGYTA	EVRPDVYSGLSVALSGCKE
	proteus	eeccccccccc	ccccccchhhhhhhcccccc
	sspro	hhhhccccccc	ccccchhhhhhhhhhhhchh
LRR9	seq.	LRCLSGFWDAV	PAYLPAVYSVCSR
	proteus	ccccccccccc	chhhhhhhhhccc
	sspro	hhhhccchhhc	hhhchhhhhhhhh
LRR10	seq.	LTTLNLSYATV	QSYDLVKLLCQCPK
	proteus	ceeeeccccccc	cchhhhhhhhhccc
	sspro	hhrcceeeeec	cchhhhhhhhhchh
LRR11	seq.	LQRLWVLDYIE	DAGLEVLASTCKD
	proteus	cceecccccccc	hhhhhhhhhcccc
	sspro	hhhhhhhhhhh	hhhhhhhhhhhhh
LRR12	seq.	LRELRVFPSEPF	VMEPNVALTEQGLVSVSMGCPK
	proteus	ceeeeecccccc	ccccccccchhhhhhhhhcchh
	sspro	hheeeecccccc	ccccccccchhhhhhhhhcchh
LRR13	seq.	LESVLYFCRQM	TNAALITIARNRPN
	proteus	hhhhccccccc	cchhhhhhhhcccc
	sspro	hhhhhhhhhhh	chhhhhhhhhcccc
LRR14	seq.	MTRFRLCIIEP	KAPDYLTLEPLDIGFGAIVEHCKD
	proteus	ccccccccccc	cccccccccccchhhhhhhhhccc
	sspro	ceeeeeeeecc	cccccccccccchhhhhhhhhcchc
LRR15	seq.	LRRLSLSGLLT	DKVFEYIGTYAKK
	proteus	ceeeeeecccc	hhhhhhhhhcccc
	sspro	cheeeccccccc	chhhhhhhhhhhh
LRR16	seq.	MEMLSVAFAGD	SDLGMHHVLSGCDS
	proteus	ccccccccccc	cchhhhhhhhcccc
	sspro	hhhhehhhccc	chhhhhhhhccccc
LRR17	seq.	LRKLEIRDCPF	GDKALLANASKLET
	proteus	ceeeeeecccc	chhhhhhhhhhccc
	sspro	ceeeeeccccc	chhhhhhhhhhhhh
LRR18	seq.	MRSLWMSSCSV	SFGACKLLGQKMPKLNVEVIDERGAPDSRPESCPVERVFIYR
	proteus	ceeeeeecccc	chhhhhhhhhhccceeeeeeccccccccccccccchhhhhhhh
	sspro	hhheeeccccc	chhhhhhhhhhcccccceeeeeccccccccccccccccceeeeeee

3B

(C) *Arabidopsis thaliana* AFB2

```
                        LxxLxLxxNxL
LRR1    seq.            RQKVFIGNCYAI NPERLLRRFPC
        proteus         cccccccccccc cchhhhhcccc
        sspro           cceeeeccchcc chhhhhhhccc
LRR2    seq.            LKSLTLKGKPHF ADFNLVPHEWGGFVLPWIEALARSRVG
        proteus         ceeeeeeccccc ccccccccccccccchhhhhhhhhccc
        sspro           ceeeeeccccch ccccccccccccccchhhhhhhhhccc
LRR3    seq.            LEELRLKRMVV  TDESLELLSRSFVN
        proteus         ceeeecccccc  cchhhhhhhhhccc
        sspro           hhhccccceee  chhhhhhhhhcccc
LRR4    seq.            FKSLVLVSCEGF TTDGLASIAANCRH
        proteus         eeeeeecccccc cchhhhhhhhcccc
        sspro           ceeeeecccccc chhhhhhhhhhccc
LRR5    seq.            LRDLDLQENEI  DDHRGQWLSCFPDTCTT
        proteus         ceeeeecccee  cccchhhhhhhhccccc
        sspro           ceeeeeccccc  cccchhhhhhhhccccc
LRR6    seq.            LVTLNFACLEG  ETNLVALERLVARSPN
        proteus         ccccccccccc  cccchhhhhhhhcccc
        sspro           eeeeeeccccc  ccchhhhhhhhhcccc
LRR7    seq.            LKSLKLNRAVPL DALARLMACAPQ
        proteus         eeeeeccccccc ccchhhhhhccc
        sspro           cheeecccccch hhhhhhhhcccc
LRR8    seq.            IVDLGVGSYEN  DPDSESYLKLMAVIKKCTS
        proteus         ccccccccccc  cccchhhhhhhhhhhcccc
        sspro           hhecccccccc  cccchhhhhhhhhhhhhhh
LRR9    seq.            LRSLSGFLEAA  PHCLSAFHPICHN
        proteus         ccccccccccc  ccccccccccccc
        sspro           heecccccccc  hhhhhhhhccccc
LRR10   seq.            LTSLNLSYAAEI HGSHLIKLIQHCKK
        proteus         ceeeeecccccc cchhhhhhhhhccc
        sspro           ceeeeecccccc chhhhhhhhhhchh
LRR11   seq.            LQRLWILDSIG  DKGLEVVASTCKE
        proteus         ceeeeeccccc  hhhhhhhhhcccc
        sspro           hhhhhhhhhhc  hhhhhhhhhcchh
LRR12   seq.            LQELRVFPSDL  LGGGNTAVTEEGLVAISAGCPK
        proteus         ceeeeeecccc  ccccccccchhhhhhhhhcccc
        sspro           hheeecccccc  ccccccccchhhhhhhhhcchh
LRR13   seq.            LHSILYFCQQM  TNAALVTVAKNCPN
        proteus         cceeecccccc  chhhhhhhhhcccc
        sspro           hhhhhhhhhcc  chhhhhhhhhhchc
LRR14   seq.            FIRFRLCILEP  NKPDHVTSQPLDEGFGAIVKACKS
        proteus         ccccceecccc  cccccccccchhhhhhhhhcccch
        sspro           hheeeeeeecc  ccccccccccchhhhhhhhhcccc
LRR15   seq.            LRRLSLSGLLT  DQVFLYIGMYANQ
        proteus         ceeeecccccc  hhhhhhhhhcccc
        sspro           cceeecccccc  hhhhhhhhhhhhc
LRR16   seq.            LEMLSIAFAGD  TDKGMLYVLNGCKK
        proteus         eeeeecccccc  chhhhhhhhhcccc
        sspro           ceeeeeccccc  chhhhhhhhhcccc
LRR17   seq.            MKKLEIRDSPF  GDTALLADVSKYET
        proteus         ceeeecccccc  cchhhhhhhhhccc
        sspro           ceeeeeccccc  cchhhhhhhhhchc
LRR18   seq.            MRSLWMSSCEV  TLSGCKRLAEKAPWL
        proteus         ceeeeeccccc  chhhhhhhhhhcccc
        sspro           cceeeeccccc  chhhhhhhhhhccce
```

3C

Figure 3 (continued on next page)

(D) Chicken CD14

		LxxLxLxxNxL	
LRR1	seq.	ASAVEFQGGDL	EKYADFPIEDDSTIDMDL
	proteus	cceeeeccccc	cccccccccccccccchh
	sspro	hheeeecccch	hhhhhhhhcccccchcch
LRR2	seq.	LGSLQIRKIIF	SDLLVPEVLLARVLRFFSYTQ
	proteus	hhhcccccеее	cccccchhhhccccccccce
	sspro	hcccceeeeee	ccccchhhhhhhhhhhhhchce
LRR3	seq.	VQEMVFKSCSF	VGRSSWADMAGRALP
	proteus	eeeeeccccee	eeccccccccccccc
	sspro	eeeeeecceee	ecccccccccccccce
LRR4	seq.	IVSLRFHNVTA	AALAGREQDLSPLSRWLGA
	proteus	ccccccccccc	ccccccccccccccccccc
	sspro	eeeeeeeecceh	hhccccchhhhhhhhhhhh
LRR5	seq.	LQELSVTASRL	AVLPCAVGQALRA
	proteus	eeeeeeecccc	ccccccccccccc
	sspro	ceeeeeeccce	eeccccchhccccch
LRR6	seq.	LRSLDLARNSL	GDDSLAPAFCQGAFTQ
	proteus	ceeeeeccccc	cccccccccccccccc
	sspro	heeecccchhhc	hhhhhhhhcccchhhe
LRR7	seq.	LRVLSLRHNNL	TSYSGVCGSVRLLGE
	proteus	ceeeeeecccc	ccccccccccccccc
	sspro	eeeeeeccccc	ccccccccchhhhhh
LRR8	seq.	LQHLDLSHNAL	TAGTPSSPCQWPAT
	proteus	ceeeecccccc	cccccccccccccc
	sspro	hhhcccccccc	ccccccccccccchh
LRR9	seq.	LRIFNLSSAGL	DHVLAPLPPA
	proteus	eeeeecccccc	cccccccccc
	sspro	eeeeecchccc	cccccccccc
LRR10	seq.	LEVLDLSSNRL	HAVDISLRS
	proteus	ccceecccccc	ccccccccc
	sspro	eeeeecccccc	ccchhhhhh
LRR11	seq.	LQALFLSRNAL	PAAPSIRGCPA
	proteus	ccceecccccc	ccccccccccc
	sspro	hhhhhhhhccc	ccccccccccc
LRR12	seq.	LHTLHLDNNLI	AELPRDEALLLEH
	proteus	ceeeecccccc	ccccccccccccc
	sspro	eeeeeeccccc	cccchhhhhhchc
LRR13	seq.	LQDVAVAGNPF	NCTCGGAGAVQALAAVGCLRQGWPQGYVCHAPARYQGVLLRDVPTSVLQC
	proteus	ceeeeeccccc	cccchhhhhhhhhhhhhhhhccccccccccccccccccccceecccccccc
	sspro	ceeeecccccc	eeechhhhhhhhhhhhheeeeccccceeeccccccccceeeeccccccccc

3D

(E) Human FSHR

		LxxLxLxxNxL	
LRR1	seq.	NRVFLCQESKV	TEIPSDLPRN
	proteus	ceeeeecccc	cccccccccc
	sspro	cceeeeecccce	eecccccccc
LRR2	seq.	AIELRFVLTKL	RVIQKGAFSGFGD
	proteus	ccceeccccc	ccccccccccccc
	sspro	ceeeeeecccc	eecchhhhccccc
LRR3	seq.	LEKIEISQNDV	LEVIEADVFSNLPK
	proteus	eeeeeeecccc	cccccccccccccc
	sspro	eeeeeeccchh	hhhcchhhhhchhh
LRR4	seq.	LHEIRIEKANNL	LYINPEAFQNLPN
	proteus	ccceeecccccc	ccccccccccccc
	sspro	hheeeecccccc	eecchhhhccccc
LRR5	seq.	LQYLLISNTGI	KHLPDVHKIHSLQ
	proteus	ceeeeeccccc	ccccccccccccc
	sspro	ceeeeeccccc	cecccccchhhhh
LRR6	seq.	KVLLDIQDNINI	HTIERNSFVGLSFE
	proteus	ccceeecccccc	cccccccccccccc
	sspro	hhhhhhccccce	eeccccchhhccchh
LRR7	seq.	SVILWLNKNGI	QEIHNCAFNGTQ
	proteus	ccceecccccc	cccccccccccc
	sspro	heeeeeccccc	cecccccccccc
LRR8	seq.	LDELNLSDNNNL	EELPNDVFHGASG
	proteus	cccccccccccc	ccchcccccccc
	sspro	hhhecccccchh	hcchhhhhcccc
LRR9	seq.	PVILDISRTRI	HSLPSYGLEN
	proteus	eeeeeeccccc	cccccccccc
	sspro	ceeeeeccccc	ccccchhhhh
LRR10	seq.	LKKLRARSTYNL	KKLPTLEKLVA
	proteus	ccchhhhccccc	ccccccccccc
	sspro	hhhhhhcccchhh	cccccchhhhhh
LRR11	seq.	LMEASLTYPSHC	CAFANWRRQISELHPI
	proteus	cccccccccccch	hhhhhhhhhhhhhhhh
	sspro	hheeecccccccc	hhhhhhhhccccccccc

3E

Figure 3 (continued on next page)

(F) Human TSHR

		LxxLxLxxNxL	
LRR1	seq.	EEDFRVTCKDI	QRIPSLPPS
	proteus	cceeeeecccc	ccccccccc
	sspro	cceeeeecccc	eeccccccc
LRR2	seq.	TQTLKLIETHL	RTIPSHAFSNLPN
	proteus	ccceeeccccc	ccccccccccccc
	sspro	ceeeeeecccc	eecchhhhhcccc
LRR3	seq.	ISRIYVSIDVTL	QQLESHSFYNLSK
	proteus	eeeeeecccccc	eeecccccccccc
	sspro	eeeeeeccccch	hecchhhhhcccc
LRR4	seq.	VTHIEIRNTRNL	TYIDPDALKELPL
	proteus	eeeeeccccccc	eeeccchhhcccc
	sspro	eeeeeeeccccc	eecchhhhhcccc
LRR5	seq.	LKFLGIFNTGL	KMFPDLTKVYSTDI
	proteus	ceeeeeeccce	eecccccccccccc
	sspro	hheeeeecccc	ceccccccccccce
LRR6	seq.	FFILEITDNPYM	TSIPVNAFQGLCNE
	proteus	ccceeeeccccc	cccccccccccccc
	sspro	eeeeeccccccc	cecchhhhhccchh
LRR7	seq.	TLTLKLYNNGF	TSVQGYAFNGTK
	proteus	cceeecccccc	cccccccccccc
	sspro	heeeeeecccc	eeeccccccccc
LRR8	seq.	LDAVYLNKNKYL	TVIDKDAFGGVYSG
	proteus	ccceeeeccccc	cccccchhhccccc
	sspro	ccheeeecccee	eecchhhhcccccc
LRR9	seq.	PSLLDVSQTSV	TALPSKGLEH
	proteus	eeeeecccccc	ccccchhhhh
	sspro	ceeeecccccc	cccchchhhh
LRR10	seq.	LKELIARNTWTL	KKLPLSLSFLHL
	proteus	hhhhhhcccccc	cccccccccccch
	sspro	hhhhehccccch	hcccccchhhhh
LRR11	seq.	LTRADLSYPSHC	CAFKNQKKIRGILESL
	proteus	cceeeeeccccc	ccccccccccccchhh
	sspro	hhhhccccccch	hhhcchhhhhcchhhh

3F

(G) Human MAPE

		LxxLxLxxNxL	
LRR1	seq.	LQVLDLRKNSH	QDFWTVWSGNRASLYSFPEPEAAQPMTKKRKVDGLSTEAEQPFIPVEV
	proteus	eeeeeccccc	cccccccccccccccccccccccccccccccccccccccccceeee
	sspro	eeeeeechccc	cccecccccccccccccccccccccccccccccccccccccccceee
LRR2	seq.	LVDLFLKEGAC	DELFSYLIEKVKRK
	proteus	eeeeeeecccc	chhhhhhhhhhhh
	sspro	eeeeeecccch	hhhhhhhhhhhhh
LRR3	seq.	KNVLRLCCKKL	KIFAMPMQDIKMILKMVQLDS
	proteus	cceeeeeeech	hhccchhhhhhhhhhhhhh
	sspro	ccceeeehhhh	ehccchhhhhhhhhcccch
LRR4	seq.	IEDLEVTCTWKL	PTLAKFSPYLGQMIN
	proteus	hcccccccccc	cccccccccchhhcc
	sspro	hheeeeecccch	hhhhhhhhhhhhhh
LRR5	seq.	LRRLLLSHIHA	SSYISPEKEEQYIAQFTSQFLSLQC
	proteus	ceeeeeeeccc	ccccchhhhhhhhhhhhhhhcccc
	sspro	hhhhhhhccc	cccccchhhhhhhhhhhhhhhhhh
LRR6	seq.	LQALYVDSLFF	LRGRLDQLLRHVMNP
	proteus	cceeeeccccc	chhhhhhhhccccc
	sspro	hhhhhhhhhhh	hhhhhhhhhhhhcc
LRR7	seq.	LETLSITNCRL	SEGDVMHLSQSPSVSQ
	proteus	ceeeeeecccc	chhhhhhhhcccccce
	sspro	cceeeeecccc	ccchhhhccccchhh
LRR8	seq.	LSVLSLSGVML	TDVSPEPLQALLERASAT
	proteus	eeeeeecceee	ccchhhhhhhhccccc
	sspro	hheeeccccee	ccchhhhhhhhhhhhhh
LRR9	seq.	LQDLVFDECGI	TDDQLLALLPSLSHCSQ
	proteus	ceeeeeccccc	chhhhhhhhhhhhhccc
	sspro	hhhhhhhccc	cchhhhhhhhhhhhhhc
LRR10	seq.	LTTLSFYGNSI	SISALQSLLQHLIG
	proteus	ceeeeeccccc	cchhhhhhhhhhh
	sspro	ceeeeeccccc	chhhhhhhhhhhhc
LRR11	seq.	LSNLTHVLYPV	PLESYEDIHGTLHLERLAYLHARLRELLCELGR
	proteus	hhccccеeccc	cccccccccccchhhhhhhhhhhhhhhhhcc
	sspro	hhhceeeeccc	cchhhcccccechhhhhhhhhhhhhhhhhcc
LRR12	seq.	PSMVWLSANPC	PHCGDRTFYDPEPILCPCFMPN
	proteus	cceeeeeeccc	cccccccccccccccccccccc
	sspro	cceeeeeeccc	cccccccccccchhhhhccccc

Figure 3 (continued on next page)

(H) Mouse Oogenesin-3

		LxxLxLxxNxL	
LRR1	seq.	LRVLKWRDEQH	DFCGIWPGSHEAEDLPELMTQKHPVQNNPDCGVKK
	proteus	eeeeeecccc	cccccccccccccccccccccccccccccccccc
	sspro	eeeeeeccccc	hhhheccccccccccccccccccccccccccccc
LRR2	seq.	ELRVTTELSVM	KGRLDDSATYLLEWAQQR
	proteus	eeeeeeeeeee	cccchhhhhhhhhhhh
	sspro	ceeeeeeeehc	ccchhhhhhhhhhhhh
LRR3	seq.	KDSIHLLCRKL	VIETLTKDTVIEIFKIVNADC
	proteus	cceeeeeeech	hhhcchhhhhhhhhhhhhh
	sspro	hcchehhhhh	ehccchhhhhhhhhcchhh
LRR4	seq.	IQELELYSLCL	EDLAFLNPYLRQMDN
	proteus	hhhhhhhccc	ccccccchhhhhh
	sspro	hhheehhhcch	hhhhhhchhhhhhh
LRR5	seq.	LLELTLDHVTD	SLSMGDSEMCEEEMITLVSQLPTFPC
	proteus	hhhhhhcccc	cccccccchhhhhhhhhhcccccc
	sspro	hhhhhhhccc	ccccccchhhhhhhhhhhhhhhhh
LRR6	seq.	LQKLCVNDVYF	IYGNLNEILRCLKKP
	proteus	eeeeeccccc	cchhhhhhhhhhhh
	sspro	hhhhhhcceee	ehcchhhhhhhccc
LRR7	seq.	LVSFCISNCEL	SQSDLDCLPYCLNIFE
	proteus	hhceeeecccc	chhhhhhhhcccccc
	sspro	ceeeeeecccc	chcchhhchhhhhhh
LRR8	seq.	LKCLYLIDIPL	NHLCLDPLGFLLESVRHT
	proteus	cceeeeccccc	ccchhhhhhhhhhhhhc
	sspro	hhhhhhhchh	ccccccchhhhhhhhh
LRR9	seq.	LECLELKSCDM	GEPQFNALLPALSQCSH
	proteus	ccccceeeecc	ccchhhccccccccc
	sspro	hhhhhhhccc	cchhhhhhhhhhhhc
LRR10	seq.	LTDVSFWENEL	SLLFLKQLLQHTSK
	proteus	eeeeeeccccc	chhhhhhhhhhhh
	sspro	ceeeeeccccc	chhhhhhhhhhcc
LRR11	seq.	LTQLSYELYPA	PLECYDDRGLILSHRLEQFCPELLDILRAKRQ
	proteus	hhccceecccc	ccccccccccchhhhhhhhhhhhhhhhhccc
	sspro	hhcceeeeccc	cchhccccchhhhhhhhhhhhhhhhhhhhhcc
LRR12	seq.	PKDIAFVTNPC	SKCGRYYVYDPKTQRFSLEDTTL
	proteus	cceeeeeeccc	cccccceeeecccccceeecccccc
	sspro	cceeeeecccc	cccccceeeeccchceeecccccc

3H

(I) *Chlamydomonas reinhardtii* Falp

		LxxLxLxxNxL	
LRR1	seq.	GGVLDLSGSGC	GPVALALLLRRLQQQQPSALTS
	proteus	cceeeccccccc	cchhhhhhhhhhhhhccccccc
	sspro	cceeeccccccc	chhhhhhhhhhhhhhcchhhhh
LRR2	seq.	LRAINLTGCEV	GPVGAAYLAAAAAAGGGGGMPAG
	proteus	ceeeeccccccc	cchhhhhhhhhhhccccccccc
	sspro	hhhhhcccccc	cchhhhhhhhhhhhccccccch
LRR3	seq.	LETACLADNGL	GPEGAEAVLCAITGAAPGGKGMAAGTGARVGGGA
	proteus	eeeeeecceee	cchhhhhhhhhhccccehhhhhhcccccccchh
	sspro	hhhhhhhcccc	ccchhhhhhhhhhccccccccceccccceecccc
LRR4	seq.	CGVLDLSLNSL	GDAGCKAVARHLPALGGS
	proteus	hhhccceccccc	chhhhhhhhhhhhhcccc
	sspro	cceeeecchhch	chhhhhhhhhhcccccc
LRR5	seq.	LGVLALSHNSI	GTDGARALAAALPTCAG
	proteus	ceeeeeeecccc	ccchhhhhhhhhhcccc
	sspro	heeeeeccccc	ccchhhhhhhhcchhhh
LRR6	seq.	LEELLLGGNQL	GDQAVRELAAAISSSTAAATAPGGSSGGLQL
	proteus	eeeeeeeeceee	chhhhhhhhhhhccccccccccceccccce
	sspro	hhhhhhcccch	hhhhhhhhhhhhhcchhhccccccchhhhh
LRR7	seq.	LAVLHLHDNPRL	AADGVAGLAAALPGLPS
	proteus	eeceeeececccc	cchhhhhhhhhhhhccc
	sspro	hhhhhcccchhh	hhhhhhhhhhhcccccc
LRR8	seq.	LTELALGRCHA	GNEGARHLSKALPSHPS
	proteus	eeeeeeeceec	chhhhhhhhhhhhhccc
	sspro	hhhhhhccccc	cchhhhhhhhhcccccc
LRR9	seq.	LQVLDLEGCKI	RAEGGAHLAAALAAMNGSPAAAAAGGGGGAAAGA
	proteus	eeeeeeeecccc	chhhhhhhhhhhhhcchhhhhcccccccccccccc
	sspro	ceeeeccccee	ecccchhhhhhhhhhhcchhhhhccccchhhhhhh
LRR10	seq.	LTRLNLARNQL	GDRGAAALAEALGRNRS
	proteus	eeeeeecceee	chhhhhhhhhhhhcccc
	sspro	hhhhhhhhhhh	cchhhhhhhhhhccccc
LRR11	seq.	LVSLDLRANGL	GLVGCRQLAGALRERNAT
	proteus	ceeeeeeecccc	cchhhhhhhhhhhccccc
	sspro	eeehchhhccc	chhhhhhhhhhhhhhhhh
LRR12	seq.	LRSLAVGGNEA	DDPVIETLEELAATQRHVTRPPP
	proteus	eeeeeeeecccee	chhhhhhhhhhhhhccccccccc
	sspro	hhhhhcccccc	cchhhhhhhhhhhhhhhcccccc

3I

Figure 3 (continued on next page)

(J) Human leiomodin-1

		LxxLxLxxNxL	
LRR1	seq.	MTEVNVNNSDC	ITNEILVRFTEALEFNTV
	proteus	eeeeeccccc	chhhhhhhhhhhhccccc
	sspro	ceeeecccccc	cchhhhhhhhhhhcccc
LRR2	seq.	VKLFALANTRA	DDHVAFAIAIMLKANKT
	proteus	eeeeeecccc	hhhhhhhhhhhhccccc
	sspro	eeeeehhhccc	cchhhhhhhhhhhhccc
LRR3	seq.	ITSLNLDSNHI	TGKGILAIFRALLQNNT
	proteus	eeeeeeccccc	hhhhhhhhhhhhcccce
	sspro	hhhccccccc	chhhhhhhhhhhhhccc
LRR4	seq.	LTELRFHNQRH	ICGGKTEMEIAKLLKENTT
	proteus	eeeeeeccccc	cchhhhhhhhhhhhhcccc
	sspro	eeeeeeccccc	ccccchhhhhhhhhhhccc
LRR5	seq.	LLKLGYHFELA	GPRMTVTNLLSRNMDKQRQKRLQEQRQAQEAKG
	proteus	eeeeeccccchh	hhhhhhhhhhhhhhhhhhhhhhhhccccccccc
	sspro	hhhhceeeecc	cchhhhhhhhhhcchhhhhhhhhhhhhhhhhcc

3J

(K) *Arabidopsis thaliana* FBL21

		LxxLxLxxNxL	
LRR1	seq.	WRKINTRDCLM	YNFDFVSMCRHIVDLSQGG
	proteus	ccccccccchh	hhhhhhhhhhhhhhccccc
	sspro	heeeecccchhh	hhhhhhhhhhhhhhccccc
LRR2	seq.	LLEINVDEHFL	SDSLLSYITDRN
	proteus	eeeeecccccc	cchhhhhhhccc
	sspro	eeeeeechccc	cchhhhhhhhhc
LRR3	seq.	LRSLGLGMCFP	RVTKLGVVNAIAKIPL
	proteus	ceeeeeeeccc	ccchhhhhhhhhhccc
	sspro	cheeeeccccc	ccccchhhhhhhhcch
LRR4	seq.	LETLEVTHSCI	KLDLKAIGHACPQ
	proteus	ceeeeeccccc	hhhhhhhhhcccc
	sspro	hhheeeccccc	hhhhhhhhhcchc
LRR5	seq.	LKTLKLNSLGR	LWPASDKYDSNVLDDMGPLECDDDALAIAESMPK
	proteus	ceeeeeccccc	cchhhhhhhhhhhhhcccccccchhhhhhhhcccc
	sspro	hhhheccccccc	ccccccceeccccccccccccchhhhhhhhhchh
LRR6	seq.	LHHLQLMANRL	TNTGLNAILDGCPH
	proteus	ceeeeeeeccc	chhhhhhhhhcccc
	sspro	hhhhhhhhhhh	hhhhhhhhhccccc
LRR7	seq.	LEHLDVRKCFRI	SLVGNLEKRCLEMIKELRRPGDSTADYPYNGV
	proteus	eeeeeeccccccc	chhhhhhhhhhhhhhhhhcccccccccccccc
	sspro	ccecchhhheee	echhhhhhhhhhhhhhhhccccccccccccccc

3K

```
(L) Ricinus communis RCOM_1091170
                  LxxLxLxxNxL
LRR1   seq.       WRSVEIHPLCDF WDLPFDLEALCRNAVDRSRGG
       proteus    cccccccccccc cccccchhhhhhhhhhhhccc
       sspro      ceeeeecccccc cccccchhhhhhhhhhhcccc
LRR2   seq.       LISISIEYFAT  DSLIKYIADRSSHL
       proteus    eeeeeeccccc  hhhhhhhhhhccc
       sspro      eeeeeeeeccc  chhhhhhhhhccc
LRR3   seq.       LKRLRLLSSYTL SDAAFSKAAKKFPL
       proteus    eeeeeeeccccc chhhhhhhhhhccc
       sspro      ceeeeecccccc chhhhhhhhhhccc
LRR4   seq.       LEELDISYCSL  STEALVGVGISCPLT
       proteus    ceeeeeccccc  chhhhhhhhhccccc
       sspro      hhhhcchhccc  chhhhhhhhccchhc
LRR5   seq.       LRSLKLNCQGY  KRPHIESNEEALAIGQWMPH
       proteus    ceeeeeccccc  cccccccchhhhhhhhcccc
       sspro      hhhhhhhcccc  ccceecccchhhhhhhhhcch
LRR6   seq.       LRYLQIFGNKL  TNDGVQAILDGCPH
       proteus    ceeeeeecccc  chhhhhhhhhcccc
       sspro      hhhhhhhcchc  chhhhhhhhccccc
LRR7   seq.       LEFLDLRQCFNV HLEGELGMLCGERIKDLRRPDDPTDDYPFNPEIPDYGSSEEDYPS
       proteus    ceeeeecccccc cchhhhhhhhhhcccccccccccccccccccccccccccccccccc 3L
       sspro      cceeeccccccce cchhhhhhhchhhhhhhcccccccccccccccccccccccccccccc
```

Figure 3. The secondary prediction of LRR domains within LRR proteins by Proteus and SSpro4.0 and the sequence alignments of the LRRs by the LRRpred. (A) *Caenorhabditis elegans* tropomodulin with 392 res. [TMOD_CAEEL]; (B) *Arabidopsis thaliana* TIR1 with 594 res. [TIR1_ARATH]; (C) Mouse CD14 with 306 res.[P10810]; (D) Human FSHR with 695 res. [FSHR_HUMAN]; (E), Human TSHR with 764 res. [TSHR_HUMAN]; (F) *A. thaliana* AFB2 with 575 res. [AFB2_ARATH]; (G) Human MAPE with 509 res. [MAPE_HUMAN]; (H) Mouse oogenesin-3 with 500 res. [Q3UWY1]; (I) *Chlamydomonas reinhardtii* Fa1p with 1,787 res. [Q9M4X9_CHLRE]; (J) Human leiomodin-1 with 600 res. [LMOD1_HUMAN]; (K) *A. thaliana* FBL21 with 304 res. [FBL21_ARATH]; (L) *Ricinus communis* RCOM_1091170 with 305 res. [B9SKT5_RICCO]. In known LRR structures, residues adopting β-strands in the HCS part are underlined by wavy lines. The conserved consensus amino acids of the highly conserved segment (HCS) of individual LRRs are highlighted with reverse-contrast. The parts of β-strand obtained by the secondary structure prediction are highlighted with a shadow. Abbreviations; h, helix; e, β-strand; c, coil.

Monocyte Differentiation Antigen CD14

CD14 cooperates with MD-2 and TLR4 to mediate the innate immune response to bacterial LPS (for review, see [27]). Amino acid sequences from nineteen species are available. PFAM identifies only 1 ~ 3 LRRs, SMART identifies confidently or plausibly 0 ~ 7 repeats, and LRRfinder identifies significant or insignificant 4 ~ 7 repeats (Table 1).

On the other hand, LRRpred indicates that mammalian CD14's all have 11 LRRs; these are consistent with those identified in the crystal structure of mouse CD14 (Figure 2C) [28]. Furthermore, CD14 from chicken [B0BL87] and zebra finch [XP_002190482] have two additional LRRs and thus 13 LRRs in total (Table 1 and Figure 3D).

TSHR and FSHR

Glycoprotein hormone receptors (GpHrs), TSHR, FSHR, and LH/CGR that belong to the rhodopsin-like G protein-coupled receptor family are receptors for thyrothropin (TSH), follicle stimulating hormone (FSH) and luteinizing hormone (LH) or human choriogonadotropic hormone (hCG), respectively (for review, see [29-31]). These GpHrs contain a large N-terminal (350 to 400 residues) ectodomain (ECD), which is an extracellular region that is responsible for high affinity and selective binding of the corresponding hormones. Here we apply LRRpred to human FSHR and TSHR whose structures are known[32,33].

PFAM identifies 2 and 0 LRRs, SMART identifies 4 and 2 repeats, LRRfinder identifies 3 and 0 LRRs, respectively (Table 1). LRRpred predicts the FSHR ECD which has 10 LRRs, while the TSHR ECD has 11 LRRs (Figures 3E and 3F). Together with multiple sequence alignments of FSHR and TSHR, it is concluded that LRRpred identifies 12 LRRs in both proteins. The structures of FSHR in complex with FSH (residues 17-268) and of TSHR (residues 22-260) reveal 10 LRRs together. Thus, LRRpred predicts one additional LRR at the C-terminus; its HCS is LMEASLTYPSHC in FSHC and LTRADLSYPSHC in TSHR (Figures 3E and 3F).

Human PRAME/MAPE/OIP4

Human MAPE [MAPE_HUMAN] [34] binds gonococcal Opa proteins [35]. Human MAPE is a member of the PRAME family and is 509 residues long. Many homologous PRAME's have been identified from human, mouse and chimpanzee [36-40]. SMART does not predict any LRRs or only one LRR in fifteen, human PRAME's with 474-509 residues and PFAM predicts only one LRR in some PRAME's; for examples, PRAMEF12 with 483 residues contains two LRRs with 320 - 342 (E-value = 6.2) and 350 - 372 residues (E-value = 6.9) and PRAME11 with 435 residues contains one LRR with 312 - 334 (E-value = 4); while LRRfinder predicts 1 ~ 7 repeats (Table 1). Moreover, the InterPro data base states that human MAPE contains 7 LRRs and that other human PRAME's contains only 3 LRRs. Finally, *Step 1* suggests strongly that all human PRAME's are LRR proteins. Taking account of irregular LRRs in the known LRR structures, human MAPE has seven plausible HCSs (IEDLEVTCTWKL, LQALYVDSLFF, LETLSITNCRL, LSVLSLSGVML, LQDLVFDE CGI, LTTLSFYGNSI, and PSMVWLSANPC).

Step 2 indicates that human MAPE contains at least seven LRRs. However, their "phasing" differs considerably from that described in InterPro.

Secondary structure prediction indicates that human MAPE contain 12 LRRs, because β-strands are preferred at positions 3 through 5 and/or its neighboring positions in twelve, plausible HCSs (Figure 3G). Multiple sequence alignment indicates that the HCS part of *LRR5* in human MAPE – LRRLLLSHIHA - corresponds to LRKLFISDGCGY in human PRAME18; the latter segment is not predicted to be a β-strand. Finally, it can be concluded that LRRpred identifies twelve LRRs in all human PRAME's (Table 1 and Figure 3G). Very recently, Wadelin *et al* [39] proposed that human PRAME contains thirteen LRRs.

Mouse Oogenesin

Mouse oogenesin is a potential oocyte-specific regulatory factor [41] that is expressed through oogenesis and early embryogenesis in the mouse [42]. Oogenesin-1 interacts with Ras and Ras-signaling proteins during early embryogenesis [43]. The family consists of more than 80 members [44] and some orthologs have been identified in rat [Q5D011]. A similarity search using oogenesin shows significant similarity with PRAME from human and mouse. Neither PFAM, SMART nor InterPro reported LRRs in oogenesin. In contrast, Dadė *et al.*

(2003) [44] proposed that mouse oogenesin-3 [Q3UWY1] contains nine LRRs. The repeat number of LRRs differs among the databases and the researchers.

Twelve mouse oogenesins were used for the present analyses. As noted, PFAM does not predict LRR at all, and SMART and LRRfinder predict only 0 ~ 3 LRRs (Table 1). On the other hand, LRRpred identifies 12 LRRs (Figure 3H). The 12 LRR repeats are equal to those in the human PRAME family.

Flagellar Autotomy Protein 1 (Fa1p)

Flagellar autotomy protein 1 (Falp) from *Chlamydomonas reinhardtii* [Q9M4X9] is essential for axonemal microtubule severing in *Chlamydomonas*; it is 1,787 residues long [45]. Falp has also been identified from *Micromonas pusilla CCMP1545* [C1N4E3] and *Micromonas sp. RCC299* [C1FHW4] which contains "Ankyrin" and "ZnF_C2HC" [46]. However, the presence of LRRs has not been reported in these three proteins in the data bases.

PFAM does not predict LRR at all in *C. reinhardtii* Falp and *M. pusilla* two Falp's (Table 1). In contrast, SMART predicts confidently 8 ~ 11 LRRs and LRRfinder predicts 7 ~ 11 LRRs. LRRpred identifies 12 LRRs (Figure 3I). Some LRRs in *C. reinhardtii* Falp are longer ; the LRRs range from 28 to 45 residues. It appears that all the LRRs have the "RI-like" motif, because the VSs assume relatively long α-helical conformation.

Leiomodin

Leiomodin that is an actin filament nucleator in muscle cells [47] is a member of the tropomodulin family and contains "WH2" domain [48]. The amino acid sequences from many vertebrate species have been identified and are ~560 residues long. LRRpred identifies five LRRs (Table 1 and Figure 3J); these are consistent with those in TMODs.

Plant FBL21

Putative F-box/LRR-repeat protein 21 (FBL21) [FBL21_ARATH] contains six LRRs in the UniProKB data base and eight LRRs by Xiao and Jang [49], but the data bases - InterPro, PFAM, SMART and PROSITE - do not identify LRR. Moreover, a large number of other homologous proteins detected by similarity search using FBL21 do not have recognizable LRRs. UniProKB recognizes five LRRs in FBL23 [FBL21_ARATH]. LRRpred indicates that FBL21 and its homologs such as *Ricinus communis* RCOM_1091170 contain seven LRRs, while only FBL23 [FBL23_ARATH] contains two additional LRRs at the N-terminus (Table 1, and Figures 3K and 3L).

Finding of a New LRR Class

LRRpred predicts that *Escherichia coli* yddk with 318 residues contains 13 LRRs; six of the 13 repeats have the consensus of LxxLxLxxNxLxxLxLxxxxx with 21 residues. The variable segment differs significantly from those of the above seven classes. Each LRR domain is a nested sequence and consists of repeats of alternating 10- and 11- residue units of LxxLxLxxNx(x/-).

LRR proteins having the IRREKO@LRR domains were identified in three steps [11]:

Step 1. Detection of LRR proteins containing the six, novel LRRs in *E-coli* yddk by using FASTA
Step 2. Identification of the IRREKO@LRRs in individual LRR proteins by LRRpred.
Step 3. Iteration of these two steps using novel LRRs in newly identified LRR proteins.

In *step 1*, we performed similarity searches using the six, novel LRRs as probes by FASTA. This procedure detected many yddK homologs from *Escherichia coli strains* and *Shigella flexneri* [Q0T447 and Q83R94] with significant similarity (E-values < 6.5×10^{-29}). In addition, two other proteins were detected with significant similarity (E-value < 3.3×10^{-9}). One is SSON_1653 that is 387 residues long [Q3Z1L5]. The other is SD1012_2081 with 163 residues [B3WXZ7].

In *step 2,* we performed multiple sequence alignment among their LRR domains of SSON_1653 and Sd1012_2081. SSON_1653 contains 14 LRRs and nine of the 12 repeats consist of LxxLxLxxNxLxxL(D/N)(L/F)xxxxx where "L" is Leu, Val, or Ile. Sd1012_2081 contains 4.5 LRRs; 3.5 of these repeats consist of LxxLxLxxNxLxxIx(I/A/F)xxaxx.

In *step 3,* the above procedures were iterated to identify other LRR proteins having this IRREKO@LRR domain.

We identified a novel LRR domain, or nested repeat observed in 134 proteins from 54 bacterial species [11]. This novel LRR domain has 21 residues with the consensus sequence of LxxLxLxxNxLxxLDLxx(N/L/Q/x)xx or LxxLxCxxNxLxxLDLxx(N/L/x)xx. This LRR domain is characterized by a nested periodicity; it consists of alternating 10- and 11- residues units of LxxLxLxxNx(x/-). We call it "IRREKO" LRR, since the Japanese word for "nested" is "IRREKO". The "IRREKO" LRR is a new class. Furthermore, LRRpred was utilized to identify LRRs in metabotropic or ionotropic glutamate receptors and IFNγ receptors [50] and human LRR proteins [51].

Detection of Non-LRR, Island Regions (IRs) Interrupting LRRs

Non-LRR, island regions (IRs) interrupting LRRs are widely distributed; they are referred to as "islands" or "loop outs" [52]. Many plant LRR proteins have IRs; they include LRR-RLKs, LRR-RLPs, insect Toll and Toll-related proteins, and Slit proteins [52-67]. Some experimental studies on the function of IRs within LRR@IR proteins have been performed [68-70].

We developed a method (called LRR@IRpred) to identify IRs utilizing LRRpred [52]. LRR@IR proteins and their IRs were identified by four steps:

Step 1. Detection of LRR@IR proteins by FASTA and BLAST .
Step 2. Identifications of all LRR units in individual LRR@IR proteins by LRRpred.
Step 3. Identification of the IRs within LRRs in individual LRR@IR proteins.
Step 4. Iteration of these three steps using LRRs and IRs in newly identified LRR@IR proteins.

In *Step 1*, the known IRs together with the N-terminal and C-terminal flanking one to three adjacent LRR units in BRI1, PSKR, Cf-9 and CLV2 are used as query sequences for BLAST and FASTA.

In *Step 2*, a new method for the recognition and alignments of LRR units in LRR@IR proteins utilizes known LRR structures and incorporates multiple sequence alignments and secondary structure predictions.

In *Step 3*, when a candidate region is not an LRR unit and it is longer than average length of the repeating unit of LRRs, it is defined as an IR.

Our recent study identified over 600 LRR@IR proteins belonging to these families. Exclusive of these, 262 LRR@IR proteins were identified and divided into 21 different protein families [52]. Most of the proteins with IRs have only a single IR. However, some LRR@IR proteins have LRRs interrupted by two to four IRs. In LRRs with a single IR, the number of repeat units in the preceding LRR block (N_1) and its number in the following block (N_2) keep usually $N_1 \gg N_2$ or $N_1 \approx N_2$, while $N_1 \ll N_2$ is very rare. Interestingly, in $N_1 \gg N_2$ there is an extreme case of $N_2 = 1$.

Features of LRRpred

LRRpred identifies irregular, non-canonical LRRs. Three residues at positions 3 to 5 in the HCS, LxxLxLxxNxL or LxxLxLxxCxxL, form a short β-strand and in canonical LRRs the short β-strands stack parallel [4-8]. In contrast, non-canonical LRRs frequently adopt longer β-strands consisting of 4-7 residues (Figures 1 and 2). For example, TIR1 has 13 non-canonical LRRs of the 18 repeats and all have long β-strands; the residues in each of the 18 parallel β-strands of the concave surface numbers are 4, 5, 5, 5, 5, 3, 3, 3, 3, 3, 5, 5, 6, 6, 4, 5, 5, and 4 (Figures 1B and 3B). In LRR proteins the β-strands on the concave surface and (mostly) helical elements on the convex surface are connected by two loops. The "ascending" loops link the C-terminal ends of the β-strands to the N termini of the helical elements on the convex; the "descending" loops link the C-terminal ends of the helices to the N termini of the following β – strands [8]. Irregular, non-canonical LRRs sometimes have various extensions on their convex, ascending or descending, sides. The longer "ascending" loops are seen in some LRRs in *A. thaliana* TIR1 (*LRR2*, *LRR8*, *LRR12*, and *LRR14*) (Figure 3B). The longer "descending" loops are also seen in three LRRs in *C. reinhardtii* Falp (*LRR2*, *LRR6*, and *LRR9*) (Figure 3I).

For the identification of LRRs PFAM and SMART use HMM based on only amino acid sequences. LRRpred, in addition to these prediction methods, utilizes protein secondary structure predictions and then uses multiple sequence alignment when an LRR protein has been identified from various species. The secondary structure predictions are performed by not one but two different programs - Proteus and SSpro4.0-. The predictions are sometimes different. If the β-strand in the HCS part of an LRR candidate is preferred by either Proteus or SSpro4.0, it is identified as an LRR. These procedures identify many irregular, non-canonical LRRs as an LRR motif. Consequently, LRRpred provides sufficient predictions that the indicated repeat number and "phasing" of LRRs are with the same, or almost so, as those revealed by structure analyses. The most structural feature of the LRR structures is parallel stacking of β-strands in the concave face. All LRRs appear to prefer strong tandem periodicity in the formation of the arc structure. We suggest that this permits them to accommodate various, irregular LRRs as canonical LRRs.

Many outstanding questions emerge from this article: why are a variety of irregular, non-canonical LRRs accommodated as canonical LRRs? Why is it possible for many LRRs to adopt longer β-strands and longer loop? These are important subjects of study in the next five years.

Several searches have developed program anda database for LRR [1-3,71-73]. We plan to refine the software of LRR pred and make it available through a public server so that the community can benefit from this expert knowledge and manual analysis, particularly in *Step 2* (Figure 1).

Conclusion

The repeat number of LRRs and their "phasing" determined by LRRpred are completely consistent with, or almost so, with those revealed by structure analyses. It is clear that these methods will provide significant improvements of annotation in the data bases. LRRpred should become a useful tool of sequence analyses of LRR proteins and contribute to understanding the structure, function and evolution of LRRs and LRR-containing proteins.

Acknowledgments

We thank Dr. Robert H. Kretsinger of the University of Virginia for his critical reading and valuable comments.

References

[1] Finn RD, Tate J, Mistry J *et al.* The Pfam protein families database. *Nucleic Acids Res*, 36(Database issue), D281-288 (2008).

[2] Letunic I, Doerks T, Bork P. SMART 6: recent updates and new developments. *Nucleic Acids Res.*, 37(Database issue), D229-232 (2009).

[3] Offord V, Coffey TJ, Werling D. LRRfinder: a web application for the identification of leucine-rich repeats and an integrative Toll-like receptor database. *Dev. Comp. Immunol*, 34(10), 1035-1041 (2010).

[4] Kobe B, Deisenhofer J. The leucine-rich repeat: a versatile binding motif. *Trends Biochem. Sci.*, 19(10), 415-421 (1994).

[5] Kobe B, Kajava AV. The leucine-rich repeat as a protein recognition motif. *Curr. Opin. Struct. Biol.*, 11(6), 725-732 (2001).

[6] Matsushima N, Enkhbayar P, Kamiya M, Osaki M, Kretsinger R. Leucine-Rich Repeats (LRRs): Structure, function, evolution and interaction with ligands. *Drug Design Reviews*, 2(4), 305-322 (2005).

[7] Matsushima N, Tachi N, Kuroki Y *et al.* Structural analysis of leucine-rich-repeat variants in proteins associated with human diseases. *Cell Mol. Life Sci.*, 62(23), 2771-2791 (2005).

[8] Bella J, Hindle KL, McEwan PA, Lovell SC. The leucine-rich repeat structure. *Cell Mol. Life Sci.*, 65(15), 2307-2333 (2008).

[9] Kajava AV. Structural diversity of leucine-rich repeat proteins. *J. Mol. Biol.*, 277(3), 519-527 (1998).

[10] Ohyanagi T, Matsushima N. Classification of tandem leucine-rich repeats within a great variety of proteins. *FASEB J.*, 11, A949 (1997).

[11] Matsushima N, Miyashita H, Mikami T, Kuroki Y. A nested leucine rich repeat (LRR) domain: the precursor of LRRs is a ten or eleven residue motif. *BMC Microbiol.*, 10, 235 (2010).

[12] Matsushima N, Tanaka T, Enkhbayar P *et al.* Comparative sequence analysis of leucine-rich repeats (LRRs) within vertebrate toll-like receptors. *BMC Genomics*, 8, 124 (2007).

[13] Kim HM, Park BS, Kim JI *et al.* Crystal structure of the TLR4-MD-2 complex with bound endotoxin antagonist Eritoran. *Cell*, 130(5), 906-917 (2007).

[14] Montgomerie S, Sundararaj S, Gallin WJ, Wishart DS. Improving the accuracy of protein secondary structure prediction using structural alignment. *BMC Bioinformatics*, 7, 301 (2006).

[15] Pollastri G, Przybylski D, Rost B, Baldi P. Improving the prediction of protein secondary structure in three and eight classes using recurrent neural networks and profiles. *Proteins*, 47(2), 228-235 (2002).

[16] Jin MS, Kim SE, Heo JY *et al.* Crystal structure of the TLR1-TLR2 heterodimer induced by binding of a tri-acylated lipopeptide. *Cell*, 130(6), 1071-1082 (2007).

[17] Bell JK, Botos I, Hall PR *et al.* The molecular structure of the Toll-like receptor 3 ligand-binding domain. *Proc. Natl. Acad. Sci. USA*, 102(31), 10976-10980 (2005).

[18] Choe J, Kelker MS, Wilson IA. Crystal structure of human toll-like receptor 3 (TLR3) ectodomain. *Science*, 309(5734), 581-585 (2005).

[19] Kostyukova AS. Tropomodulin/tropomyosin interactions regulate actin pointed end dynamics. *Adv. Exp. Med. Biol.*, 644, 283-292 (2008).

[20] Littlefield RS, Fowler VM. Thin filament length regulation in striated muscle sarcomeres: pointed-end dynamics go beyond a nebulin ruler. *Semin Cell Dev. Biol.*, 19(6), 511-519 (2008).

[21] Krieger I, Kostyukova A, Yamashita A, Nitanai Y, Maeda Y. Crystal structure of the C-terminal half of tropomodulin and structural basis of actin filament pointed-end capping. *Biophys. J.*, 83(5), 2716-2725 (2002).

[22] Lu S, Symersky J, Li S *et al.* Structural genomics of Caenorhabditis elegans: crystal structure of the tropomodulin C-terminal domain. *Proteins*, 56(2), 384-386 (2004).

[23] Dharmasiri N, Dharmasiri S, Estelle M. The F-box protein TIR1 is an auxin receptor. *Nature*, 435(7041), 441-445 (2005).

[24] Kepinski S, Leyser O. The Arabidopsis F-box protein TIR1 is an auxin receptor. *Nature*, 435(7041), 446-451 (2005).

[25] Tan X, Calderon-Villalobos LI, Sharon M *et al.* Mechanism of auxin perception by the TIR1 ubiquitin ligase. *Nature*, 446(7136), 640-645 (2007).

[26] Hayashi K, Tan X, Zheng N *et al.* Small-molecule agonists and antagonists of F-box protein-substrate interactions in auxin perception and signaling. *Proc. Natl. Acad. Sci. USA*, 105(14), 5632-5637 (2008).
[27] Mollnes TE, Christiansen D, Brekke OL, Espevik T. Hypothesis: combined inhibition of complement and CD14 as treatment regimen to attenuate the inflammatory response. *Adv. Exp. Med. Biol.*, 632, 253-263 (2008).
[28] Kim JI, Lee CJ, Jin MS *et al.* Crystal structure of CD14 and its implications for lipopolysaccharide signaling. *J. Biol. Chem.*, 280(12), 11347-11351 (2005).
[29] Vassart G, Pardo L, Costagliola S. A molecular dissection of the glycoprotein hormone receptors. *Trends Biochem. Sci.*, 29(3), 119-126 (2004).
[30] Bogerd J. Ligand-selective determinants in gonadotropin receptors. *Mol. Cell Endocrinol*, 260-262, 144-152 (2007).
[31] Caltabiano G, Campillo M, De Leener A *et al.* The specificity of binding of glycoprotein hormones to their receptors. *Cell Mol. Life Sci*, 65(16), 2484-2492 (2008).
[32] Fan QR, Hendrickson WA. Structure of human follicle-stimulating hormone in complex with its receptor. *Nature*, 433(7023), 269-277 (2005).
[33] Sanders J, Chirgadze DY, Sanders P *et al.* Crystal structure of the TSH receptor in complex with a thyroid-stimulating autoantibody. *Thyroid*, 17(5), 395-410 (2007).
[34] Ikeda H, Lethe B, Lehmann F *et al.* Characterization of an antigen that is recognized on a melanoma showing partial HLA loss by CTL expressing an NK inhibitory receptor. *Immunity*, 6(2), 199-208 (1997).
[35] Williams JM, Chen GC, Zhu L, Rest RF. Using the yeast two-hybrid system to identify human epithelial cell proteins that bind gonococcal Opa proteins: intracellular gonococci bind pyruvate kinase via their Opa proteins and require host pyruvate for growth. *Mol. Microbiol*, 27(1), 171-186 (1998).
[36] Gregory SG, Barlow KF, McLay KE *et al.* The DNA sequence and biological annotation of human chromosome 1. *Nature*, 441(7091), 315-321 (2006).
[37] Carninci P, Kasukawa T, Katayama S *et al.* The transcriptional landscape of the mammalian genome. *Science*, 309(5740), 1559-1563 (2005).
[38] Okazaki Y, Furuno M, Kasukawa T *et al.* Analysis of the mouse transcriptome based on functional annotation of 60,770 full-length cDNAs. *Nature*, 420(6915), 563-573 (2002).
[39] Wadelin F, Fulton J, McEwan PA, Spriggs KA, Emsley J, Heery DM. Leucine-rich repeat protein PRAME: expression, potential functions and clinical implications for leukaemia. *Mol. Cancer*, 9, 226).
[40] Wadelin F, Fulton J, McEwan PA, Spriggs KA, Emsley J, Heery DM. Leucine-rich repeat protein PRAME: expression, potential functions and clinical implications for leukaemia. *Mol Cancer*, 9, 226 (2010).
[41] Song JL, Wessel GM. How to make an egg: transcriptional regulation in oocytes. *Differentiation*, 73(1), 1-17 (2005).
[42] Minami N, Aizawa A, Ihara R, Miyamoto M, Ohashi A, Imai H. Oogenesin is a novel mouse protein expressed in oocytes and early cleavage-stage embryos. *Biol. Reprod.*, 69(5), 1736-1742 (2003).
[43] Tsukamoto S, Ihara R, Aizawa A *et al.* Oog1, an oocyte-specific protein, interacts with Ras and Ras-signaling proteins during early embryogenesis. *Biochem. Biophys. Res. Commun.*, 343(4), 1105-1112 (2006).

[44] Dade S, Callebaut I, Mermillod P, Monget P. Identification of a new expanding family of genes characterized by atypical LRR domains. Localization of a cluster preferentially expressed in oocyte. *FEBS Lett.*, 555(3), 533-538 (2003).

[45] Finst RJ, Kim PJ, Griffis ER, Quarmby LM. Fa1p is a 171 kDa protein essential for axonemal microtubule severing in Chlamydomonas. *J. Cell Sci.*, 113 (Pt 11), 1963-1971 (2000).

[46] Worden AZ, Lee JH, Mock T *et al.* Green evolution and dynamic adaptations revealed by genomes of the marine picoeukaryotes Micromonas. *Science*, 324(5924), 268-272 (2009).

[47] Chereau D, Boczkowska M, Skwarek-Maruszewska A *et al.* Leiomodin is an actin filament nucleator in muscle cells. *Science*, 320(5873), 239-243 (2008).

[48] Conley CA, Fritz-Six KL, Almenar-Queralt A, Fowler VM. Leiomodins: larger members of the tropomodulin (Tmod) gene family. *Genomics*, 73(2), 127-139 (2001).

[49] Xiao W, Jang J. F-box proteins in Arabidopsis. *Trends Plant Sci.*, 5(11), 454-457 (2000).

[50] Tarakanov AO, Fuxe KG, Agnati LF, Goncharova LB. Possible role of receptor heteromers in multiple sclerosis. *J. Neural. Transm*, 116(8), 989-994 (2009).

[51] Ng ACY, Eisenberg JM, Heath RJW *et al.* Human leucine-rich repeat proteins: a genome-wide bioinformatic categorization and functional analysis in innate immunity. *Proc. Natl. Acad. Sci. USA*, 108, 4631-4638 (2011).

[52] Matsushima N, Mikami T, Tanaka T, Miyashita H, Yamada K, Kuroki Y. Analyses of non-leucine-rich repeat (non-LRR) regions intervening between LRRs in proteins. *Biochim. Biophys. Acta*, 1790(10), 1217-1237 (2009).

[53] Torii KU. Leucine-rich repeat receptor kinases in plants: structure, function, and signal transduction pathways. *Int. Rev. Cytol.*, 234, 1-46 (2004).

[54] van der Hoorn RA, Wulff BB, Rivas S *et al.* Structure-function analysis of cf-9, a receptor-like protein with extracytoplasmic leucine-rich repeats. *Plant Cell*, 17(3), 1000-1015 (2005).

[55] Fritz-Laylin LK, Krishnamurthy N, Tor M, Sjolander KV, Jones JD. Phylogenomic analysis of the receptor-like proteins of rice and Arabidopsis. *Plant Physiol.*, 138(2), 611-623 (2005).

[56] Wang G, Ellendorff U, Kemp B *et al.* A genome-wide functional investigation into the roles of receptor-like proteins in Arabidopsis. *Plant Physiol.*, 147(2), 503-517 (2008).

[57] Li J, Chory J. A putative leucine-rich repeat receptor kinase involved in brassinosteroid signal transduction. *Cell*, 90(5), 929-938 (1997).

[58] Nomura T, Bishop GJ, Kaneta T, Reid JB, Chory J, Yokota T. The LKA gene is a BRASSINOSTEROID INSENSITIVE 1 homolog of pea. *Plant J.*, 36(3), 291-300 (2003).

[59] Scheer JM, Ryan CA, Jr. The systemin receptor SR160 from Lycopersicon peruvianum is a member of the LRR receptor kinase family. *Proc. Natl. Acad. Sci. USA*, 99(14), 9585-9590 (2002).

[60] Montoya T, Nomura T, Farrar K, Kaneta T, Yokota T, Bishop GJ. Cloning the tomato curl3 gene highlights the putative dual role of the leucine-rich repeat receptor kinase tBRI1/SR160 in plant steroid hormone and peptide hormone signaling. *Plant Cell*, 14(12), 3163-3176 (2002).

[61] Matsubayashi Y, Ogawa M, Morita A, Sakagami Y. An LRR receptor kinase involved in perception of a peptide plant hormone, phytosulfokine. *Science*, 296(5572), 1470-1472 (2002).

[62] Jones DA, Thomas CM, Hammond-Kosack KE, Balint-Kurti PJ, Jones JD. Isolation of the tomato Cf-9 gene for resistance to Cladosporium fulvum by transposon tagging. *Science*, 266(5186), 789-793 (1994).

[63] Kruijt M, Kip DJ, Joosten MH, Brandwagt BF, de Wit PJ. The Cf-4 and Cf-9 resistance genes against Cladosporium fulvum are conserved in wild tomato species. *Mol. Plant Microbe Interact*, 18(9), 1011-1021 (2005).

[64] Jeong S, Trotochaud AE, Clark SE. The Arabidopsis CLAVATA2 gene encodes a receptor-like protein required for the stability of the CLAVATA1 receptor-like kinase. *Plant Cell*, 11(10), 1925-1934 (1999).

[65] Taguchi-Shiobara F, Yuan Z, Hake S, Jackson D. The fasciated ear2 gene encodes a leucine-rich repeat receptor-like protein that regulates shoot meristem proliferation in maize. *Genes Dev.*, 15(20), 2755-2766 (2001).

[66] de Kock, J.D. M, Brandwagt *et al.* The tomato Orion locus comprises a unique class of Hcr9 genes. *Molecular Breeding*, 15, 409-422 (2005).

[67] Chow B, McCourt P. Plant hormone receptors: perception is everything. *Genes Dev.*, 20(15), 1998-2008 (2006).

[68] Kinoshita T, Cano-Delgado A, Seto H *et al.* Binding of brassinosteroids to the extracellular domain of plant receptor kinase BRI1. *Nature*, 433(7022), 167-171 (2005).

[69] Shinohara H, Ogawa M, Sakagami Y, Matsubayashi Y. Identification of ligand binding site of phytosulfokine receptor by on-column photoaffinity labeling. *J. Biol. Chem.*, 282(1), 124-131 (2007).

[70] Wang G, Long Y, Thomma BP, de Wit PJ, Angenent GC, Fiers M. Functional analyses of the CLAVATA2-like proteins and their domains that contribute to CLAVATA2 specificity. *Plant Physiol.*, 152(1), 320-331).

[71] Enkhbayar P, Kamiya M, Osaki M, Matsumoto T, Matsushima N. Structural principles of leucine-rich repeat (LRR) proteins. *Proteins*, 54(3), 394-403 (2004).

[72] Hindle KL, Bella J, Lovell SC. Quantitative analysis and prediction of curvature in leucine-rich repeat proteins. *Proteins*, 77(2), 342-358 (2009).

[73] Wei T, Gong J, Jamitzky F, Heckl WM, Stark RW, Rossle SC. LRRML: a conformational database and an XML description of leucine-rich repeats (LRRs). *BMC Struct. Biol*, 8, 47 (2008).

In: Bioinformatics
Editor: Renu Tuteja

ISBN: 978-1-62100-913-9

Chapter IV

MicroRNA Profiling Involved in Human Tumorigenesis Using Bioinformatics Tools

Minal Garg*
Department of Biochemistry, University of Lucknow,
U.P., India

Abstract

MicroRNAs (miRNAs) are the class of 21-23 nucleotide small non coding RNAs (ncRNAs), play fundamental roles in the regulation of gene expression in many biological processes including development, proliferation, apoptosis, stress response, and fat metabolism. Mutations of miRNAs, the dysfunction of miRNA biogenesis, misregulation of miRNAs or disruption of their target sites have been implicated in various human diseases including cancer. This chapter focuses on the application of various computational tools used to identify putative miRNAs throughout the human genome and their associations with cancer associated genomic regions (CAGRs) and fragile sites (FRAs). Bioinformatics studies are being increasingly used in order to identify miRNA expression profile and discover specific molecular signatures for cancer diagnosis, prognosis and response to treatment modalities. The implementation of new algorithms and tools specifically focused on characterization of miRNA functions would enable physicians to gain insight into the heterogeneity of the disease and will help in designing the important therapeutic strategies using anti-miRNAs and antagomiRs in the treatment of human cancer.

Keywords: miRNAs, Cancer associated genomic regions, fragile sites, Bioinformatics tools, miRNA profiling and cancer

---

* Corresponding Author: Dr. Minal Garg*, Ph.D., Assistant Professor, Department of Biochemistry, University of Lucknow, Lucknow – 226007, U.P., INDIA. e-mail: minal14@yahoo.com. Phone no.: 91-09335820857.

1. Introduction

MicroRNAs (miRNAs) are the 21-23 nucleotide short non coding RNA sequences, many of them are evolutionary conserved among distantly related organisms and some are expressed in tissue-specific or developmental stage-specific manner. miRNAs are normally expressed as polycistronic transcripts. They regulate mRNA function at post transcriptional as well as post translational level and play key role in various fundamental biological processes such as cell cycle, cell growth and differentiation, apoptosis and embryo development, cardiac and immune system function [1]. miRNAs were discovered way back in 1993 by Victor Ambros, Rosalind Lee and Rhonda Feinbaum, during a study of the gene *lin-14* in *Caenorhabditis elegans* development. Partial binding of 61 nucleotide precursor from lin-4 gene matured to a 22 nucleotide to complementary sequences in the 3' UTR of the lin-14 mRNA inhibited the translation of lin-14 mRNA [2]. Over a brief span of seven years in 2000, second miRNA was characterized as let-7, which repressed lin-41, lin-14, lin-28, lin-42, and daf-12 expression during developmental stage transitions in *C. elegans* [3]. This is followed by the identification of many more. Based on computational and or experimental evidences, around 700 miRNAs hairpin sequences are currently known to be contained in the miRNA registry (miRBase, release 12.0) of which 92% have been experimentally verified [4]. Study by MiRanda et al., 2006 provides a recent estimate of total number of miRNA genes in the human genome in the range of approximately 55,000 and still many of them are yet to be experimentally verified [5]. More than 5300 human genes are supposed to be regulated by miRNA which accounts to 30% of all the genes and around 60% of protein non coding genes.

1.1. Biogenesis of miRNA

Much progress has been made towards an understanding of the mechanisms involved in RNA mediated gene silencing. Several studies have come up that have shown large number of miRNA-disease associations including cancer; cardiovascular disease; schizophrenia; renal function disorders; Tourette's syndrome; psoriasis; primary muscular disorders; Fragile-X mental retardation syndrome; chronic hepatitis; polycythemia vera; AIDS; and obesity [6-17]. In order to discover the links between the mechanisms of RNA mediated silencing and human diseases, it is important to discuss the miRNA biogenesis pathway (Figure 1).

Micro-RNAs are either produced from their own genes or from introns. The miRNA genes are usually transcribed by RNA polymerase II while those with upstream Alu sequences, transfer RNAs (tRNAs), and mammalian wide interspersed repeat (MWIR) promoter units by RNA polymerase III [18]. This is followed by capping, polyadenylation, and splicing to form the primary miRNA (pri-miRNA) having hundreds or thousands of nucleotides and one or more miRNA stem loops [19]. Hairpin loop structures of around 70 nucleotides in length are cleaved with the enzyme Drosha, RNA III endonuclease and DGCR8 (double stranded RNA binding protein) that together form a microprocessor complex or Pasha in invertebrates to form resulting hairpin, known as a pre-miRNA (precursor-miRNA) which has 5' phosphate and 2 nucleotide 3' overhang [20]. Pre-miRNAs that are spliced directly out of introns are known as miRtrons.

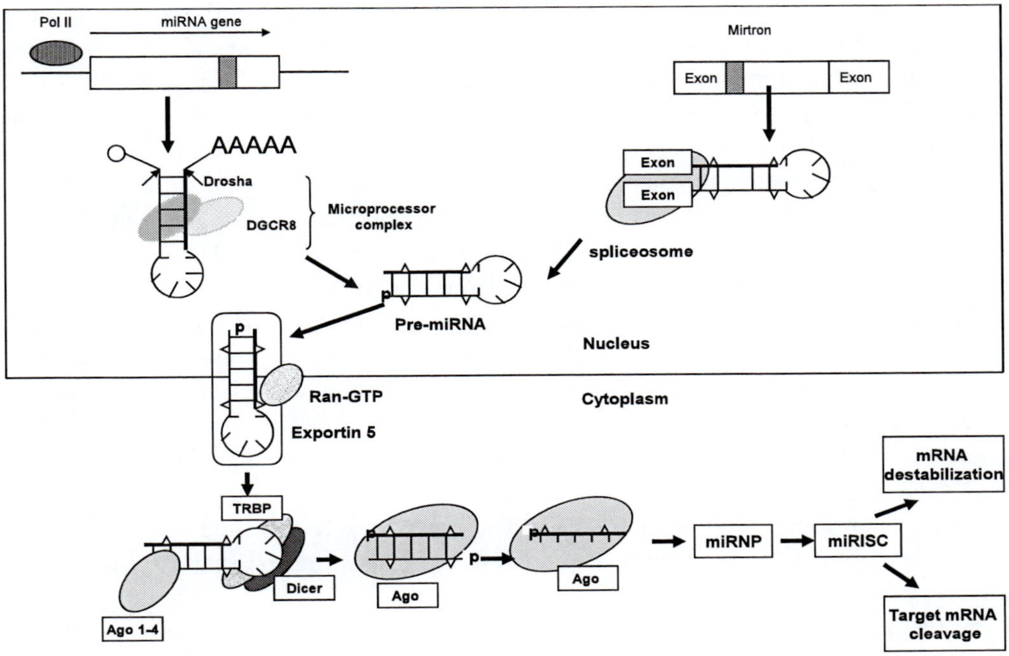

Figure 1. MicroRNAs are produced from either their own genes or from introns. Large primary transcripts known as Pri-miRNA are processed inside the nucleus and is mediated by Drosha, along with DGCR8, together form the microprocessor complex. The strands of hairpin stem at sites near the base of primary stem loop are cleaved, thereby a 60-70 nucleotide pre-**miRNA having 5' phosphates and 2 nucleotide 3' overhangs is released.** Pre-miRNA can be spliced directly out of introns known as miRtrons. Exportin 5 mediates pre-miRNA transfer to cytoplasm in an energy dependent manner where Dicer along with RBP partner TRBP cleaves it to form 21-23 nucleotide mature miRNAs. This is followed by incorporation of it into Ago 1-4 forming miRNPs also known as miRISC where miRNA and its mRNA target interact. Gene silencing may occur via direct mRNA degradation by cleavage with Ago 2 or preventing mRNA from being translated.

Processed pre-miRNA are then exported from the nucleus, an Ran (*RA*s-related *N*uclear protein) -GTP dependent process via nucleocytoplasmic shuttle Exportin-5 [21]. Pre-miRNA then subsequently cleaved by another RNA III endonuclease known as Dicer in cytoplasm in partnership with its TRBP [human immunodeficiency virus (HIV) transactivating response RNA (TAR) binding protein], an RNA binding protein to form a final product of 21-23 **nucleotide miRNA with 5'phosphates** and a **2-nucleotide 3' overhangs [22]. Either strand of** the duplex mature miRNAs are incorporated into the members of argonaute (Ago) protein family, Ago 1–4, forming miRNPs (microRNA ribonucleoprotein complex) along with other proteins such as GW182 and known as miRISC (miRNA-induced silencing complex). miRNPs either direct miRNA-mediated destabilization through interaction with CCR4 (C-C chemokine receptor type 4), or translational repression of miRNAs through interaction with ribosomes without perfectly complementary mRNA targets [23-24]. Mutations of miRNAs, the dysfunction of miRNA biogenesis, misregulation of miRNAs or disruption of their target sites especially near the cancer associated regions may have oncogenic consequences [25].

In order to evaluate the miRNAs as potential therapeutic targets and novel biomarkers for cancer detection, it is important to understand their regulatory functions throughout the human genome. Location of miRNAs at or near cancer associated regions, epigenetic regulation of miRNA expression and abnormalities in miRNA processing genes and proteins strongly correlate abnormalities in miRNA expression with cancer [26]. miRNAs and their linkage to cancer can be studied by identifying its genomic localization, transcriptional regulation, processing steps and post-transcriptional and post-translational modification. The involvement of miRNAs in the pathogenesis of cancer in humans was not known until a study came up in 2002 by Calin and colleagues. About 52.5% of miRNA genes are located in cancer associated genomic regions (CAGRs) or fragile sites (FRAs). The study by Calin et al., reports the association of miR-15 and miR-16 encoded on chromosome 13q14 in B cells with the prognosis in patients with chronic lymphocytic leukemia (CLL) [27]. Some other examples include amplification of miR-17-92 cluster in lymphoma, translocation of miR-17-92 in T cell acute lymphoblastic leukemia (T-ALL) and amplification of miR-26a in glioblastoma [28-30]. Table 1 lists some of the common miRNAs associated with human malignancies. MicroRNAs function as normal miRNAs but upon alterations could become tumor suppressor miRNAs like let-7, miR-16 families or oncogenic miRNAs like miR-21 depending upon their location in CAGRs. They affect protein coding targets and cancer development (Figure 2). The reports on miRNA and cancer are exponentially increasing in pubmed and is close to more than 3500 articles.

miRNAs are the important targets of tumour suppressor and oncogenes and therefore, contribute to the initiation and progression of human cancer. In order to understand the mechanisms for the pathogenesis of cancer, early detection, accurate prediction of an individual's response to a specific treatment and forecast the outcomes of cancer after therapy, it is important to analyse the diagnostic, prognostic and predictive values of miRNAs as genetic signatures in human cancer.

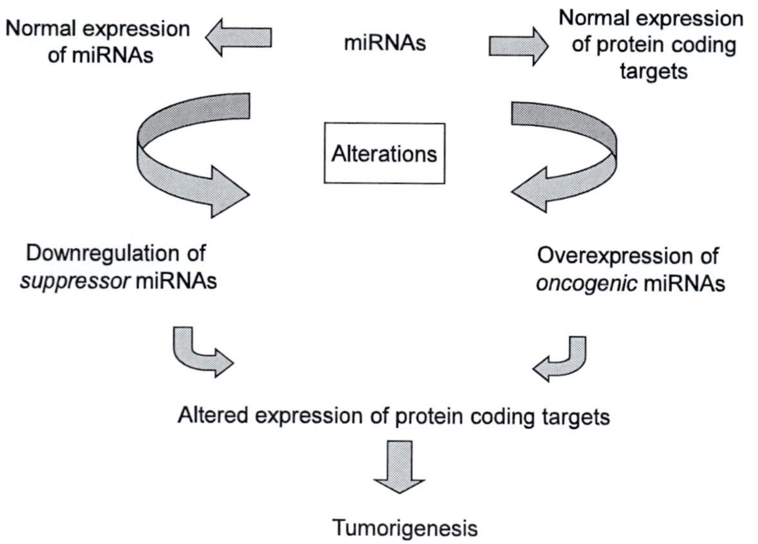

Figure 2. Micro-RNAs or ncRNAs, depending upon their location in cancer associated genomic regions (CAGRs) and or fragile sites (FRAs), could be classified as oncogenic and or suppressor miRNAs are derived from normal miRNAs upon genomic alterations. These miRNAs are either highly expressed or downregulated and affect expression of protein coding targets thus play important role in cancer pathogenesis.

Table 1. miRNAs, their chromosomal locations and association with malignancies

miRNA	Chromosomal location	Type of chromosomal instability	Tissue type specificity	Malignancy
15a/16-1 cluster	13q14.2	Deletion	Ubiquitous	Chronic B cell lymphocytic leukemia (B-CLL) [48], Multiple myeloma [49], Mantle cell lymphoma [49], Pituitary adenoma [48], prostate [48], Pancreatic [50]
miR-17-92 family	7, 13, X	Amplification *Translocation*	Ubiquitous	Lymphoma [28], Leukaemia [51], multiple myeloma, lung [47], colon [47],
miR-21	17q23.1	Amplification	Ubiquitous	medulloblastoma [52], breast [47], prostate [53], renal cell [54], *T cell acute lymphoblastic leukaemia (T-ALL) [30]*
miR-200/141 family	1,12	Amplification Deletion/ Amplification	Epithelial specific	Ovarian [55], pancreatic, prostate [56-57], Liver [58], Gastric [59], cervical [59], head and neck [59], colorectal [59], glioblastoma [59], Breast [59]
miR-34a/b/c	1p36.22(a) 11q23.1(b) 11q23.1(c)	*Deletion*	Ubiquitous	Endometrial endometrioid, breast [55], prostate [57], Lung [60], ovarian [61], *Anaplastic thyroid carcinomas [62], Gastric [63]*
		Deletion	Ubiquitous	Gastric [64], Pancreatic [65], Glioma [66], Leukaemia [67], prostate [57], Breast [55], Lung [68], Ovarian [69], Osteosarcoma [70]
miR *let-7* family	3, 9, 11, 19, 21, 22	Deletion	Haematopoietic system	Lung [71](Takamizawa et al., 2004), Breast [55], prostate [57], Gastric [72], Colon [72], Acute myeloid leukaemia [73], Melanoma [72], Lymphoma [74]
miR-155	21q21.3	Amplification		Pediatric Burkitt lymphoma [75], Breast [47], lung [47], colon [47], pancreatic adenocarcinoma [76]

1.2. Methods for miRNA Profiling

In order to discover novel approaches for cancer therapies as well as individualization of therapy that has significantly and positively impacted survival outcomes, it is important to identify specific signatures associated with diagnosis, prognosis, and response to treatment of human tumors. Large fraction of genomic ultra conserved regions (UCRs) encoding miRNAs have been identified to be frequently located in or near cancer associated genomic regions with altered expression in human cancers. The main methods for miRNA profiling are based on experimental and computational approaches [4]. miRNA databases that collect and analyse large amount of genetic information need to be integrated with biomedical and clinical information. This review further emphasizes on the application of bioinformatics as an important tool to correlate the information contained in miRNA databases with clinical parameters in order to find their significance in cancer diagnosis and prognosis.

Identification of genes responsible for cancer predisposition is primarily required for better diagnosis. Currently available experimental approaches for miRNA profiling require high quality preserved fresh-frozen samples or formalin-fixed paraffin-embedded tissues (FFPE). RNA and protein extracted from tissues samples can be analysed for miRNA expression which clearly shows fold changes among healthy and cancerous tissues using extensive microarray or quantitative real time PCR. Microarray expression data set can be used to predict the diagnostic categories of cancerous or multiclass tumour samples. Among these, *in situ* hybridization (ISH) allows sensitive detection of miRNA in heterogenous tissues. These methods are lower throughput and require standardization [31]. Inefficiencies are also due to wrong selection of genomic region for investigation, or difficulties in isolating certain miRNAs by cloning due to low expression, stability, tissue specificity and technical failures of the cloning procedure. Besides this, tracing large family lineages for locating abnormal chromosomal gene region is difficult as well as time consuming. All these problems make these approaches complex and suboptimal [4].

Advancements in biological-computational approaches and necessity to retrieve, share and understand vast amount of data, have led to development of genome databases, providing an open source of genetic information. Recent reports indicate that integration of experimental knowledge and computational methods with effective miRNA (features) and tissues (samples) can enhance the prediction accuracy by increasing intra class and decreasing inter class similarity among tumour versus non tumour samples [32]. This chapter discusses some of the commonly used bioinformatics tools to identify putative miRNAs throughout the human genome, association with cancer genomic regions, their potential involvement as diagnostic and therapeutic markers in human malignancies.

1.2.1. Identification of miRNAs Genes

Putative miRNAs are identified via the use of either supervised algorithms that are trained on known miRNA general biological features like sequence composition, secondary structure and species conservation or un-supervised algorithms such as alignment or conservation. Other multiple biological features such as free energy, paired bases, loop length and stem conservation can also be used to predict novel miRNAs [4]. Recently, freely available prediction tool, SSCprofiler at the web service http://www.imbb.forth.gr/SSC profiler.html, where *Profile* HMMs (Hidden Markov Models), statistical methods for multiple sequence alignments simultaneously considers sequence, structure and conservation features of miRNA genes. All the three features used to train HMMs and eight filtering parameters including Hairpin = 1; **Bulges ≤16; Loops ≤32; Asymmetry ≤13; Bulges-loops ≤37; Hairpin length ≤16; Folding min energy ≤−25.44 kcal/mol; Conservation ≥25% of nucleotides** conserved, SSCprofiler achieves a very high performance accuracy of 88.95% sensitivity and 84.16% specificity on a large set of human miRNA genes. Sequential consideration of sequence, structure and conservation at the nucleotide level offer an important advantage to SSCprofiler as compared to other existing supervised machine-learning algorithms including MiRRim, BayesMiRNAfind (Bayes classifier) and TripletSVM (SVM classifier) in order to identify novel miRNAs [4]. Due to imbalance between the number of positive (178 human pre-miRNAs) and negative (5395 random sequences from tRNA, rRNA, and mRNA genes) samples, SVM classifier model obtained high specificity (91%) and low sensitivity (71%). Compared with similar methods, MiRFinder has been observed as high throughput miRNA prediction tool which is based on search for hairpin candidates and exclusion of the nonrobust structures based on the parameters analysed by the SVM including the local secondary structure differences of the stem region of miRNA and the secondary structures of pre-miRNAs [33]. miPred, another algorithm is based on contiguous structure-sequence composition, minimum of free energy of the secondary structure. Table 2 summarizes some of the commonly used methods for finding human miRNA and their comparative testing in terms of percent sensitivity and percent specificity. These methods rely on genome annotation and sufficient number of characterized miRNAs as training samples to reduce the number of predicted putative miRNAs.

Table 2. Commonly used algorithms for finding miRNA genes in human genome

Alogrithm	Classifiers	Sensitivity	Specificity	Web server	Reference
MiR-abela	SVM	71%	97%	http://www.miRz.unibas.ch/cgi/pred_miRNA_genes.cgi	[77]
Triplet-SVM (Support vector machine)	SVM	93.3%	88.1%	http://bioinfo.au.tsinghua.edu.cn/miRnasvm/	[78]
miPred	SVM	84.55	97.97	http://web.bii.a-star.edu.sg/_stanley/Publications	[79]
Microprocessor SVM	SVM	95	90	https://demo1.interagon.com/miRNA/	[80]
SSCProfiler	HMM	88.95	84.16	http://www.imbb.forth.gr/SSCprofiler.html	[4]
miPred	RF	89.3	93.21	http://www.bioinf.seu.edu.cn/miRNA/	[81]
MiRFinder	SVM	99.6	70	http://www.bioinformatics.org/miRfinder	[33]

1.2.2. miRNAs and Cancer Associated Genomic Regions (CAGRs)

Among thousands of miRNA genes, also known as UCRs, only a small proportion of them are found to be implicated in human tumorigenesis. Computational technologies have allowed the geneticist to determine the location of miRNAs and/or UCRs in fragile sites, in regions involved in cancers, or in tumor susceptibility loci [6, 34, 35]. To identify the precise location of miRNA, BLAST search is done against human genome and BLAST jobs are then submitted automatically to genome resources at the NCBI (http://www.ncbi.nlm.nih.gov) using various modules like LWP, HTML, and HTTP, PERL and BIOPERL. New database can be established having combined information regarding the positions of fragile sites and cancer associated regions with the positions of known and predicted miRNAs throughout the human genome. These informations are directly extracted manually from published work and then compare it with already compiled in GenBank (http://www.ncbi.nlm.nih.gov) and the miRNA Registry (http://www.microrna.sanger.ac.uk). According to the miRBase of Sanger Institute approximately 700 miRNAs are found in human and about 50% of miRNA genes are localized in cancer associated genomic regions (CAGR) or fragile sites as well as in minimal regions of loss of heterozygosity, minimal regions of amplification, or common breakpoint regions [6].

1.2.3. Predicting miRNA Targets and Cancer

The untranslated regions (3'UTRs) of transcripts having binding sites for them lead to the identification of miRNA targets. TarBase, a manually curated database of experimentally supported miRNA targets, have led to identification of only small proportion of human miRNA targets which are also confirmed experimentally. The efficacy of use of computational tools in predicting miRNA targets is largely supported by the relatively high degree of miRNA complementarity to experimentally determined binding sites. There are three types of target sites including (a) 5'-dominant canonical having good or perfect complementarity at both the 5' and 3' ends of the miRNA, with a characteristic bulge in the middle; (b) Dominant seed sites have perfect seed 5' complementarity to the miRNA but poor 3' complementarity. (c) Compensatory sites have a mismatch or wobble in the 5' seed region but compensate through excellent complementarity at the 3' end. Since these target sites differ in level of complementarity of miRNA sequences to the site sequences, therefore, the tools are designed to look for sequence complementarity and/or for favorable miRNA– target duplex thermodynamics. MiRanda algorithm developed for miRNA target prediction uses dynamic programming to search for optimal sequence complementarities between a set of mature miRNAs and a given mRNA [36]. RNAhybrid, although similar to an RNA secondary structure prediction algorithm, but determines the most favourable hybridization sites between the two sequences. Since an evolutionarily conserved target sequences are more likely to be true targets, TargetScanS and TargetScanHuman 5.1 (the latest revised version) scores target sites based on the conservation of the target sequences between five genomes (human, mouse, rat, dog and chicken) and in around 24 species respectively [37-39]. RNA22, another computational tool does not rely upon cross-species conservation but is able to recover most of the known target sites with validation of some of its new predictions [39]. NBmiRTar, another target prediction method, not require sequence conservation but generates a model from sequence and miRNA–mRNA duplex information derived from validated target sequences and artificially generated negative examples.

It produces fewer false-positive predictions and fewer target candidates to be tested than MiRanda [39]. The predicted targets component of miRecords is an integration of predicted miRNA targets produced by 11 established miRNA target prediction programs including DIANA-microT, MicroInspector, miRanda, MiRTarget2, miTarget, NBmiRTar, PicTar, PITA, RNA22, RNAhybrid, and TargetScan/TargertScanS [40]. Table 3 mentions some of the commonly used computational tools and methods used by them to predict miRNA targets.

Table 3. Commonly used algorithms for predicting miRNA targets in human genome

Alogrithm	Description	Web server	Reference
TargetScan	Based on seed complementarity, thermodynamic free energy of binding, conservation over different species	targetscan.org	[82]
TargetScanS	Based on seed complementarity, thermodynamic free energy of binding, conservation over different species	targetscan.org	[37]
TargerScanHuman 5.1	Based on complementarity search for 8mer and 7mer sites that match the seed region of each miRNA.	www.targetscan.org/	[38]
MiRanda	Based on sequence complementarity, free energy of the RNA-RNA duplex, extent of conservation	http://www.microrna.org/	[36]
PicTar	Based on seed complementarity, thermodynamics and a combinatorial prediction for common targets in sets of coexpressed miRNAs	http://pictar.bio.nyu.edu/	[83]
RNAhybrid	Based on the finding the minimum free energy hybridization of a long and a short RNA where the short sequence is hybridized to the best fitting part of the long one	http://bibiserv.techfak.uni-bielefeld.de/rnahybrid/	[84]
DIANA microT	Based on thermodynamics	http://diana.pcbi.upenn.edu/	[85]
miRecords	Based on an integration of predicted miRNA targets produced by 11 established miRNA target prediction programs	http://miRecords.biolead.org/	[40]
TargetMiner	Based on miRNA–mRNA structural interactions and seed-site conservation	www.isical.ac.in/~bioinfo_miu	[86]

Due to the requirements for evolutionary conservation, many of the tools suffer from significant false positive and false negative rates while none of these algorithms utilize the experimentally validated binding sites as training set and provide a genome wide map of predictions that utilizes post-transcriptional regulation within a broad framework of relevant target site prediction features [41]. Recently an online source is developed called ExprTargetDB (http://www.scandb.org/apps/microrna/) by integrating computational algorithm and miRNA expression data generated experimentally, provide a reference dataset of miRNA-mRNA relationships on the HapMap lymphoblastoid cell line samples [41].

Extensive PubMed (http://www.pubmed.gov) search for the overlapping of the reported genomic alterations in human cancers and the regions containing miRNAs identify the significant dysregulation of approximately 200 miRNAs in various cancer malignancies. Table 1 lists some of them.

Cancer-miRNA network has been recently developed for detecting hub miRNAs dysregulated in many cancer types and identifying cancer specific miRNAs. Cancer-miRNA network is being generated by mining the literature of experimentally verified cancer miRNA relationships. Several cancer-miRNA modules have been identified from the network which are based on a computational approach (Association rule mining (ARM)) to mine associations between cancer types and miRNAs that are found to have a number of common predicted target onco/tumor suppressor genes. About 67% of the cancer types have at least two neighboring miRNAs (located within 50 kb of genomic location) showing either downregulation or upregulation and is found to be statistically significant ($p < 10^{-7}$, Randomization test) [42].

1.2.4. miRNAs Expression Levels and Their Biological Characteristics

miRNAs profiling has allowed the identification of specific signatures associated with cancer diagnosis, prognosis and patient's survival in response to therapeutic modalities. In order to validate the involvement of miRNAs, their functions, characteristics, correlations and associations to cancer through targeting proto oncogenes or tumor suppressor genes, various statistical methods are used. In one of the studies by Gaur et al., 2007, data for expression profiles of 241 human miRNAs in normal tissues and the NCI-60 panel of human tumor-derived cell lines were statistically analyzed with leave-one-out sensitivity analyses, Student's t tests, and Spearman rank correlation analysis [43]. The experimentally normalized delta Ct values for the miRNA profiles were used to evaluate the NCI-60 cell lines by agglomerative hierarchical clustering where subclusters reflected miRNA expression as a fundamental tissue characteristic. The expression levels of 241 different miRNAs were further characterized by multiscale bootstrap analysis and identified four highly significant clusters of cell lines, derived from hematologic, colon, melanoma, and CNS tumors. Leave-one-out sensitivity analysis narrows down the expression of miRNA types to identify the tissue-specific clustering of cell lines, thus validating the roles of molecular signatures in mediating tissue-specific differentiation and/or tumorigenesis. Spearman rank correlations then further used to identify not only the association of miRNAs but also their pattern in cell lines. Expression levels of miRNAs can be correlated with cell proliferation indices where elevated expression with longer doubling times, an indicator of candidate antiproliferative RNAs, whereas expression with rapid cell cycling represent a set of potentially acting enhancers or proliferation [43].

Several reports are coming up identifying miRNA expression profiles as diagnostic and prognostic markers of cancer. The cumulative 5-year survival rate of glioma patients is observed as 51.54% in the low miR-182-expression group, whereas only 7.23% in the high miR-182-expression group as well as multivariate Cox regression analysis identifies miR-182 as an independent valuable marker of glioma progression and its high expression is associated with poor overall survival in patients with malignant glioma [44]. Experimental data containing miRNA expression profiles of head and neck squamous cell carcinoma (HNSCC) tumor and adjacent normal tissue were analysed using univariate and multivariable statistical models. Low levels of miR-205 are found to be significantly associated with loco-regional recurrence independent of disease severity at diagnosis and treatment while low levels of miR-205 and hsa-let-7d together in HNSCC tumors are associated with poor head and neck cancer survival [45]. In stage I lung adenocarcinomas, high miR-155 and low hsa-let-7a-2 expression correlated with poor survival by univariate analysis as well as multivariate analysis for miR-155, confirmed by real-time RT-PCR analysis of precursor miRNAs and cross-validated with an independent set of adenocarcinomas [46]. From a large-scale miRnome analysis on 540 samples including lung, breast, stomach, prostate, colon, and pancreatic tumors, sets of miRNAs such as miR-17-5p, miR-20a, miR-21, miR-92, miR-106a, and miR-155 are found to be are extensively involved in cancer pathogenesis of solid tumors [47].

Bioinformatics and statistical tools are highly utilized to crack genome-wide expression profiling of miRNAs and reveal associations between the ncRNAs and CAGRs and/ or FRAs. Several ncRNAs are shown to affect tumor development and progression and therefore, targeting suppressor or oncogenic miRNAs may provide an important therapeutic strategy for treatment of human cancer. The implementation of new algorithms and tools specifically focused on identifying miRNAs and their functions in cancer development may lead to important findings regarding miRNAs as tumor biomarkers and therapeutic targets.

1.3. Conclusions

Reports are coming up regarding the implications of high percentage of miRNAs in cancer due to its close proximity to chromosomal breakpoints, cancer associated genomic regions and or fragile sites and dysregulated expression levels in many malignancies. Differential expression of miRNAs promotes carcinogenesis either by amplifying the expression of proto-oncogenes or by reducing the expression of tumor suppressor genes. Various algorithms and statistical methods are recently being applied and provide a concise biological picture of cancer pathways and genes regulated by novel miRNAs. New bioinformatics and statistical methods are needed for very efficient genome wide search for characterisation of miRNAs and their functional interactions associated with the emergence of cancerous phenotype. These types of studies could help in selecting the best candidates for using miRNAs as potential therapeutic targets in cancer.

Competing Interests

The author declares no competing interest.

References

[1] Kwak PB, Iwasaki S, Tomari Y. The microRNA pathway and cancer. *Cancer Sci.*, 2010; 101(11): 2309-15.

[2] Lee RC, Feinbaum RL, Ambros V. The C. elegans heterochronic gene lin-4 encodes small RNAs with antisense complementarity to lin-14. *Cell,* 1993; 75 (5): 843–54.

[3] Reinhart BJ, Slack FJ, Basson M, Pasquinelli AE, Bettinger JC, Rougvie AE, Horvitz HR, Ruvkun G. The 21-nucleotide let-7 RNA regulates developmental timing in Caenorhabditis elegans. *Nature,* 2000; 403 (6772): 901–6.

[4] Oulas A, Boutla A, Gkirtzou K, Reczko M, Kalantidis K, Poirazi P. Prediction of novel microRNA genes in cancer-associated genomic regions—a combined computational and experimental approach. *Nucleic Acids Res*, 2009; 37(10): 3276-87.

[5] MiRanda KC, Huynh T, Tay Y, Ang YS, Tam WL, Thomson AM, Lim B, Rigoutsos I. A pattern-based method for the identification of MicroRNA binding sites and their corresponding heteroduplexes. *Cell,* 2006; 126: 1203–1217.

[6] Calin G, Sevignani C, Dumitru C, Hyslop T, Noch E, Yendamuri S, Shimizu M, Rattan S, Bullrich F, Negrini M, Croce C. Human microRNA genes are frequently located at fragile sites and genomic regions involved in cancers. *Proc. Natl. Acad. Sci. USA*, 2004; 101: 2999-3004.

[7] Latronico MV, Catalucci D, Condorelli G. Emerging role of microRNAs in cardiovascular biology. *Circ. Res.*, 2007; 101(12):1225-36.

[8] Perkins DO, Jeffries CD, Jarskog LF, Thomson JM, Woods K, Newman MA, Parker JS, Jin J, Hammond SM. microRNA expression in the prefrontal cortex of individuals with schizophrenia and schizoaffective disorder. *Genome Biol*, 2007; 8(2): R27.

[9] Karolina DS, Wintour EM, Bertram J, Jeyaseelan K. Riboregulators in kidney development and function. *Biochimie*, 2010; 92(3): 217-25.

[10] Esau CC, Monia BP. Therapeutic potential for microRNAs. *Adv. Drug Deliv Rev*, 2007; 59(2-3):101-14.

[11] Sonkoly E, Wei T, Janson PC, Sääf A, Lundeberg L, Tengvall-Linder M, Norstedt G, Alenius H, Homey B, Scheynius A, Ståhle M, Pivarcsi A. MicroRNAs: novel regulators involved in the pathogenesis of psoriasis? *PLoS One*, 2007; 2(7): e610.

[12] Chen JF, Callis TE, Wang DZ. microRNAs and muscle disorders. *J. Cell Sci*, 2009; 122(Pt 1):13-20.

[13] Fiore R, Schratt G. MicroRNAs in synapse development: tiny molecules to remember. *Expert Opin. Biol. Ther*, 2007; 7(12):1823-31.

[14] Chen Y, Cheng G, Mahato RI. RNAi for treating hepatitis B viral infection. *Pharm. Res.,* 2008; 25(1): 72-86.

[15] Bruchova H, Yoon D, Agarwal AM, Mendell J, Prchal JT. Regulated expression of microRNAs in normal and polycythemia vera erythropoiesis. *Exp. Hematol*, 2007; 35(11): 1657-67.

[16] Bagasra O, Prilliman KR. RNA interference: the molecular immune system. *J. Mol Histol.*, 2004; 35(6): 545-53.

[17] Heneghan HM, Miller N, Kerin MJ. Role of microRNAs in obesity and the metabolic syndrome. Obes Rev, 2010; 11(5): 354-61.

[18] Faller M, Guo F. MicroRNA biogenesis: there's more than one way to skin a cat. Biochim Biophys Acta, 2008; 1779 (11): 663–667.

[19] Lee Y, Kim M, Han J, Yeom KH, Lee S, Baek SH, Kim VN. MicroRNA genes are transcribed by RNA polymerase II. *EMBO J.*.2004; 23 (20): 4051–60.

[20] Gregory R., Chendrimada T, Shiekhattar R. MicroRNA biogenesis: isolation and characterization of the microprocessor complex. *Methods in molecular biology*, 2006; 342: 33–47.

[21] Murchison, E.; Hannon, G. MiRNAs on the move: miRNA biogenesis and the RNAi machinery. *Current opinion in cell biology*, 2004; 16 (3): 223–229.

[22] Lund E, Dahlberg J. Substrate selectivity of exportin 5 and Dicer in the biogenesis of microRNAs. *Cold Spring Harbor symposia on quantitative biology*, 2006; 71: 59–66.

[23] Schwarz D, Zamore P. Why do miRNAs live in the miRNP? *Genes and Development*, 2002; 16 (9):1025–1031.

[24] Pratt A, MacRae I. The RNA-induced silencing complex: a versatile gene-silencing machine. *The Journal of biological chemistry*, 2009; 284 (27): 17897–17901.

[25] Gregory RI and Shiekhattar R. MicroRNA biogenesis and cancer. *Cancer Res*, 2005; 65: 3509-3512.

[26] Rossi S, Sevignani C, Nnadi SC, Siracusa LD, Calin GA. Cancer-associated genomic regions (CAGRs) and noncoding RNAs: bioinformatics and therapeutic implications. *Mamm Genome*, 2008; 19: 526–540.

[27] Asslaber D, Hofbauer JP, Greil R, Egle A. MicroRNAs as biomarkers for the diagnosis and prognosis of human cancer. *Journal of Nucleic Acids Investigation*, 2010; 1:e14.

[28] Tagawa H, Seto M. A microRNA cluster as a target of genomic amplification in malignant lymphoma. *Leukemia*, 2005; 19: 2013–2016.

[29] Huse JT, Brennan C, Hambardzumyan D, et al. The PTENregulating microRNA miR-26a is amplified in high-grade glioma and facilitates gliomagenesis in vivo. *Genes Dev,* 2009; 23: 1327–1337.

[30] Mavrakis KJ, Wolfe AL, Oricchio E, et al. Genome-wide RNAmediated interference screen identifies miR-19 targets in Notchinduced T-cell acute lymphoblastic leukaemia. *Nat. Cell Biol.*, 2010; 12: 372–379.

[31] Farazi TA, Spitzer JI, Morozov P and Tuschl T. miRNAs in human cancer. *J. Pathol,* 2011; 223: 102–115.

[32] Mitra R, Bandyopadhyay S, Maulik U and Zhang MQ. SFSSClass: an integrated approach for miRNA based tumor classification. *BMC Bioinformatics*, 2010; 11(Suppl 1): S22.

[33] Huang TH, Fan B, Rothschild MF, Hu ZL, Li K et al. MiRFinder: an improved approach and software implementation for genome-wide fast microRNA precursor scans. *BMC Bioinformatics*, 2007; 8: 341.

[34] Makunin IV, Pheasant M, Simons C, Mattick JS. Orthologous microRNA genes are located in cancer-associated genomic regions in human and mouse. *PLoS ONE*, 2007; 2(11): e1133.

[35] Sevignani C, Calin GA, Nnadi SC, Shimizu M, Davuluri RV et al. MicroRNA genes are frequently located near mouse cancer susceptibility loci. *Proc. Natl. Acad. Sci. USA*, 2007; 104: 8017–8022.

[36] Enright AJ, John B, Gaul U, Tuschl T, Sander C, Marks DS. MicroRNA targets in Drosophila. *Genome Biol*, 2003; 5(1): R1.

[37] Lewis B P, Burge C B and Bartel D P. Conserved seed pairing, often flanked by adenosines, indicates that thousands of human genes are microRNA targets. *Cell,* 2005; 120, 15-20.

[38] Friedman RC, Kai-How Farh K, Burge CB, Bartel DP. Most Mammalian mRNAs Are Conserved Targets of MicroRNAs. *Genome Research* 2009; 19:92-105.

[39] Garzon R, Fabbri M, Cimmino A, Calin GA, Croce CM. MicroRNA expression and function in cancer. *Trends Mol. Med.*, 2006; 12(12): 580-7.

[40] Yousef M, Showe L, Showe M. A study of microRNAs in silico and in vivo: bioinformatics approaches to microRNA discovery and target identification. *FEBS J.*, 2009; 276(8): 2150-6.

[41] Xiao F, Zuo Z, Cai G, Kang S, Gao X, Li T miRecords: an integrated resource for microRNA-target interactions. *Nucleic Acids Res*, 2009; 37: D105-D110.

[42] Gamazon ER, Im HK, Duan S, Lussier YA, Cox NJ, Dolan ME, Zhang W. ExprTarget: An Integrative Approach to Predicting Human MicroRNA Targets. *PlOS ONE*, 2010; 5(10): e13534.

[43] Bandyopadhyay S, Mitra R, Maulik U, Zhang MQ. Development of the human cancer microRNA network. *Silence,* 2010; 1(1): 6.

[44] Gaur A, Jewell DA, Liang Y, Ridzon D, Moore JH et al. Characterization of microRNA expression levels and their biological correlates in human cancer cell lines. *Cancer Res,* 2007; 67: 2456–2468.

[45] Jiang L, Mao P, Song L, Wu J, Huang J, Lin C, Yuan J, Qu L, Cheng SY, Li J. miR-182 as a prognostic marker for glioma progression and patient survival. *Am. J. Pathol.,* 2010; 177(1): 29-38.

[46] Childs G, Fazzari M, Kung G, Kawachi N, Brandwein-Gensler M, McLemore M, Chen Q, Burk RD, Smith RV, Prystowsky MB, Belbin TJ, Schlecht NF. Low-level expression of microRNAs let-7d and miR-205 are prognostic markers of head and neck squamous cell carcinoma. *Am. J. Pathol.* 2009; 174(3): 736-45.

[47] Yanaihara N, Caplen N, Bowman E, Seike M, Kumamoto K, Yi M, Stephens RM, Okamoto A, Yokota J, Tanaka T, Calin GA, Liu CG, Croce CM, Harris CC. Unique microRNA molecular profiles in lung cancer diagnosis and prognosis. *Cancer Cell,* 2006; 9(3): 189-98.

[48] Volinia S, Calin GA, Liu CG, Ambs S, Cimmino A, Petrocca F, Visone R, Iorio M, Roldo C, Ferracin M, Prueitt RL, Yanaihara N, Lanza G, Scarpa A, Vecchione A, Negrini M, Harris CC, Croce CM. A microRNA expression signature of human solid tumors defines cancer gene targets. *Proc. Natl. Acad. Sci. USA*, 2006; 103(7): 2257-61.

[49] Aqeilan RI, Calin GA and Croce CM. miR-15a and miR-16-1 in cancer: discovery, function and future perspectives. *Cell Death and Differentiation*, 2010; 17: 215-220.

[50] Lerner M, Harada M, Lovén J, Castro J, Davis Z, Oscier D, Henriksson M, Sangfelt O, Grandér D, Corcoran MM. DLEU2, frequently deleted in malignancy, functions as a critical host gene of the cell cycle inhibitory microRNAs miR-15a and miR-16-1. *Exp. Cell Res.,* 2009; 315(17): 2941-52.

[51] Finnerty JR, Wang WX, Hébert SS, Wilfred BR, Mao G, Nelson PT. The miR-15/107 group of microRNA genes: evolutionary biology, cellular functions, and roles in human diseases. *J. Mol. Biol.* 2010; 402: 491–509.

[52] Cui JW, Li YJ, Sarkar A, Brown J, Tan YH, Premyslova M, Michaud C, Iscove N, Wang GJ, Ben-David Y. Retroviral insertional activation of the Fli-3 locus in erythroleukemias encoding a cluster of microRNAs that convert Epo-induced differentiation to proliferation. *Blood*, 2007; 110 (7): 2631–40.

[53] Northcott PA, Fernandez-L A, Hagan JP, Ellison DW, Grajkowska W, Gillespie Y, Grundy R, Van Meter T, Rutka JT, Croce CM, Kenney AM, Taylor MD. The miR-17/92 polycistron is up-regulated in sonic hedgehog-driven medulloblastomas and induced by N-myc in sonic hedgehog-treated cerebellar neural precursors. *Cancer Research*, 2009; 69(8): 3249-55.

[54] Poliseno L, Salmena L, Riccardi L, et al. Identification of the miR-106b-25 microRNA cluster as a proto-oncogenic PTEN targeting intron that cooperates with its host gene MCM7 in transformation. *Sci. Signal*, 2010; 3: ra29.

[55] Chow TF, Mankaruos M, Scorilas A, Youssef Y, Girgis A, Mossad S, Metias S, Rofael Y, Honey RJ, Stewart R, Pace KT, Yousef GM. The miR-17-92 cluster is over expressed in and has an oncogenic effect on renal cell carcinoma. *J. Urol.*, 2010; 183(2): 743-51.

[56] O'Day E, Lal A. MicroRNAs and their target gene networks in breast cancer. *Breast Cancer Res,* 2010; 12(2): 201.

[57] Ribas J, Lupold SE. The transcriptional regulation of miR-21, its multiple transcripts, and their implication in prostate cancer. *Cell Cycle*, 2010; 9(5): 923-9.

[58] Pang Y, Young CY, Yuan H. MicroRNAs and prostate cancer. *Acta Biochim. Biophys. Sin (Shanghai),* 2010; 42(6): 363-9.

[59] Gramantieri L, Fornari F, Callegari E, Sabbioni S, Lanza G, Croce CM, Bolondi L, Negrini M. MicroRNA involvement in hepatocellular carcinoma. *J. Cell Mol. Med.,* 2008; 12(6A): 2189-204.

[60] Jazbutyte V, Thum T. MicroRNA-21: from cancer to cardiovascular disease. *Curr. Drug Targets*, 2010; 11: 926–935.

[61] Roybal JD, Zang Y, Ahn YH, Yang Y, Gibbons DL, Baird BN, Alvarez CA, Thilaganathan N, Saintigny P, Liu D, Heymach JV, Creighton CJ, Kurie JM. miR-200 Inhibits Lung Adenocarcinoma Cell Invasion and Metastasis by Targeting Flt1/VEGFR1. *Mol. Cancer Res*, 2011; 9(1): 25-35.

[62] Mezzanzanica D, Bagnoli M, De Cecco L, Valeri B, Canevari S. Role of microRNAs in ovarian cancer pathogenesis and potential clinical implications. *Int. J. Biochem. Cell Biol,* 2010; 42(8):1262-72.

[63] Braun J, Hoang-Vu C, Dralle H, Hüttelmaier S. Downregulation of microRNAs directs the EMT and invasive potential of anaplastic thyroid carcinomas. Oncogene, 2010; 29(29): 4237-44.

[64] Shinozaki A, Sakatani T, Ushiku T, Hino R, Isogai M, Ishikawa S, Uozaki H, Takada K, Fukayama M. Downregulation of microRNA-200 in EBV-associated gastric carcinoma. *Cancer Res.*, 2010; 70(11):4719-27.

[65] Qing Ji, Xinbao Hao, Min Zhang, Wenhua Tang, Yang Meng, Ling Li, Debing Xiang, Jeffrey T. DeSano, Guido T. Bommer, Daiming Fan, Eric R. Fearon, Theodore S. Lawrence, Liang Xu. MicroRNA miR-34 Inhibits Human Pancreatic Cancer Tumor-Initiating Cells. *PLoS ONE*, 2009; 4(8): e6816.

[66] Qing Ji, Hao X, Meng Y, Zhang M, DeSano J, Fan D and Xu L. Restoration of tumor suppressor miR-34 inhibits human p53-mutant gastric cancer tumorspheres. *BMC Cancer,* 2008; 8: 266.

[67] Luana S, Sunb L, Huang F. MicroRNA-34a: A Novel Tumor Suppressor in p53-mutant Glioma Cell Line U251. *Archives of Medical research*, 2010; 41(2): 67-74.

[68] Merkel O, Asslaber D, Piñón JD, Egle A, Greil R. Interdependent regulation of p53 and miR-34a in chronic lymphocytic leukemia. *Cell Cycle*, 2010; 9(14): 2764-8.

[69] Cheng J, Zhou L, Xie QF, Xie HY, Wei XY, Gao F, Xing CY, Xu X, Li LJ, Zheng SS. The impact of miR-34a on protein output in hepatocellular carcinoma HepG2 cells. *Proteomics,* 2010; 10(8): 1557-72.

[70] Corney DC, Hwang CI, Matoso A, Vogt M, Flesken-Nikitin A, Godwin AK, Kamat AA, Sood AK, Ellenson LH, Hermeking H, Nikitin AY. Frequent downregulation of miR-34 family in human ovarian cancers. *Clin. Cancer Res.*, 2010; 16(4): 1119-28.

[71] He C, Xiong J, Xu X, Lu W, Liu L, Xiao D, Wang D. Functional elucidation of MiR-34 in osteosarcoma cells and primary tumor samples. *Biochem. Biophys. Res. Commun*, 2009; 388(1): 35-40.

[72] Takamizawa J, Konishi H, Yanagisawa K, et al. Reduced expression of the let-7 microRNAs in human lung cancers in association with shortened postoperative survival. *Cancer Res.*, 2004; 64: 3753–6.

[73] Peter ME. Let-7 and miR-200 microRNAs: guardians against pluripotency and cancer progression. *Cell Cycle*, 2009; 8: 843–852.

[74] Cammarata G, Augugliaro L, Salemi D, Agueli C, La Rosa M, Dagnino L, Civiletto G, Messana F, Marfia A, Bica MG, Cascio L, Floridia PM, Mineo AM, Russo M, Fabbiano F, Santoro A. Differential expression of specific microRNA and their targets in acute myeloid leukemia. *Am. J. Hematol*, 2010; 85(5): 331-9.

[75] Roush S, Slack FJ. The let-7 family of microRNAs. *Trends Cell Biol*, 2008; 18: 505–516.

[76] Metzler M, Wilda M, Busch K, Viehmann S, Borkhardt A. High expression of precursor microRNA-155/BIC RNA in children with Burkitt lymphoma. *Genes Chromosomes Cancer*, 2004; 39: 167–9.

[77] Ryu JK, Hong SM, Karikari CA, Hruban RH, Goggins MG, Maitra A. Aberrant MicroRNA-155 expression is an early event in the multistep progression of pancreatic adenocarcinoma. *Pancreatology*. 2010; 10(1): 66-73.

[78] Sewer A, Paul N, Landgraf P, Aravin A, Pfeffer S, Brownstein MJ, Tuschl T, van Nimwegen E, Zavolan M. Identification of clustered microRNAs using an ab initio prediction method. *BMC Bioinformatics*, 2005; 6, 267.

[79] Xue C, Li F, He T, Liu G-P, Li Y and Zhang X. Classification of real and pseudo microRNA precursors using local structure-sequence features and support vector machine. *BMC Bioinformatics*, 2005; 6, 310.

[80] Ng KL, Mishra SK. De novo SVM classification of precursor microRNAs from genomic pseudo hairpins using global and intrinsic folding measures. *Bioinformatics,* 2007; 23: 1321–1330.

[81] Helvik SA, Snove O Jr, Saetrom P. Reliable prediction of Drosha processing sites improves microRNA gene prediction. *Bioinformatics,* 2007; 23: 142–149.

[82] Jiang P, Wu H, Wang W, Ma W, Sun X et al. MiPred: classification of real and pseudo microRNA precursors using random forest prediction model with combined features. *Nucleic Acids Res.*, 2007; 35: W339–W344.

[83] Lewis B P, Shih I-h, Jones-Rhoades M W, Bartel D P and Burge C B. Prediction of mammalian microRNA targets. *Cell,* 2003; 115: 787-798.

[84] Krek A, Grun D, Poy MN, Wolf R, Rosenberg L, et al. Combinatorial microRNA target predictions. *Nat. Genet*, 2005; 37: 495–500.

[85] Rehmsmeier M, Steffen P, Hochsmann M, Giegerich R. Fast and effective prediction of microRNA/target duplexes. *RNA*, 2004; 10: 1507–1517.

[86] Kiriakidou, M, Nelson PT, Kouranov A, Fitziev P, Bouyioukos C, Mourelatos Z, Hatzigeorgiou A. A combined computational–experimental approach predicts human microRNA targets. *Genes Dev*, 2004; 18: 1165–1178.

[87] Bandyopadhyay S, Mitra R. TargetMiner: microRNA target prediction with systematic identification of tissue-specific negative examples. *Bioinformatics*, 2009; 25: 2625-2631.

In: Bioinformatics
Editor: Renu Tuteja
ISBN: 978-1-62100-913-9

Chapter V

Human Bio-Banks and Human Genetic Databases – The Bioethical Dimensions

G. Gandhi
Department of Human Genetics, Guru Nanak Dev University,
Amritsar, India

Abstract

Bio-banks (bio-repositories or tissue banks) are collections of biological material and as they have to be tagged they contain associated vital demographic and personal information, ethnic background and health details. Human genetic research databases are resources for genetic research and include information generated from analysis of biological samples. Biological collections are made for both, research and non-research purposes and have applications useful for diagnostics, therapeutics, treatment, forensics, transplantation, transfusion, public health surveillance, quality assurance purposes or as teaching materials. Such endeavours require establishment, harmonization and sharing of resources viz. data, biological samples and information derived from their analyses. However, as health records are extremely personal and sensitive, there is a plethora of bioethical concerns: type of informed consent, re-consent, individual and group consent, genetic privacy, genetic discrimination, stigmatization, benefit-sharing, distributive and compensatory justice, solidarity, etc. Concerns arise about compromises in donor protection from uncontrolled use of samples and data, potential donors may be pressurized into taking unreasonable risks or divulge personal information. Donors and their genetic relatives or entire population groups may not be protected from genetic discrimination and stigmatization. Though pseudonymization and anonymization can give data protection, national and international cooperation and networks require transfer of samples and data from bio-banks to third parties. Hence, donor privacy in sample and data storage requires organizational measures that ensure that the code and the encrypted data are stored in compliance with legal requirements applicable to the handling of personal data. Any breach in confidentiality of research results associating diseases with genetic endowment may be perceived as stigmatizing by those concerned and in their social environment. Revelation of their genetic status, may lead to insurance refusal, higher premium and/ denial of employment.

Genetic analyses of samples may also generate information applicable to a donor's relatives. The relevance of results to specific ethnicities may require group consent in addition to the consent of individual donors. Deceased persons' samples and data also require consent by kin in accordance with the deceased's wishes, expressed or presumed, during his/her lifetime. National bio-banks for "stocking" samples and data from populations for genetic epidemiology studies also require stringent sample storage and data protection measures. Although individual and social utility of bio-banks cannot be undermined, some concerns about establishment of longitudinal follow-up of databases and usage of stored samples for different purpose in future, would require a re-consent or should it be *carte blanche*? Mining of databases by companies for gene patenting, genetic medicines and ethnic-specific drug-development are other issues requiring good governance. DNA profiling and DNA forensics raise concerns about availability of genetic information in government and police databases, querying about access and privacy as well as ownership /control of our data and samples. Therefore, bio-banks and genetic research databases require regular and independent ethics review and compliance with all relevant legislation, guidelines and standards to ensure that the bioethical issues and concerns are looked after and to ensure genetic privacy and donor samples' and data protection.

Introduction

A bio-bank is a collection of human tissue samples and information about participants stored for a period of time for use for research studies. Bio-banking includes the storage of DNA, tissue or cell collections and/or genetic and personal information. Demographic details include name, sex, address, age, ethnicity and, depending on the reasons for collection, other records include health care data (disease history, treatments and therapies received), lifestyle information (dietary pattern, physical activity, socioeconomic status, family background, use, frequency and types of alcohol consumption, tobacco usage), and in case of genetic diseases and genetic variant analyses, genealogy, and pedigree charts. These collections and the associated databases with biological resources are vital for varied scientific purposes. In fact, bio-banks have the potential to accelerate the development of personalized diagnostics and therapeutics which can be enhanced through scientific sharing and co-operation by way of sharing/providing bio-specimens to pharmaceutical, clinical, academic and biosciences research and to medical and research organizations. The Organisation for Economic Co-operation and Development [1] in its Guidelines for Human Biobanks and Genetic Research Databases has defined bio-banks as structured resources (human biological materials and/or information generated from their analysis as well as extensive associated information) useful for genetic research. In fact, the Time Magazine in 2009 listed Biobanking among the TOP 10 ideas changing the world [c.f.2].

The term 'Bio-bank' is often used for both registries and bio-repositories although a 'registry´ is the collection of clinical data, demographics and contact information while a 'bio-repository´ is the collection and storage of human blood or tissue samples [3]. Samples of blood, saliva, skin biopsies, hair, or even whole organs of the deceased may be collected and cell lines established for backups so that re-collection need not be done. Genetic data derived from analysing this biological material and then stored on a computer, forms a genetic database. In this digital age, data are therefore easier to collect, store and analyse. Concerns regarding discrimination, privacy and the mis-appropriate use of samples exist [c.f.4] despite confidentiality and security arrangements of bio-material and genetic information protected by computer firewalls and alarm systems [3]. The International Society for Biological and Environmental Repositories has provided information and guidance on the safe and effective management of repositories of specimens in the document: Best Practices for Repositories: Collection, Storage, Retrieval and Distribution of Human Biological Materials for Research [5].

Types and Purposes of Bio-Banks

The systematic and organized collection of cells, tissues, blood or DNA samples are stored and maybe analysed and/retrieved later for analysis. These maybe governmental or private.This practice has been going on for research and diagnostic purposes as university- and/or hospital-based working collections [6]. However, the population-level bio-banks for health monitoring can be stratified as those required for medical treatment and diagnostics (diagnostic bio-banks and treatment bio-banks/clinical and therapeutic bio-banks) and the research bio-banks [c.f.7]. In the latter case, the genomic data rather than the sample become more important. Other bio-banks include DNA databanks for forensic and military use [8] required for crime scene sample DNA profiling and for casualty identification, respectively. Other uses include for the identification of those killed/unidentified minors/ family separations in the aftermath of mass disasters. There may also be information exchange among the forensic, military and medical DNA databanks raising various controversies [8]. The FBI Laboratory's Combined DNA Index System (CODIS) and the National DNA Index System (NDIS) for law enforcement purposes use DNA profiling while more than 30 countries are using the CODIS software for their own database initiatives [9].

Deceased personal DNA Banking- With preference of cremation rather than traditional methods of burial even in the west, DNA evidence is lost and so a need to store DNA from the elderly and affected members of the family with reduced life expectancy has been proposed with a view that descendants, who may wish to ascertain about any serious medical conditions in future, can access these [c.f. 10] and this could be easily carried out with the cooperation of funeral directors. Rather the ability to store DNA from deceased persons was also there in 1998, and presently, this is on the increase as a number of companies worldwide have provisions for DNA retrieval from those recently deceased and where samples can be taken before or after embalmment. Blood spots are dried on specially-coated cards and stored while skin samples after embalmment are used to retrieve the DNA [11]. Will not the DNA retrieval from the dead be against their written or vocal wishes? And taken without consent?

The Million Veteran Program- It has been proposed to enroll one million veterans over the next five to seven years for the Million Veteran Program which aims to be one of the largest databases of its kind in the United States as an effort to consolidate genetic, military exposure, health, and lifestyle information in a single database [12].

Genetic Genealogy Databases-Commercial enterprises have facilities for DNA testing for genetic genealogy and the leader is Family Tree DNA which started in 2000 and has the world's largest Y-DNA (205,868 Y-DNA records) and mtDNA (127,444 mtDNA records) databases and with a total of 333,312 records as on May 10, 2011 [13].

Family Members' DNA- Home DNA Storage-After the tragedy of 9/11, a trend to store the family's DNA at home has been in circulation for identification/finding those lost. Online instructions using a DNA storage kit are available for banking the DNA of children in own freezer [14].

Surreptitious sampling-Surreptitious sampling is sample collection from known individuals (rather than from crime scenes) by swabbing cups and glasses, and as the individual is ignorant about it, there is no informed consent [15]. Therefore privacy challenges arise as genetic information can be revealed from material that is unconsciously left behind, such as cells left on computer keys or saliva left on a drinking glass [16]. Native American mtDNA was obtained from chewed betel quids and from menstrual (probably) blood on aprons available from these artifacts in museum and university collections and used for genetic characterization [17]. If ancient DNA can be extracted from artifacts, skeletons and feces, any bio-material presently is also amenable to DNA amplification and analysis without consent- from chewing gum, cigarette butts, spits, hair, etc.

DNA 'espionage'-Imagine if one's DNA and/or data fell into wrong hands? Instructions to US diplomats in Africa, Middle East and the UN to collect biometric information on 'key civilian and military officials' including 'fingerprints, facial images, DNA, and iris scans' were in the news sometime back [18].

Other Purposes-Human bio-collections can have a wide array of potential [19] and contentious uses other than medical such as towards genetic enhancement, for genetic basis of mental and behavioural processes, racial and ethnic profiling, etc.

Therefore , it is a matter of observation that biobanks can vary in size, degree of access, the status of the institution(s) in charge of their constitution and/or management (public, private companies, private not-for-profit organizations), the range of possible uses and the extent to which samples can be traced back to their donor [c.f.20]. Therefore the type, size and purpose of bio-banks/ collections of biospecimens may be large-scale, multi-research, linked population-based collections or smaller-scale, multi-research collections or collections for specific research or those collections created originally for non-research purposes. The National Health and Medical Research Council of Australia in the Biobanks Information Paper has delibrated on the establishment, management and governance of bio-banks and stated that the information provided has relevance also to other collections of biospecimens in use for health and medical research [21] and that regardless of size, bio-banks inherently involve some risk due to the sensitive status of the data they contain and also pose legal, ethical and regulatory challenges.

The Ethics of DNA Banking

The advantages to DNA banking are logically and ethically correct for disease diagnostics and therapeutics, drug discovery, personalized, predictive and preventive medicine, forensics, and in identification of missing persons, in tracing relatives, in casualty identification during warfare or mass disasters. The disadvantages of DNA banking include a violation of privacy and issues related to confidentiality and security of stored samples, access to samples and derived and/or personal and demographic data. In criminal and military databases, the period for retention of samples and their access to authority and at which level, are other concerns. Hence safeguards become necessary to prevent genetic material or data falling into wrong hands.

Since genetic information has led to stigmatization and/or discrimination because of misunderstanding and mis-interpretation of genetic information, vigilance is needed in allowing DNA collections by governments; in case of sample collections for petty offenses, the resulting DNA databases may be used as sources for social control requiring collection policies so that DNA testing is reserved for only those situations demanding it so that the potential governmental access to the citizen's genetic information is limited [22], and for DNA databanks and genomic research for forensic and military uses,that individual liberties are not curtailed [8].

Though DNA analysis for forensic purposes is popular where direct evidence is lacking, yet a number of legal, social and ethical controversies exist for the collection and retention of DNA samples. In blanket coverage databanks there are concerns about familial searching and over representation of minority ethnic groups, racial profiling and the very action of providing samples to law-enforcement authorities [23].

An important carry-over from biological material from mass disasters, archived samples and samples of the deceased and of vulnerable persons is its secondary use. This should be done without compromising the dignity, autonomy and rights of victims and their families and only permit specific research for improving methods for DNA identification and increasing public safety alongwith feedback from the public [24].

DNA Banking Concerns

Genetic data are different from other analyses as the genetic data are personal, permanent and shared by relatives. The potential for discrimination in insurance, employment and social standing and refusal in schools are important concerns [4]. As with bioinformatics it became possible to reveal the identity of a person from pooled data, controversies in respect of genetic privacy were raised following which the US National Institutes of Health removed the genome data from genome-wide association studies from off the web [25]. The possibility of tracing the person (from whom sample and data were obtained) also depends on how the sample and data are tagged and linked to their donor identity in the database [20].

An ethical framework respecting privacy and right to informed consent are the required hallmarks for all DNA testing. Prenatal and Pre-implantation diagnoses, useful in clinical genetics, can determine the genetic condition of the unborn and provide for non-directive genetic counseling for available options. The availability of genetic information of any individual has a scope for stigmatizing those with genetic dispositions, and pressurize them to either enhancement or annihilation. Also with the possibility of databases being breached, laws are required. Legislation to prohibit genetic discrimination exists in most European countries, in the US and Australia. Rather, in view of the constitution and functioning, bio-banks face different ethical and juridical regulations [26].

Other quandaries are who should/should not be tested? Different regulations are *en force* in different nations. Which cases require DNA testing and why? The rules are different worldwide. In the UK, all arrested persons must submit to a DNA test. However, is it appropriate to reveal the genetic profile of a person who is convicted of a minor offense? The different states in the US can decide regarding DNA testing. And if someone is arrested but not yet convicted, is it appropriate that the person be required to undertake DNA testing (arrestee DNA sampling)? Also is testing DNA on everyone before being convicted ethical? This procedure does raise concerns about genetic privacy.

Recently under the Protection of Freedoms Bill, the British government required the removal of the genetic profiles of an estimated one million innocent people from the National DNA Database, following public pressure and European Court of Human Rights' ruling [27]; the database was set- up by the police to collect DNA routinely on all arrests, ten years or older with retention of all data to age 100, irrespective of conviction or offense.

Concerns about Biomaterial, Data Collections and Genetic Databases

DNA banking is appropriate in situations when techniques currently available have limitations or where genetic testing is not feasible. However in the mandatory newborn screening, the retention of Guthrie cards for future research requires appropriate justification and safeguards as it poses a potential threat to privacy and human rights while the irrevocable consent from the mother circumvents withdrawal and is unethical [28]. Therefore, sample collections and retention and access to genetic information requirements need to be specified. Information should be also provided for the collection of samples from adults and those incapacitated, with or without informed consent, re-consent/ blanket consent, period of sample retention, future testing and analyses, access to samples and derived information to family and third parties, codification and anonymisation, and speculative searching of derived profiles. Some other important considerations include that of retention and access to genetic information to which authority (local, national, international). In cases where guilt has not been established, there are controversies about breach of privacy regarding retention of biological samples and the associated profile data and speculative searching against new samples. Also is testing DNA on everyone before being convicted ethical? On the other hand the conviction of a person with murder 15 years after committing the crime was feasible by a DNA test of saliva from a cigarette butt [29] despite the challenge of privacy violation regarding DNA testing on a tossed cigarette butt.

Bioethical Dimensions

The principles of Bioethics - autonomy, beneficence, non-maleficence and justice should be applied for all biomaterial collections (where DNA testing is to be carried out presently or in future) and for genomic data. Informed consent and informed choice are of paramount importance to accommodate solidarity and equity in the ethical framework [30]. There is a need for harmonization of guidelines and appropriate legislation as bio-banking facilitates exchange across institutes and nations. DNA banking and storage and associated genomic analyses for the varied purposes (as described above) by the government or private companies for commercial or other usage must comply with such an ethical framework in order to pave the way for reaping the benefits of genomics and bioinformatics in a rightful manner. There is a dire need to protect personal and public interests on one hand and have gainful, maximal research utility from these repositories on the other.

The Human Genome Organisation (HUGO) Ethics Committee in its Statement on DNA Sampling: Control and Access, 1998[31] had addressed several ethical issues pertinent to sample collection and sharing in genetic resources and emphasized the need for international standardization of the ethical requirements for the control and access of DNA samples and information. In the Statement on Human Genomic Databases, 2002[32], the committee stated the purpose of such databases and gave a number of recommendations including upholding genetic privacy, and protection from dicrimination and stigmatization.

The Universal Declaration on the Human Genome and Human Rights emphasizes the protection of the human genome and prevention of genetic discrimination, and any use of genetic information that would be contrary to human dignity and human rights [33]. The International Declaration on Human Genetic Data provides information for the collection, use and storage of human genetic data and includes informed consent in genetics, confidentiality of genetic data, prevention of genetic discrimination, anonymization of personal genetic information, the right not to know, genetic counselling, solidarity and benefit sharing [34]. The Universal Declaration on Bioethics and Human Rights is a comprehensive framework of principles for biomedical activities [35]. The OECD Guidelines also state that appropriate measures be taken to avoid discrimination against or stigmatisation of a person, family or group (Best Practice 1.4) and that the risks involved relate not only to disclosure of an individual's information, but the potential for discrimination or stigmatisation resulting from research that may, for example, identify a correlation between a specific heritage and a particular disease (see OECD Guidelines, Annotation 10) [1].

Consent

Consent was established [c.f.21] in The Nuremberg Code (1947) as part of of modern medical ethics and in biomedical research in the Declaration of Helsinki (World Medical Association 1964). This principle is related to the principles of autonomy, human rights and respect for human dignity [36].

Consent is an ongoing process and has three important elements: disclosure and comprehension, capacity to understand, and voluntariness without coercion or manipulation [37]. As future research and risks involved are not apparent, *a priori* consent cannot be obtained and broad or blanket consent has been suggested also but this empowers the keeper or the researcher [38].

An authorization model for future indescribable uses of DNA data collected has been proposed wherein the participants are able to exercise some control over future uses of genetic material by giving permission and discretion to researchers and hence preserving autonomy [39]. The Genetic Alliance Registry and Bio-Bank, a repository for the standardized collection, storage and distribution of biological samples and clinical data for research purposes, was launched in 2004 [3] and states on its website that it allows for ethical re-contact and robust protections in the context of the communities served by its seven advocacy organizations.

Informed Consent and Informed Choice

This is required so that the participants are provided details about the reason, usage and storage of and access to biomaterial and analyzed results. On understanding and if in agreement, the participants can give their written consent and only then, sample and information should be taken. This is not legally binding and the participant can withdraw at any time, which requires that the sample and the primary/secondary information need to be destroyed.

Role of Institutional Review Boards/ Institutional Ethics Committees- The purpose of such bodies is to approve and establish rules for the research study before the start of the study so that the participants are neither exploited nor compromised physically, mentally or psychologically. The type of security measures and protection and confidentiality of personal data and samples, the issue of access to samples and nature and type of research possible, need to be explicit. As per the international guidelines, the participants can withdraw their consent without penalty or explanation [c.f. 21]. However this may not be feasible if samples and/or information have been distributed, anonymisation of these has occurred or if the softwares do not have the possibilty. There also may be a need for re-consent for which ethical review could be required [c.f.21].

Confidentiality

Genetic data have predictive potential and have relevance for the individual, the family and the population group or community. The disclosure of personal information or access to data without approval can lead to risk of discrimination for the person and to the sub-group, of stigmatisation.

Privacy, in the context of collection of information, and confidentiality, as related to the disclosure of personal data to third parties, requires protection. The UNESCO International Declaration on Human Genetic Data (Article 14) [34] and HUGO Statement on Human Genomic Databases (Recommendation 4c) [32] have sressed on the protection of privacy of individuals and the confidentiality of their genetic data.

Therefore genetic material and data require confidentiality to be maintained. In fact, confidentiality in genetic bio-banking needs to cover aspects of individual privacy, community privacy, data protection, access to and the sharing of data [40]. This can be feasible by de-linking/removing personal identifiers through codification or anonymization but has drawbacks in terms of providing any future therapeutical aid to the appropriate participant and/or to the genetic relatives while pseudo-anonymisation (the reversible dissociation procedure for separating the identifiable data of the source from the other data of medical-scientific interest) is another method but it also grants a limited protection of the data [c.f. 37].

Personal data include information relating to an identifiable person by reference to an identification number or to one or more specific factors of his physical, physiological, mental, economic, cultural or social identity and the sensitive personal data are those that reveal race or ethnic origin, political opinions, religious or philosophical beliefs, and membership of a trade union, health, or sex life [c.f.37].

Security

As the sample-collections have a dual nature, being biological material as well as a source of information [41], their security is required. The American Society of Human Genetics has a policy statement on "DNA Banking and DNA Analysis: Points to Consider"[42] and under Minimal Standards for Quality Assurance for DNA Banks has stated that a DNA bank should exist apart from other DNA work and have secure, alarm-equipped storage facilities; other adages include codification of samples, limited access, quality assurance, ownership, consent, etc. Further as health data are any data concerning the past, present or future physical or mental well-being of a living or deceased human being and the medical data in a described situation include also the information related to the abuse of alcohol, drugs consumption as well as genetic data, there is a need for a higher level of security, and the sensitive personal data, about the ethnical origin also have to be subjected to reinforced protection [c.f. 41]. A balance needs to be maintained whereby access to bio-collections and/or data is to authorised individuals without undue restrictions and this may require legislation [c.f.21].

Data Protection

This has implications for collection, storage and use of genetic information by employers, insurers, researchers, and other individuals and organizations [43]. The insurance aspect has ramifications in the developing world and as also applicable to America, where medical insurance is not so common compared to European countries with socialised medical systems.

The Euro-Bio-bank [44] since 2001 is a network of 12 biological banks with ~ 155,000 biological samples and associated data from patients with rare diseases. Though national guidelines and legislations exist for the collection, distribution and use of human biological materials so that there is protection of the rights of patients, families, care-givers, scientists and the society, yet the harmonization of legal and ethical matters require further action [41]. The binding legal network in Europe is the EC Directive 95/46/EC16 [45] which regulates the processing of the personal data and international exchange while the EC countries also have implemented a version of this directive at national levels [c.f.7].

An international consortium, The Public Population Project in Genomics (P^3G) [46] with membership from more than 40 countries (including India) facilitates collaboration between population genomics researchers besides harmonizing the design of biobank infrastructures and research projects and ascertains that basic ethical and legal guidelines are followed and that security measures exist for the protection of genomic data in about its 123 listed bio-banks [c.f.47].

Genetic Privacy

The concept of privacy is based on principles of human dignity and respect for individual freedom and includes the individual's informational privacy as a fundamental human right as embodied in different international organizations [c.f.21].

Privacy, as a dimension of bioethics, is bound to become more prominent with rapid deposition of personal information in archived government or commercial databases and the concurrent chances of its abuse [48] since the testing of DNA has the potential to reveal insights into many intimate aspects of a person and his/her families besides disease-predisposition/ susceptibility, birth legitimacy and may be predispositions to certain behaviours and sexual orientation increasing reasons for genetic discrimination by government, insurers, employers, schools, banks, and others [49]. When companies declare bankruptcy or when there is a take-over, the questions of genetic privacy and access to genetic information and biomaterial get raised. Informed consent and appropriate safeguards for future research and commercial activities besides consumer privacy and security of data are needed. The Icelander deCode Genetics, launched in 2007 and the first to offer direct-to-consumer personal genomics, faced bankruptcy in 2009 [50] and this raises the ownership and access issues regarding genetic information and DNA samples. The New deCode after the sale to Saga Investments, has plans for whole genome sequencing of 2500 individuals from its Icelandic database by mid-2011 for rare variants contributing to common diseases and traits without re-contacting these individuals for consent because their original consent agreements cover whole genome sequencing [51].

The European Convention on Human Rights and Biomedicine (EU Convention) was passed in 1997 by the Council of Europe to legislatively proscribe genetic discrimination [52]. Article 10 of the EU Convention specifically upholds human dignity and respect for privacy [c.f.7]. In the USA, the Genetic Privacy Act, 1995 was drafted as part of ELSI (Ethical, legal and social implications) of the Human Genome Project and after a debate of 13 years in the Congress, The Genetic Information Nondiscrimination Act (GINA) was passed in 2008; the legislation protects 'genetic rights' by preventing discrimination from health insurers and employers. However, three states in the US have introduced Genetic Data Protection Bills since it is felt that the existing protections under GINA do not provide protection of genetic data from misuse [53].

The Indian Scenario- Privacy protection is not a great personal concern in the Indian sub-continent as not many are aware or knowledgeable about it. The recent census data collection, besides information about the number of members in a household, had a multitude of other queries pertaining to personal information and health status and data on materialistic attributes. There were appeals by the government as advertisements in national dailies and in the television channels about providing accurate data to the collection team. In fact the Indian Census has amassed information on demography, economic activity, literacy and education, housing and household amenities, urbanization, fertility and mortality, scheduled castes and scheduled tribes, language, religion, migration, disability and many other socio-cultural and demographic data since 1872 [54].

Privacy is also a key concern with respect to the Multipurpose National Identity Card (MNIC) or Unique Identification card (UID Card) scheme as all of an individual's personal information will be stored in one database. The unique identification number will have associated identifying biometric data and photographs by 2011. However the biometrics are not reliable enough, and the linking of the databases can cause misuse of personal information and therefore potentially violate privacy rights [55].

In fact, India has not paid any attention to data protection, data security and privacy laws though India is a signatory to the International Covenant on Civil and Political Rights, 1966 whose Article17 provides for the 'right of privacy' [c.f.56]. In the absence of specific legislation, data protection in India is achieved through the enforcement of privacy and property rights. Privacy rights are enforced under the Indian Constitution and the Information Technology Act, 2000 [57] and as interpreted by the courts that the right to privacy is enshrined in the Article 21 of the Constitution of India [58]. However, the Information Technology (IT) Act requires amendment for conferring strong data protection to individuals and organizations [c.f.56] as the IT Act is not data or privacy protection legislation *per se* but deals with the issue of data protection and privacy in a piecemeal fashion. Rather the proposed amendments to the Information Technology Act, 2000 (21 of 2000) still lack a real legal framework for data protection and privacy [59].The amendment - The Information Technology (Reasonable security practices and procedures and sensitive personal data or information) Rules, 2011 [60] has included DNA data under biometrics among the technologies that measure and analyse human body characteristics for authentication purposes. In sensitive personal data, information relating to physical, physiological and mental health condition, sexual orientation, medical records and history as well as biometrics, has been included. Organizations undertaking genomic research and having bio-material collections and databases should be members of the Data Security Council of India which is a self regulatory organization to establish, popularize, monitor and enforce privacy and data protection standards [61].

However as early as 1997 among 11 international biotechnology organizations, the All India Biotech Association had endorsed the Biotechnology Industry Organization's policy statement regarding genetic privacy, calling for strong protections against the misuse of personal medical information including data derived from genetic diagnostic tests [62]. In harnessing the benefits of genetic knowledge and setting of guidelines, the pluralistic nature of the Indian society, which is multicultural, multi-religious and multi-ethnic, has to be kept in mind. The Indian Council of Medical Research (ICMR, 2000) in its guidelines on biomedical research [63] has stated the creation of independent, multidisciplinary and pluralist ethics committees to oversee genomic research.

Bio-Banking in India

The ICMR has Draft Ethical Guidelines for Bio-banking in India [64] as well as Guidance on Transfer of Human Biological Material for research/ commercial purposes with biological materials specified as blood and/or blood components; dried blood spots and faecal occult blood; medical or clinical wastes considered under the 'infectious substance' category [65]. The Department of Biotechnology, Government of India also has guidelines for DNA and Cell-line Banking emphasizing on the principles of bioethics to be upheld including statements on benefit-sharing from commercialization and the roles of ethics committees overseeing consent and approval of research and the National Bioethics committee / or Department of Biotechnology, Government of India, on transfer of material in India and abroad [66]. In order to prevent exploitation and protect the rights of participants, the three main requirements which the ICMR guidelines emphasize are individual informed consent, approval of the Institutional Ethics Committee and the Repository Ethics Committee in bio-banking [67]. The pending Biomedical Research Human Subjects Promotion and Regulation

Bill has strictures on biomedical research with suitable deterrents for an ethical framework on the likes of similar laws in force in about 35 countries [68].

Examples of Bio-banks with sample sizes more than 200,000 include the UK Bio-bank (500,000+,40–69 y), Estonian Genome Project (~1 million), Icelandic deCode Bio-bank (250,000), Kadoorie Study of Chronic Diseases in China (500,000,35–74 y) and The Indian National Bio-bank with a proposed size of ~2–3 million(18+ y) [69]. The national Indian Bio-bank has been initiated to assess the underlying risk factors and correlates of disease. It is a collaborative effort of ICMR, Registrar General of India and the University of Toronto. This national repository has stored biospecimens and health data at various research centres (universities, public and private hospitals, and research institutes) but lacks well-defined and stringent regulatory mechanisms with the main concern of minimising the gap between the bioethical guidelines and protocols and the practice at the ground level [40].

The global directory of bio-banks, tissue banks , and bio-repositories includes five listings from India viz. Bangalore Brain Bank, Narayana Hrudayalaya Tissue Bank and Stem Cell Research Center, Tata Memorial Hospital Tissue Bank and TCG Life Sciences Bio-bank and Dhruv Dental Stem Cell Bank [70]. However two bio-banks find prominence [7]: The Women's Health Bio-bank Initiative was started with the collaboration of UK in 2004 to study gene regulation and female fertility in order to design effective contraceptive measures and to study the interaction between genotype and environmental factors for a range of debilitating conditions with a provision for storage of samples for future analyte testing and for follow-up of participants for up to ten years [71]. The other is The Indian Genome Variation (IGV) Database Project, initiated in 2004 by a network of six governmental laboratories and a private firm to map SNPs in 15,000 individuals from different sub-populations for predictive medicine, has ethical and social ramifications in terms of result outcomes which can be interpreted to reveal sensitive personal data and cause destability for the population group and those closely-related to it [72].

Asia's first and the world's second bank having DNA identification system (DIS) was established in Uttar Pradesh (which other states also want to follow) under public- private partnership and it was envisaged that each of the members would get a microchip- based DNA card containing general information, anthropological and medical information, biometrics DNA and fingerprint-details as well as very rare personal and confidential information [73].

The Human DNA Repository of Indian Population (HDRIP) was started in 2002 for studying endogamous sub-populations to establish a genomic DNA Bank of Indian sub-population for forensic and medico-diagnostic studies. Housed in it are more than 8,000 DNA samples and the genotype profiles from individuals belonging to 107 population groups [74].

Conclusion

The scientific potential of bio-banks for studying "genomic variation"as well as the "normal" distribution of health and disease and in achieving the medical paradigm shift from 'cure' to 'prevention' is undoubtedly immense [c.f. 2]. However the DNA data link with personal information as well as analysis, retention and storage, access and third parties, and future use and re-use of bio-material, pose a myriad of complications. Regulations to prevent misuse need to be addressed and safeguards implemented to prevent discriminatory consequences. The principles of bioethics should be applied for all biomaterial collections and for genomic data. As per the EuroBio-bank [41], it is recommended that the sample collections require that the following information be provided before obtaining consent viz. purpose of the research, expected gains, further sampling, right of withdrawal of the consent and the effects of doing so (destruction or anonymisation of the sample), the retention of samples, place of storage (banking) and the outcome for the sample on research completion by dissociation and/or destruction; Also the confidentiality of the collected data, access to personal data, possibility of disclosure of the information from the genetic analysis, right to disclosure and advice on the implications of disclosure to relatives. Other clarifications include information about any unexpected discoveries, right to grant donations and benefit-sharing from commercialization of product from the sample. Some points which may not be clear at the time of consent include the type of analysis and future innovative procedures possible. Therefore some flexibility in consent process is required though committed to the protection of the person's rights. This can be overseen by the Ethics Review committee. Genetic data of personal nature may be transmitted to third parties only with the grant of an express and written consent from the subject. The transfer of samples from one place to another also requires protection and security of tissues and cells.

Requirements in different regions for sample collections for DNA pharmacogentic studies from all subjects in clinical trials were perused from the survey results from 40 countries (including India) and it was felt that understanding regional requirements can assist in the harmonization of regulations for genetic sample collections [75].

The two other bioethical principles which need to be emphasized for non-commercialization are altruism (no compensation if the sample is gift to science for the benefit of society at large) and that no commercial transactions should be there involving the human body and/or its parts based on the principle that the human being should not to be considered as a means but as an end (*res extracomercium)* [41]. A charitable-trust model for bio-banks to maximize the alturuistic value of samples donated as a gift for research purposes had been earlier proposed [76]. 'Genetic alturism' also finds mention in the Universal Declaration on the Human Genome and Human Rights [33]. The wish to contribute to medically relevant research may be the probable reason that individuals participate in biobanking which has increasing commercial remuneration. However altruism should be protected by prior informed consent before commercialisation.

On the other hand, benefit- sharing has been mentioned also in various international guidelines. The UNESCO's International Declaration on Human Genetic Data states that benefits from the use of human genetic data and biological samples should be shared with the society as a whole and the international community (Article 19a) and that special assistance may be provided to the persons and groups who have taken part in the research [c.f.21]. The consent process therefore should also clear this prior to obtaining the biospecimen.

The return of information to research subjects has also been delibrated upon and though private benefits may not occur, yet research results or incidental findings may be of interest but communicating these findings require clinical expertise and genetic counselling and hence need to be incorporated in policy making [77]. Genetic counsellors can have other roles in biobanking by educating the participants for informed decision-making [78] . It has been envisaged that the release of individual results to participants could be problematic (OECD Guidelines,Annotation 46) as the research results are unlikely to be scientifically validated, or meaningful to participants, and could potentially be harmful in the absence of informed interpretation, counselling and support especially when no therapeutic action is available [c.f.21].

Cultural sensitivity also needs fundamental consideration in biobanking. The OECD and ISBER Guidelines emphasize that sample and data collection should take into consideration the known cultural heritage and/or religious beliefs of participants and their representative groups as they consider that certain types of biological material have a special status, particularly after post-mortem removal and as deserving of special treatment in its use and disposal [c.f.21]. This has special significance for biobanking in the Indian sub-continent.

Hence it becomes apparent that bio-banks are emerging as longitudinal population studies, designed to serve as long term research platforms providing avenues for understanding the complex interaction between genes and environment and their impact on health or on the disease process [2]. However as different types of biobanks may trigger different bioethical issues,it is therefore proposed that a combination of statutes, guidelines and legislation besides the personal ethics of all stakeholders can assist in maintaining genetic privacy and at the same time, maximal utility from what bio-banking has to offer. A step in this direction is a model of Biobank Policy/ Oversight [2].

References

[1] OECD-Organisation for Economic Co-operation and Development (2009). *Guidelines for Human Biobanks and Genetic Research Databases.* Available from: www.oecd.org/sti/biotechnology/hbgrd.

[2] Caulfield T. and Knoppers B.M. (2010). *Consent, Privacy and Research Biobanks. Policy Brief No. 1.* Available from: http://www.genomecanada.ca/medias/pdf/en/GPS-Policy-Directions-Brief.pdf.

[3] Genetic Alliance Registry and BioBank. [2011-05-10].Available from: http://www.gabiobank.org/.

[4] Ikeda, W. (2008). The public's attitudes towards the use of Genetic information for medical purposes and its related factors in Japan. *Community Genetics,* 11, 8-25.

[5] International Society for Biological and Environmental Repositories, ISBER (2008). Best Practices for Repositories: Collection, Storage, Retrieval and Distribution of Human Biological Materials for Research. *Cell Preservation Technology,* 6, 5-58.

[6] Cambon-Thomsen, A., Hirtzlin, I., Prèaubert, N., Dubreuil, C. Duchier, J., Jansen, B. *et al,.* (2003). An Empirical Survey on Biobanking of Human Genetic Material and Data in Six EU Countries. In B.M. Knoppers (Ed.), *Populations and Genetics: Legal and Socio-Ethical Perspectives* (141-168), Leiden: Martinus Nijhoff.

[7] Jansen,B. (2009). Biobanking and Genetic testing: A comparison between European countries and India. *Journal International de bioethique,* 20, 57-68.

[8] Williams, R. (2008). DNA Databanks and Genomics Research: Forensic and Military (Ab) uses. *Encyclopedia of Life Sciences, DOI: 10.1002/9780470015902.a0005859.*

[9] CODIS-Combined DNA Index System. CODIS Brochure [2011-05-15] Available from: http://www.fbi.gov/about-us/lab/codis/codis_brochure.

[10] Bettinger, B.T. (2007). 10 DNA Myths Busted. Available from:

http://www.thegeneticgenealogist.com/wpcontent/uploads/10%20DNA%20Myths%20 Busted.PDF.

[11] The Genetic Genealogist (2007). DNA from the Dead: DNA Banking is Legal, but is it Ethical? Part I. Available from: http://www.thegeneticgenealogist.com/.

[12] The Million Veteran Program - A partnership between VA and Veterans. [2011-05-10]. Available from: http://www.research.va.gov/MVP/veterans.cfm#1.

[13] Genetic Genealogy Databases. [2011-05-10]. Available from: http://www.familytreedna.com/Default.aspx.

[14] Lei. Saving Family Members' DNA. (2005). Available from: http:// blisstree.com/feel/saving-family-members-dna/?utm_source=blisstree andutm_medium =webandutm_ campaign=b5hubs_migration.

[15] Economist Debates (2009). *The ethics of DNA databasing*. Available from: www.economist.com/debate /days/view/286/CommentKey:257098.

[16] Jaswal, P. S. and Jolly, S. (2006). Genetic Privacy and Legislative Response- A Critical Analysis. *NALSAR Law Review*, 3(1), 7-19.

[17] LeBlanc, S. A., Cobb Kreisman, L. S., Kemp, B., Smiley, F. E., Carlyle, S. W., Dhody, A. N., and Benjamin, T. (2007). Quids and Aprons: Ancient DNA from Artifacts from the American Southwest. *Journal of Field Archaeology,* 32(20), 161-175.

[18] Guy, S. WikiLeaks uncovers DNA 'espionage'. (2010). Available from: http://www.bionews.org.uk/page_83482.asp.

[19] Gene Watch UK. (2001).*Who Benefits from Gene Banking? Human Bio-collections: Briefing Number*. [2001-04-14]. Available from: http://www.genewatch.org/ uploads/f03c6d66a9b354535738483c1c3d49e4/brief14.pdf.

[20] Cambon-Thomsen, A. (2004). The social and ethical issues of post-genomic human biobanks. *Nature Reviews Genetics,* 5, 866-873.

[21] National Health and Medical Research Council (NHMRC) of Australia (2010). *Biobanks Information Paper 2010.* Available from: http://www.nhmrc.gov.au/ _files_nhmrc/file/your_health/egenetics/practioners/biobanks_information_paper.pdf.

[22] Sankar, P. (1997). The proliferation and risks of government DNA databases. *Am. J. public Hlth*, 87, 336-337.

[23] Michael, K. (2010). The Legal, Social And Ethical Controversy of the Collection and Storage of Finger Print Profiles and DNA Samples in Forensic Science. Available from: http://works.bepress.com/kmichael/191.

[24] Knoppers, B. M., Saginur, M. and Cash, H. (2006). Ethical issues in secondary uses of human biological materials from mass disasters. *Journal of Law, Medicine and Ethics*. 352-365.

[25] National Institutes of Health, NIH. (2008). *Modifications to Genome-Wide Association Studies (GWAS) Data Access.* Available from: http://gwas.nih.gov/03policy2.html.

[26] Cambon-Thomsen, A. Rial-Sebbag, E. and Duchier J. (2003). Ethical and Legal aspects of biological sample banks: synthesis, practical questions and proposals. *Rev. Epidemiol. Santé Pbl, 51:* 121-126.

[27] Wallace, H. (2011).Police DNA database to be rolled back in Britain. *Biopolitical Times*. Available from: http://www.biopoliticaltimes.org/article.php?id=5623.

[28] GeneWatch UK (2006). *Keeping blood spots from newborn babies*. Available from: www.genewatch.org/.../Blood_spots_from_newborn_babies.doc.

[29] Walsh, D. (2010). California ruling backs police use of DNA from discarded cigarette. *The Sacramento Bee*. Available from: http://www.geneticsandsociety.org/article.php?id=5480

[30] Chadwick, R. and Berg, K. (2001). Solidarity and equity: new ethical frameworks for genetic databases. *Nature Review Genetics, 2,* 318-321.

[31] Human Genome Organisation (HUGO) Ethics Committee (1998).*Statement on DNA Sampling: Control and Access.* Available from: http://www.hugo-international.org/img/dna_1998.pdf.

[32] Human Genome Organisation (HUGO) Ethics Committee (2002). *Statement on human genomic databases*. Available from: http://www.hugo-international.org/img/genomic_2002.pdf.

[33] UNESCO (1998).*Universal Declaration on the Human Genome and Human Rights.* Available from: http://www.unesco.org/new/en/social-and-human-sciences/themes/bioethics/human-genome-and-human-rights/.

[34] UNESCO (2003). *International Declaration on Human Genetic Data.* Available from: http://portal.unesco.org/en/ev.php-.URL_ID=17720 and URL_ DO= DO_TOPIC and URL_SECTION=201.html.

[35] UNESCO (2005) .*Universal Declaration on Bioethics and Human Rights*. Available from: 2005http://unesdoc.unesco.org/images/0014/001428/142825e.pdf#page=80.

[36] UNESCO (2008). *Report of the International Bioethics Committee of UNESCO On Consent*. Available at: http://unesdoc. unesco.org/images/ 0017/001781/178124e.pdf.

[37] Bernstein, M, Bampoe, J. and Daar, A. S. (2004). Ethical issues in molecular medicine of relevance to surgeons. *Can J. Surg*, 47(6), 414-421.

[38] Anderlik, M. (2003). Commercial biobanks and genetic research: ethical and legal issues. *Am J Pharmacogenomics*, 3, 203-215.

[39] Caulfield, T., Upshur, R. E. and Daar, A. (2003). DNA databanks and consent: a suggested policy option involving an authorization model. *BMC Med Ethics*, 4(1), E1.

[40] Patra, P. K. and Sleeboom-Faulkner, M. (2009). The Indian Genomic Biobank Initiative and Emerging Bioethical Issues: a Community-based Perspective. In: M. Sleeboom-Faulkner (ed.) *Human Genetic Biobanks in Asia: Politics of trust and scientific advancement.* (pp.151-167). London: Routledge.

[41] Martín Uranga, A., Martín Arribas, M. C., Jaeger, C. and Posadas, M. (2005).Outstanding ethical-legal issues on biobanks. An overview on the regulations of the Member States of the Eurobiobank project. *Rev Derecho Genoma Hum*, 22, 103-114.

[42] ASHG. *Ad Hoc Committee on DNA Technology*, American Society of Human Genetics, (1988). Available from: http://www.ashg.org/pages/statement_ajhg42.shtml.

[43] Crosbie, D. (2000). *Report to the Human Genetics Commission. Protection of Genetic Information: An International Comparison*. Available from: http://www.hgc.gov.uk/UploadDocs/DocPub/Document/international_regulations.pdf.

[44] The Euro-Biobank. [2011-05-11].Available from: http://www.eurobiobank.org/

[45] Directive 95/46/EC of the European Parliament and of the Council of 24 October 1998 on the protection of individuals with regard to the processing of personal data and on the free movement of such data. Available from: http://www.cdt.org/privacy/eudirective/EU_Directive_html.(c.f.7).

[46] The Public Population Project in Genomics (P³G). [2011-05-11] Available from: http://www.p3gconsortium.org/index.cfm.

[47] Periera, N. (2008). Futuristic Biobanks. *Current Science, 95(3)*, 301.

[48] Cook, M. (2010). Editorial.*BioEdge,* [2010-12-11]. Available from: http://www.bioedge.org/index.php/tools/newsletter/.

[49] Human Genome Project Information. *DNA Forensics*. Available from: http://www.ornl.gov/sci/techresources/Human_Genome/elsi/forensics.shtml#5.

[50] Roberts, M. (2009). Possible setback for genetic medicine as deCODE Genetics faces bankruptcy. Available from: http://www.bionews.org.uk/page_13754.asp.

[51] Vorhaus, D. (2010). Meet the New deCODE, Same as the Old deCODE? Available from: http://www.genomicslawreport.com/index.php/2010/01/25/meet-the-new-decode-same-as-the-old-decode/.

[52] European Convention on Human Rights and Biomedicine. (1997). CETS No. 164. Available from: http://conventions.coe.int/treaty/en/treaties/html/164.html.

[53] Anonymous. (2011). Available from: http://www.genomeweb.com/dxpgx/vermont-california-follow-massachusetts-introducing-genetic-data-protection-bill.

[54] Census of India Organisation (2011). *Why Census is important?* [2011-05-09]. Available from: http://censusindia.gov.in/2011-Common/why_imp.html.

[55] UIDAI Strategy Overview: *Creating a unique identity number for every resident in India.* Planning Comission, Government of India. 2010. Available from: http://uidai.gov.in.

[56] Dalal, P. (2010). Privacy And E-Surveillance Issues Must Be Resolved First By UIDAI. [2010-01-11]. Available from: hrpic.blogspot.com/.../privacy-and-e-surveillance-issues-must.html.

[57] Majmudar and Co., International Lawyers, India. *Data Protection in India.* [2011-05-13]. Available from: http://www.majmudarindia.com/pdf/Data%20Protection%20in%20India.pdf.

[58] Nair, L. R. (2005). Does India need a separate data protection law? (Published in the *World Data Protection Report* Volume 5 Number 12, December 2005).

[59] Sharma, V. (2011). White Paper on Privacy Protection in India. Available from: http://www.iamai.in/Upload/IStandard/White%20Paper%20on%20Privacy.%202007.pdf.

[60] The Information Technology (Reasonable security practices and procedures and sensitive personal data or information) Rules, 2011-Amendment of The Information Technology Act,2000(21 of 2000) [To be published in THE GAZETTE OF INDIA, EXTRAORDINARY, Part II, Section 3, Sub-section (i) of dated the 11th April 2011]Government of India MINISTRY OF COMMUNICATIONS AND INFORMATION TECHNOLOGY (Department of Information Technology).[2011-05-11]. Available from: http://www.dsci.in/sites/default/files/Rules%20for%20sections%2043A%20and%2079%20ºf%20the%20IT%20(Amendment)%20Act,%202008.pdf.

[61] DSCI (Data Security Council of India). (2008). A Self Regulatory initiative in Data Security and Privacy Protection. Available from: www.dsci.in.

[62] Biotechnology Industry Organization (BIO) (1997). International Biotechnology Community Supports Genetic Privacy - BIO Seeks White House Support. Available from: http://www.bio.org/news/pressreleases/newsitem.asp?id=1997_0605_01.

[63] Indian Council of Medical Research (ICMR). (2000). *Ethical Guidelines for Biomedical Research on Human Subjects*. Available from: http://www.icmr.nic.in/vsicmr/ethical/pdf.
[64] Indian Council of Medical Research (ICMR). [2011-05-13].*Draft Ethical Guidelines for Bio-banking in India.* Available from: http://icmr.nic.in/bioethics/bioethics%20cell/guidelines.htm.
[65] Indian Council of Medical Research (ICMR). [2011-05-13]. *Guidance on Transfer of Human Biological Material for research/commercial purposes.*Available from: http://www.icmr.nic.in/ihd/ihd.htm.
[66] Department of Biotechnology (DBT), Government of India. (2002). *Ethical policies on human genome, genetic research and services*. Available from: http://dbtindia.nic.in/policy/polimain.html.
[67] Indian Council of Medical Research (ICMR). (2006). *Ethical guidelines for Biomedical research on Human Participants*. Available from: http://www.icmr.nic.in/ethical_guidelines.pdf.
[68] Anonymous (2010). Law on biomedical research soon. Available from: http://www.deccanherald.com/content/73452/law-biomedical-research-soon.html.
[69] Sgaier, S. K., Jha, P., Mony, P., Kurpad, A., Lakshmi, V., Kumar, R. and Ganguly, N. K. (2007). Biobanks in Developing Countries: Needs and Feasibility. *Science,* 318, 1074-1075.
[70] The Global Directory of Bio-banks. [2011-05-13]. Available from: http://www.specimencentral.com/bio-bank-directory.
[71] **Kennedy, S. (2004).Women's health** biobank in India. *Indian J Med Res,* 120,131-132.
[72] Indian Genome Variation Consortium. (2005). The Indian Genome Variation database (IGVdb): a project overview. *Hum Genet,* 118(1), 1-11.
[73] Pandey, A. Asia's 1st human DNA bank in UP. [2008-06-15]. Available from: http://www.rediff.com/news/2008/jun/15bank.htm.
[74] Anonymous (2002). DNA Repository dedicated to nation. *Financial Daily,The Hindu*, Available from: blonnet.com/2002/06/10/03hdline.htm.
[75] Ricci, D. S., Broderick, E. D., Tchelet, A., Hong, F., Mayevsky, S., Mohr, D. M., Schaffer, M. E., Warner, A. W., Hakkulinen, P. and Snapir, A. (2011). Global requirements for DNA sample collections: results of a survey of 204 Ethics Committees in 40 countries. *Clinical Pharmacology and Therapeutics,* 89(4), 554-561.
[76] Winickoff, D. E., Winicoff, R. N. In reply to: Otten, J., Wyle, H. R. and Phelps, G. D. (2004). The Charitable Trust as a Model for Genomic Biobanks. *The New Eng Journal of Medicine*, 350(1), 85-86.
[77] Hansson, M. G. (2009). Ethics and biobanks. *British Journal of Cancer,* 100, 8-12.
[78] Hawkins, AK. (2010). Biobanks: importance, implications and opportunities for genetic counselors. *J. Genet Counsel, 19,* 423-429.

In: Bioinformatics
Editor: Renu Tuteja

ISBN: 978-1-62100-913-9

Chapter VI

Application of Next-Generation DNA Sequencing in Medical Discovery

***Dilyara Cheranova*[1], *Li Qin Zhang*[1], *Daniel P. Heruth*[1] *and Shui Qing Ye*[1, 2 *]**
[1]Department of Pediatrics
[2]Department of Biomedical and Health Informatics,
Children's Mercy Hospitals and Clinics,
University of Missouri School of Medicine,
Kansas City, MO, US

Abstract

Over the past few years, next-generation DNA sequencing has emerged as a new and powerful technology allowing a much more comprehensive view of biological systems at the genetic level and thus more medical discoveries. Not long ago, the Sanger method for DNA sequencing, considered as first-generation DNA sequencing, was the workhorse in the completion of the human genome project. The Sanger method was often the best choice for de novo sequencing, resequencing, genotyping, and fragment analysis. However, the high cost and low speed of the Sanger method presented a hurdle to sequence large numbers of bases as needed in the era of personalized medicine. To address these drawbacks, many companies and researchers have successfully developed "next-generation DNA sequencing" methods, i.e., any new sequencing method with lower cost and higher speed that is not based on the Sanger method. This chapter will first introduce next generation DNA sequencing technology and then synopsize its application into medical discoveries. The objective of this chapter is to pitch readers to be acquainted with next generation DNA sequencing technology and to help scientists and physician-scientists to realize the potential of next-generation sequencing technology and its application to medical delivery and discovery.

[*] Address correspondence to: Shui Qing Ye, MD, PhD, Department of Pediatrics. Children's Mercy Hospitals and Clinics. University of Missouri-School of Medicine. 2401 Gillham Road, PRC/4th FL. Kansas City, MO 64108. Email: sqye@cmh.edu.

Introduction

DNA sequencing is the process of determining the order of the nucleotide bases adenine, guanine, cytosine, and thymine—in a molecule of DNA. The advent of DNA sequencing has significantly accelerated medical research and discovery since knowledge of DNA sequences is indispensable for basic medical research, specifically, for the understanding of human evolution, the causation of disease, and the interplay between an individual's genes and the environmental factors involved in defining the human condition. The completion of the human genome project has been a significant milestone in the history of medicine and biology by deciphering the order of the 3 billion units of DNA that go into making a human genome, as well as to identify all of the genes located in this vast amount of data. The information garnered from the human genome project has the potential to forever transform healthcare by fueling the hope of genome-based medicine, frequently called personalized medicine, which is the future of healthcare. Although it is a great feat, the $3-billion human genome project has taken more than 13 years for the completion of a reference human genome sequence using a DNA technology based on the chain terminator method or Sanger method, now considered as a first generation DNA sequencing. Both the cost and speed in the first generation DNA sequencing are prohibitive to sequence everyone's entire genome, a prerequisite to realize personalized medicine. Over the past several years, next-generation DNA sequencing technologies have been developed, which reduce the costs of DNA sequencing by several orders of magnitude and dramatically increase the speed of sequencing. This chapter will first introduce next generation DNA sequencing technology and then synopsize its application into medical discoveries. The objective of this chapter is to pitch readers to be acquainted with next generation DNA sequencing technology and to help scientists and physician-scientists to realize the potential of next-generation sequencing, and its application to medical delivery and discovery.

Technology of Next-Generation DNA Sequencing

Two hallmarks of next-generation DNA sequencing technology that distinquish it from first generation DNA sequencing technology are faster speed and lower cost. At least three technical advances have made the development of next-generation DNA sequencing technology possible or practical to realize [1]. First, general progress in technology across disparate fields, including microscopy, surface chemistry, nucleotide biochemistry, polymerase engineering, computation, data storage and others, have provided building blocks or foundations for the production of next generation DNA sequencing platforms. Second, the availability of whole genome assemblies for *Homo sapiens* and other model organisms provide references against which short reads, typical of most next generation DNA sequencing platforms, can be mapped or aligned. Third, a growing variety of molecular methods have been developed, whereby a broad range of biological phenomena can be assessed to elucidate the role and functions of any gene in health and disease, thus increasing demand of gene sequence information by high-throughput DNA sequencing (e.g., genetic variation, RNA expression, protein-DNA interactions and chromosome conformation) [2]. Over the past few years, several platforms of next generation DNA sequencing technologies have emerged as new and more powerful strategies for DNA sequencing, replacing the first generation DNA sequencing technology based on the Sanger method as a preferred technology for high throughput DNA sequencing tasks.

454 Pyrosequencing

The 454 pyrosequencing system was originated by 454 Life Sciences (Branford, CT, USA). 454 Life Sciences was founded in June, 2000, by Jonathan Rothberg, Ph.D. and it was conceived while he was on paternity leave and wanted a way to sequence the genome of his new born son who had been admitted to a pediatric intensive care unit. In late March, 2007, Roche Diagnostics purchased 454 Life Sciences. The Genome Sequencer FLX with GS FLX Titanium series reagents is 454's current system.

In the 454 system, a parallelized version of the pyrosequencing method amplifies DNA inside water droplets in an oil solution (emulsion PCR), with each droplet containing a single DNA template attached to a single primer-coated bead that then forms a clonal colony. Emulsion PCR avoids the preferential amplification of short templates and chimeric products even at low concentrations of DNA [3].The sequencing machine contains many picoliter-volume wells with each containing a single bead and sequencing enzymes. Pyrosequencing uses luciferase to generate light for detection of the individual nucleotides added to the nascent DNA, and the combined data are used to generate sequence read-outs [4].

In October, 2005, 454 Life Sciences released the Genome Sequencer 20, the first next-generation sequencing system on the market. In the same year, 454 Life Sciences was awarded the Wall Street Journal's Gold Medal for Innovation. In May 2007, Project "Jim", a project initiated by Dr. Rothberg and 454 Life Sciences to determine the first sequence of an individual was completed. The result of the project, the complete genome sequence of James Dewey Watson, was handed to Dr. Watson at a ceremony taking place at Baylor College of Medicine. Advantages of the current 454 pyrosequencing system are longer reads and fast run times and thus it may be the method of choice for certain applications where long read-lengths are critical (e.g., *de novo* assembly and metagenomics). The drawbacks of this system are high reagent cost and high error rates in homopolymer repeats [2, 5].

Illumina (Solexa) Sequencing

Illumina was founded in April of 1998 by David Walt, Ph.D., CW Group (Larry Bock), John Stuelpnagel, D.V.M., Anthony Czarnik, Ph.D., and Mark Chee, Ph.D. and headquartered in San Diego, CA, USA. On Nov. 1, 2006, Illumina acquired Solexa. Solexa kept its name after its merge in 2004 with Lynx Therapeutics, which developed the first of the "next-generation" sequencing technologies, Massively Parallel Signature Sequencing (MPSS), in the 1990s. Although MPSS became obsolete, the essential properties of the MPSS output are typical of later "next-gen" data types, including hundreds of thousands of short DNA sequences. Illumina's current sequencing platforms include HiSeq 2000, HiSeq 1000, HiScanSQ, Genome Analyzer IIX, and MiSeq. Among them, the HiScanSQ system integrates the power and resolution of next-generation sequencing with the high-throughput capacity of microarray-based experiments and thus offers unlimited possibilities for genomic research.

Solexa, now part of Illumina, developed a sequencing technology based on sequencing by synthesis. DNA molecules are first attached to primers on a slide and amplified so that local clonal colonies are formed (bridge PCR). Bridge PCR has both amplification and detection coupled to a single solid phase that would both circumvent carryover of contaminants and increase the throughput of detection (6). Four types of ddNTPs are added, and non-incorporated nucleotides are washed away. Unlike pyrosequencing, the DNA can only be extended one nucleotide at a time. A camera takes images of the fluorescently labeled nucleotides and then the dye along with the terminal 3' blocker is chemically removed from the DNA, allowing the next cycle [7]. Figure 1 illustrates the steps of sequencing by synthesis technology.

The first Solexa sequencer, the Genome Analyzer, was launched in 2006. It could only sequence 1Gb in a single run. Through refinements and optimization, the 2010 generation of Illumina sequencing instruments generates 200G of bases per run and up to 25 Gb per day (2×100 bp read length, two billion paired-end reads/run).

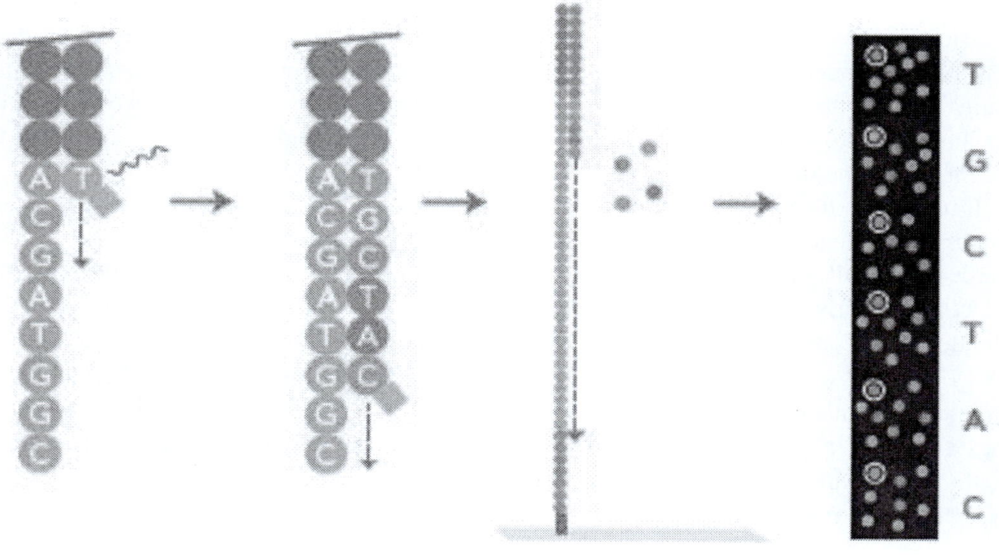

Figure 1. Sequencing by synthesis (SBS) technology uses four fluorescentlylabeled nucleotides to sequence the tens of millions of clusters on the flow cell surface in paralle. During each sequencing cycle, a single labeled deoxynucleoside triphosphate (dNTP) is added to the nucleic acid chain. The nucleotide label serves as a terminator for polymerization, so after each dNTP incorporation, the fluorescent dye is imaged to identify the base and then enzymatically cleaved to allow incorporation of the next nucleotide. Since all four reversible terminator-bound dNTPs (A, C, T, G) are present as single, separate molecules, natural competition minimizes incorporation bias. Base calls are made directly from signal intensity measurements during each cycle, which greatly reduces raw error rates compared to other technologies. The end result is highly accurate base-by-base sequencing that eliminates sequence-context specific errors, enabling robust base calling across the genome, including repetitive sequence regions and within homopolymers. (Courtesy of Illumina, Inc., San Diego, CA, USA, http://www.illumina.com/Documents/products/brochures/brochure_genome_analyzer.pdf).

This is a 200-fold increase in throughput in less than four years. Illumina's sequencing platforms are currently the most widely used platforms in the field. A drawback of this system is the low multiplexing capability of samples.

Solid Sequencing

SOLiD (Sequencing by Oligonucleotide Ligation and Detection) is a next-generation sequencing technology developed by Life Technologies and has been commercially available since 2008. This platform has its origins in the system described by Dr. Jay Shendure and colleagues in 2005 and in work by McKernan and colleagues at Agencourt Personal Genomics (Beverly, MA, USA), which was incorporated into Applied Biosystems (Foster City, CA, USA) in 2006, now part of Life Technologies [2]. 5500xl SOLiD™ System is its current platform.

SOLiD technology employs sequencing by ligation. After the preparation of a fragment or mate-paired library, the DNA is amplified by emulsion PCR. The resulting bead, each containing only copies of the same DNA molecule, are deposited on a glass slide. In sequencing by ligation, primers hybridize to the complementary adapter sequence on the template beads. A set of four fluorescently labeled di-base probes compete for ligation to the sequencing primer. Specificity of the di-base probe is achieved by interrogating every 1st and 2nd base in each ligation reaction. Multiple cycles of ligation, detection and cleavage are performed with the number of cycles determining the eventual read length.

Solid system's two-base encoding provides inherent error correction but requires longer run time.

Helicos' True Single Molecule Sequencing (tSMS)™

HeliScope™ Single Molecule Sequencer was marketed by Helicos Biosciences (www.helicosbio.com), which was headquartered in Cambridge, Massachusetts and co-founded in 2003 by life science entrepreneur Stanley Lapidus, Stephen Quake, D. Phil. a professor of Bioengineering and Applied Physics at Stanford University, and Noubar Afeyan, Ph.D. managing partner of Flagship Ventures.

tSMS Technology is dependent on a sequencing-by-synthesis process. Fluorescently-labeled nucleotides – the building blocks of DNA: C, G, A or T – are added one at a time and incorporated into the growing complementary strand by a DNA polymerase enzyme, based on the sequence of the template. Unused nucleotides are washed away. Upon illumination with a laser, the incorporated nucleotides emit light that is detected by the HeliScope™ Single Molecule Sequencer [8].

Helicos offers the first universal genetic analysis platform that does not require amplification. Pursuing a single molecule sequencing strategy simplifies the DNA sample preparation process, avoids PCR-induced bias and errors, simplifies data analysis, and tolerates degraded samples. A drawback is high error rates compared with other reversible terminator chemistries. Dr. Quake reported in Nature Biotechnology [9] in Sept., 2009 that he had sequenced his own entire genome, named "patient zero" in one week for under $50,000 using a single machine.

Single Molecule Real Time (SMRT™) Sequencing

Single Molecule Real Time (SMRT™) Sequencing is a parallelized single molecule DNA sequencing by synthesis technology developed by Pacific Biosciences (PacBio), which is located in Menlo Park, California. PacBio was founded in 2004 with the goal of developing a Single Molecule Real Time approach for nucleic acid sequencing. This concept was initially developed at Cornell University in the Laboratories of Watt Webb, Sc.D. and Harold Craighead, Ph.D. The PacBio *RS* system is its current platform to conduct, monitor and analyze single molecule, realtime (SMRT™) sequencing reactions.

PacBio 's SMRT technology is built upon two key innovations: a. The SMRT chip, which enables observation of individual fluorophores against a dense background of labeled nucleotides by maintaining a high signal-to-noise ratio; and b. Phospholinked nucleotides, which produce a completely natural DNA strand through fast, accurate, and processive DNA synthesis.

The advantages of the SMRT sequencing are long reads, short runs, and high quality sequence data at lower cost. However, it has the highest error rates compared with other NGS chemistries.

Semiconductor Sequencing

Semiconductor sequencing was invented and developed by H. Esfandyarpour, Ph. D. at Stanford Genome Technology Center. Dr. Esfandyarpour developed the technology during his Ph.D. work, culminating in his thesis entitled "Electronic Gene Sequencing: A novel method for DNA sequencing based on direct heat or pH measurement" at the Stanford Genome Technology Center and the Center for Integrated Systems at Stanford University. It was commercialized by Ion Torrent Systems Inc. and was released in February 2010. Ion Torrent has marketed their machine as a rapid, compact and economical sequencer that can be utilized in a large number of laboratories as a bench top machine [10]. Life Technologies acquired Ion Torrent in August, 2010.

Semiconductor sequencing is a method of DNA sequencing based on the detection of hydrogen ions that are released during the polymerization of DNA. In its Ion Personal Genome Machine (PGM™) sequencer, microwells on a semiconductor chip, which contain a single-stranded template DNA molecule to be sequenced and one DNA polymerase, are sequentially flooded with unmodified A, C, G or T dNTP.

If an introduced dNTP is complementary to the next unpaired nucleotide on the template strand it is incorporated into the growing complementary strand by the DNA polymerase. If the introduced dNTP is not complementary there is no incorporation and no biochemical reaction. The hydrogen ion that is released in the reaction changes the pH of the solution, which is detected by a hypersensitive ion sensor. The unattached dNTP molecules are washed out before the next cycle when a different dNTP species is introduced.

The major benefits of ion semiconductor sequencing are rapid sequencing speed and low upfront operating costs. This has been enabled by the avoidance of modified nucleotides and optical measurements. A major limitation of the system is the difficulty to enumerate long repeats. Also, it suffers from a short read length and lower throughput.

Application of Next-Generation DNA Sequencing

The unprecidently high throughput and the dramatically reduced cost make the Next-Generation DNA Sequencing platforms practical and useful for many applications. These include genomic applications for variant discovery by resequencing targeted regions of interest or whole genomes, *de novo* assemblies of bacterial and lower eukaryotic genomes, and species classification and/or gene discovery by metagenomics studies; epigenome applications by identifying epigenetic marks and chromatin structure changes using seq-based methods (ChIP–seq, methyl–seq and DNase–seq); and transcriptomics applications by cataloguing the transcriptomes of cells, tissues and organisms (RNA–seq).

Genomic Application

Sequencing Whole Genome

The genetic blueprints, or genomes, of any two humans are more than 99 percent identical at the genetic level. However, it is important to understand the small fraction of genetic material that varies among people because it can help explain individual differences in susceptibility to disease,response to drugs or reaction to environmental factors. Thus, knowing the full genome DNA sequence and cataloguing single nucleotide variants and structural variants of each person are a prerequisite for personalized medicine. The 1000 Genomes Project (http://www.1000genomes.org/) is leading this effort. This project was launched on Jan. 22, 2008 by an international research consortium, which receives major support from the Wellcome Trust Sanger Institute in Hinxton, England, the Beijing Genomics Institute, Shenzhen (BGI Shenzhen) in China and the National Human Genome Research Institute (NHGRI), part of the National Institutes of Health (NIH). The sequencing work has been carried out at the Sanger Institute, BGI Shenzhen and NHGRI's Large-Scale Sequencing Network, which includes the Broad Institute of MIT and Harvard; the Washington University Genome Sequencing Center at the Washington University School of Medicine in St. Louis; and the Human Genome Sequencing Center at the Baylor College of Medicine in Houston. Three firms that have pioneered development of next generation DNA sequencing technologies: 454 Life Sciences (a Roche company, Branford, Conn), Applied Biosystems (an Applera Corp. business, Foster City, CA) and Illumina Inc. (San Diego, CA), have also joined this international effort.

The scientific goals of the 1000 Genomes Project, actually including 2500 human genomes in its modified plan, are to produce a catalog of variants that are present at 1 percent or greater frequency in the human population across most of the genome, and down to 0.5 percent or lower within genes. Among the populations whose DNA will be sequenced in the 1000 Genomes Project are: Yoruba in Ibadan, Nigeria; Japanese in Tokyo; Chinese in Beijing; Utah residents with ancestry from northern and western Europe; Luhya in Webuye, Kenya; Maasai in Kinyawa, Kenya; Toscani in Italy; Gujarati Indians in Houston; Chinese in metropolitan Denver; people of Mexican ancestry in Los Angeles; and people of African ancestry in the southwestern United States.

Joining this fray is the Personal Genome Project (PGP, http://www.personal genomes. org/), which was founded by George M. Church, Ph.D., Professor of Genetics and Director of the Center for Computational Genetics at Harvard Medical School. The PGP aims to enroll 100,000 informed participants from the general public. Currently, More than 1,000 individuals have enrolled in the PGP and volunteered to share their DNA sequences, medical information, and other personal information with the research community and the general public. PGP is committed to making research data from the Personal Genome Project freely available to the public to promote discovery and to advance science.

The 1000 Genomes Project has released the data sets for the pilot projects and for more than 1000 samples for the full-scale project. This type of effort using next generation DNA sequencing technology has contributed or will continuously contribute to speed up the cataloguing of potential etiologically and functionally implicated rare and common single nucleotide variants or structural variants to human diseases and traits. The rare variants that have a severe effect, such as the variants responsible for causing cystic fibrosis and Huntington's disease, which typically affect fewer than one in 1,000 people, researchers often must spend years on studies involving affected families. The common variants that have weak effects, perhaps increasing risk of a common condition by 25 percent or less, which affects common diseases such as diabetes and heart disease. On March 25, 2011, based on HuGENavigator (http://hugenavigator.net), there are 818 published genome wide association studies (GWAS), which reported a total of 4898 GWAS hits associated with 484 diseases and traits. With this information as well as those obtained from ongoing and expanding individual genome sequencing, genomics based medicine is beginning to be realized in the clinic.

Targeted DNA Sequencing

Routine whole genome sequencing of large numbers of individuals is still not feasible partly due to the high cost associated with the technique. At present, it is necessary to use an alternative approach, in which certain regions of the genome, such as the "exome", are targeted, enriched and sequenced, which requires ~5% as much sequencing as a whole genome [11]. Exome sequencing is a prime example of targeted DNA sequencing in the application of next generation DNA technology.

Exons are short, functionally important sequences of DNA which represent the regions in genes that are translated into protein and untranslated region flanking them (UTR). UTRs are usually not included in exome studies. In total there are about 180,000 exons found in the human genome. These protein coding regions constitute about 1% of the human genome which translates to about 30 megabases (Mb) in length [11]. It is estimated that the protein coding regions of the human genome constitute about 85% of the disease-causing mutations [12]. The National Heart, Lung, and Blood Institute (NHLBI) and National Human Genome Research Institute (NHGRI) have sponsored an Exome Project. The goal of this project is to develop cost-effective, high-throughput sequencing of the protein coding regions of the human genome for application in well-phenotyped populations. Three groups (Broad Institute, Harvard Medical School and University of Washington) were funded to test and implement approaches in four key areas - sample preparation, target capture, sequencing, and data management and analysis - to generate an integrated resequencing pipeline with the potential to reduce the cost of exome analysis. The goal of exome sequencing is to identify the functional variation that is responsible for both Mendelian and common diseases without the high costs associated with whole-genome sequencing while maintaining high coverage in sequence depth. Some of the data generated by the Exome Project have been deposited into the NCBI's dbGaP database (http://www.ncbi.nlm.nih.gov/gap/?term=exome).

This approach has already produced successful outcomes. One example is that a mutation in the XIAP gene identified by whole exome sequencing is linked to inflammatory bowel disease. Nic Volker, a 6-year-old Wisconsin boy, developed inflammatory bowel disease shortly before his second birthday. Multiple intestinal fistulas occurred, making it impossible for him to eat normally. Despite numerous tests and more than 100 surgeries, doctors could not make a correct diagnosis. Then, researchers at the Medical College of Wisconsin carried out whole-exome sequencing, examining the protein-coding regions of every gene in Nic's genome. Ultimately, they identified a mutation in his XIAP gene. XIAP mutations were not previously associated with bowel symptoms, but had been linked to a severe blood disorder that is curable through bone marrow transplantation.

The medical team raised the possibility of a transplant, which would not have been considered without a firm diagnosis. It was performed in July 2010, using stem cells from the cord blood of a matched, healthy donor. Seven months later, Nic appears to be on the road to recovery. While he is still on immunosuppressants, doctors report the new stem cells are stably engrafted, blood counts are good, and there has been no return of bowel disease. Now Nic can finally eat solid foods [13, 14].

Metagenomics

Metagenomics refers to the application of modern genomics techniques to the study of communities of microbial organisms directly in their natural environments, bypassing the need for isolation and lab cultivation of individual species [15]. Traditional microbiology and microbial genome sequencing rely upon cultivated clonal cultures. This relatively new field of genetic research enables studies of organisms that are not easily cultured in a laboratory as well as studies of organisms in their natural environment.

Metagenomic sequencing is being used to characterize the microbial communities from 15-18 body sites from at least 250 individuals. This is part of the Human Microbiome initiative with primary goals to determine if there is a core human microbiome, to understand the changes in the human microbiome that can be correlated with human health, and to develop new technological and bioinformatics tools to support these goals. Based on the collection of a metagenomics analysis server (MG-RAST, http://metagenomics.anl.gov/), there are 1866 metagenome sequences available as of March 27, 2011.

Beginning in 2003, Craig Venter, Ph.D. leader of the privately-funded parallel of the Human Genome Project, has led the Global Ocean Sampling Expedition, circumnavigating the globe and collecting metagenomic samples throughout with the goal of assessing the genetic diversity in marine microbial communities and to understand their role in nature's fundamental processes. The Global Ocean Sampling Expedition datasets were submitted to both NCBI and Community Cyberinfrastructure for Advanced Marine Microbial Ecology Research and Analysis (CAMERA), a new online resource for marine metagenomics funded by the Gordon and Betty Moore Foundation, developed by JCVI and hosted by UC San

Diego's Division of the California Institute for Telecommunications and Information Technology. Many new species of bacteria have been found in this endeavor [17, 18].

Epigenomics Applications

The mapping of the human epigenome is seen by many scientists as the logical follow-up to the completion of the Human Genome Project. This epigenomic information will be important in understanding how the function of the genetic sequence is implemented and regulated. Since the epigenome is less stable than the genome, it is thought to be important in gene-environment interactions.

Epigenomic mapping is inherently more complex than genome sequencing, since the epigenome is much more variable than the genome. While each of us has only one genome, our epigenome varies with age, differs between tissues, is altered by environmental factors, and shows aberrations in diseases. So mapping a person's epigenome at different ages, in different tissue types and disease states, would yield valuable information on the mechanisms leading to aging and disease.

ChIP-Sequencing (ChIP-Seq). ChIP-Sequencing, also known as ChIP-Seq, is a procedure that combines chromatin immunoprecipitation (ChIP) with massively parallel DNA sequencing or next generation DNA sequencing to identify the cistrome of DNA-associated proteins. It can be used to precisely map global binding sites for any protein of interest.

In ChIP-Seq, specific DNA sites in direct physical interaction with transcription factors and other proteins can be isolated by chromatin immunoprecipitation. ChIP produces a library of target DNA sites bound to a target *in vivo*. Massively parallel sequence analyses are used in conjunction with whole-genome sequence databases to analyze the interaction pattern of any protein with DNA [19], or the pattern of any epigenetic chromatin modifications. Compared to ChIP-chip which requires a hybridization array and is thus restricted to a fixed number of probes, ChIP-Seq can provide a complete spectrum of DNA sites interacted with a particular protein.

Determining how proteins interact with DNA to regulate gene expression is essential for fully understanding many biological processes and disease states. This epigenetic information is complementary to genotype and expression analysis. Robertson et al. used ChIP-seq to map STAT1 target regions in interferon-gamma stimulated and unstimulated human HeLa S3 cells (20). They found 41,582 and 11,004 sites, respectively. Before this report, only 34 loci were known to contain STAT1 interferon-responsive binding sites. Although it cannot exclude that they may contain some noise (or false) signals, this report would open a much wider door for further experimentation on the role of STAT1 in the regulation of gene expression landscape [20]. Using ChIP-Seq, Schmid et al discovered that yeast genes seem to have a minimal nucleosome-free promoter region of 150bp in which RNA polymerase can initiate transcription [21].

Methylation sequencing (Methyl–seq). Methylation sequencing (Methyl–seq) most often refers to the sequencing of bisulfite-treated genomic DNA to determine methylation status at CpG dinucleotides. Treatment of DNA with bisulfite converts cytosine residues to uracil, but leaves 5-methylcytosine residues unaffected. Thus, bisulfite treatment introduces specific changes in the DNA sequence which depend on the methylation status of individual cytosine residues. The global pattern of this type of DNA sequence change can be detected by next generation DNA sequencing technology, yielding single- nucleotide resolution information about the methylation status of a segment of DNA.

DNA methylation was the first discovered epigenetic mark, and remains the most studied. In animals it predominantly involves the addition of a methyl to the carbon-5 position of cytosine residues of the dinucleotide CpG, and is implicated in repression of transcriptional activity. Hereby DNA methylation is involved in the regulation of many cellular processes, including: protecting the genome from parasitic DNA, chromosome stability, chromatin structure, X chromosome inactivation, genomic imprinting, embryonic development and human tumor.

Next generation DNA sequencing has spurred advances in DNA methylation analysis. NGSmethDB (http://bioinfo2.ugr.es/NGSmethDB/gbrowse/), a database for next-generation sequencing single-cytosine-resolution DNA methylation data [22], has been available and useful for experimental researchers, as well as for bioinformaticians, who might use the data as input for further research.

DNase-seq. DNase-seq is a high-throughput method that identifies DNase I hypersensitive sites across the whole genome by capturing DNase-digested fragments and sequencing them by next-generation sequencing. In a single experiment, DNase-seq can identify most active regulatory regions from potentially any cell type, from any species with a sequenced genome [23]. Mapping DNase I hypersensitive (HS) sites has historically been a valuable tool. This method utilizes DNase I to selectively digest nucleosome-depleted DNA (presumably by transcription factors), whereas DNA regions tightly wrapped in nucleosome and higher-order structures are more resistant. The traditional low-throughput method for identifying DNase I HS sites uses Southern blots.

Identification of active gene regulatory elements including promoters, enhancers, silencers, insulators, and locus control regions is a key to understanding transcriptional control governing biological processes such as cell-type specificity, differentiation, development, proliferation, and response to the environment.

Transcriptomics Application

RNA-seq is an increasingly popular application of next-generation DNA sequencing into transcriptomics analysis. The characterization of gene expression in cells via measurement of mRNA levels has long been of interest to researchers. Measuring mRNA concentration levels is a useful tool in determining how the transcriptional machinery of the cell is affected in the presence of external signals (e.g. drug treatment), or how cells differ between a healthy state and a diseased state.

RNA-seq refers to the use of next-generation DNA sequencing technologies to sequence cDNA in order to get information about a sample's RNA content. Compared to the gene expression profiling using DNA microarray technology, RNA-Seq has the advantages to generate an unprecedented global view of the transcriptome and its organization for a number of species and cell types. Before the advent of RNA-Seq, it was known that a much greater than expected fraction of the yeast, *Drosophila melanogaster* and human genomes are transcribed, and for yeast and humans a number of distinct isoforms have been found for many genes. However, the starts and ends of most transcripts and exons have not been precisely resolved and the extent of spliced heterogeneity remained poorly understood. RNA-

Seq, with its deep coverage and base-scale resolution, has revealed many novel transcribed regions and splicing isoforms of known genes, and has mapped 5′ and 3′ boundaries for many genes [24].

The next big challenge for RNA-Seq is to target more complex transcriptomes to identify and track the expression changes of rare RNA isoforms from all genes. Technologies that will advance achievement of this goal are strand-specific sequencing and the use of longer reads to increase coverage and depth. As the cost of sequencing continues to fall, RNA-Seq is expected to replace microarrays for many applications that involve determining the structure and dynamics of the transcriptome.

Conclusion

Over last 10 years, progress in genomics has been monumental. The completion of human genome project has opened the door to the era of genomic medicine. Next generation DNA sequencing has sped up our pace to navigate a course from the base pairs of the human genome sequence to the bedside of patients. However, there are still several hurdles to overcome before the full benefit of personalized medicine can be realized. Although the sequencing speed is continuously increasing and the sequencing cost is continuously decreasing with the advent and the refinement of next (or third) generation DNA sequencing technology, the capacity and cost of whole genome sequencing are still prohibitive to be applied to every individual.

The quality control of different sequencing platforms is lacking. No uniform phenotyping standard of patients in particular disease categories makes some discovered SNPs less valuable. The majority of reported SNPs are without validation in larger patient populations; and lack mechanistic and functional characterization. Although many hurdles are ahead, we should be able to conquer some of these challenges in the coming decade or beyond [24]. We should remain unabashed about the ultimate impact of genomic medicine, which will be to transform the health care of our children and our children's children [25].

References

[1] Shendure J, Mitra RD, Varma C, Church GM. Advanced sequencing technologies: methods and goals. *Nat. Rev. Genet.* 5, 335–344 (2004).

[2] Shendure J, Ji H. Next-generation DNA sequencing. *Nat. Biotechnol.* 26, 1135–1145 (2008).

[3] Williams R, Peisajovich SG, Miller OJ, Magdassi S, Tawfik DS, Griffiths AD. Amplification of complex gene libraries by emulsion PCR. *Nat. Methods.* 3(7): 545-550 (2006).

[4] Margulies M, Egholm M, Altman WE, Attiya S, Bader JS, Bemben LA, Berka J, Braverman MS, Chen YJ, Chen Z, Dewell SB, Du L, Fierro JM, Gomes XV, Godwin BC, He W, Helgesen S, Ho CH, Irzyk GP, Jando SC, Alenquer MLI, Jarvie TP, Jirage KB, Kim JB, Knight JR, Lanza JR, Leamon JH, Lefkowitz SM, Lei M, Li[1] J, Lohman[1] KL, Lu H, Makhijani VB, McDade KE, McKenna MP, Eugene W. Myers, Nickerson E,

Nobile JR, Plant R, Puc BP, Ronan MT, Roth GT, Sarkis GJ, Simons JF, Simpson JW, Srinivasan M, Tartaro KR, Tomasz A, Vogt KA, Volkmer JA, Wang SH, Wang Y, Weiner MP, Yu P, Begley RF, Rothberg JM. Genome sequencing in microfabricated high-density picolitre reactors. *Nature* 437 (7057): 376–80 (2005).

[5] Metzker ML. Sequencing technologies — the next generation. *Nat. Rev. Genet.* 11, 31-46 (2010).

[6] Oroskar AA, Rasmussen SE, Rasmussen HN, Rasmussen SR, Sullivan BM, Johansson A. Detection of immobilized amplicons by ELISA-like techniques. *Clin. Chem.* 42(9):1547-55 (1996).

[7] Mardis ER. Next-generation DNA sequencing methods. *Annu. Rev. Genomics Hum. Genet.* 9: 387–402 (2008).

[8] Harris TD, Buzby PB, Babcock H, Beer E, Bowers J, Braslavsky I, Causey M, Colonell J, DiMeo J, Efcavitch J.W, Giladi E, Gill J, Healy J, Jarosz M, Lapen D, Moulton K, Quake SR, Steinmann K, Thayer E, Tyurina A, Ward R, Weiss H, and Xie Z. Single-Molecule DNA Sequencing of a Viral Genome. *Science,* 320(5872): 106-109. (2008).

[9] Pushkarev D, Neff NF Quake SR. Single-molecule sequencing of an individual human genome. *Nature Biotechnology*, 27(9):847-850 (2009).

[10] Pennisi, E. Semiconductors inspire new sequencing technologies. *Science* 327(5970): 1190 (2010).

[11] Sarah B. Ng, Emily H. Turner, Peggy D. Robertson, Steven D. Flygare, Abigail W. Bigham, Choli Lee1, Tristan Shaffer, Michelle Wong, Arindam Bhattacharjee, Evan E. Eichler, Michael Bamshad, Deborah A. Nickerson and Jay Shendure. "Targeted capture and massively parallel sequencing of 12 human exomes". *Nature*, 461, 272-276 (2009).

[12] Murim Choia, Ute I. Scholla, Weizhen Jia, Tiewen Liua, Irina R. Tikhonovab, Paul Zumbob, Ahmet Nayirc, Ayşin Bakkaloğlud, Seza Özend, Sami Sanjade, Carol Nelson-Williamsa, Anita Farhia, Shrikant Maneb and Richard P. Lifton. Genetic diagnosis by whole exome capture and massively parallel DNA sequencing. *PNAS*. 106: 19096-19101 (2010).

[13] Worthey EA, Mayer AN, Syverson GD, Helbling D, Bonacci BB, Decker B, Serpe JM, Dasu T, Tschannen MR, Veith RL, Basehore MJ, Broeckel U, Tomita-Mitchell A, Arca MJ, Casper JT, Margolis DA, Bick DP, Hessner MJ, Routes JM, Verbsky JW, Jacob HJ, Dimmock DP. Making a definitive diagnosis: Successful clinical application of whole exome sequencing in a child with intractable inflammatory bowel disease. *Genet Med.* Mar;13(3):255-62 (2011).

[14] Collins FS. Faces of the Genomes. *Science.* 331 (6017): 546 (2011).

[15] Chen K, Pachter L. Bioinformatics for whole-genome shotgun sequencing of microbial communities. *PLoS Comp. Biol.* 1 (2): 24. (2005).

[16] Nelson KE and White BA. Metagenomics and Its Applications to the Study of the Human Microbiome. *Metagenomics: Theory, Methods and Applications*. Caister Academic Press. ISBN 978-1-904455-54-7 (2010).

[17] Venter JC, Remington K, Heidelberg JF, Halpern AL, Rusch D, Eisen JA, Wu D, Paulsen I, Nelson KE, Nelson W, Fouts DE, Levy S, Knap AH, Lomas MW, Nealson K, White O, Peterson J, Hoffman J, Parsons R, Baden-Tillson H, Pfannkoch C, Rogers Y-H, Smith HO. Environmental Genome Shotgun Sequencing of the Sargasso Sea. *Science*. 304 (5667): 66–7 (2004).

[18] Johnson DS, Mortazavi A, Myers RM Wold B. Genome-wide mapping of in vivo protein-DNA interactions. *Science* 316: 1497-1502 (2007).

[19] Robertson G, Hirst M, Bainbridge M, Bilenky M, Zhao Y, Zeng T, Euskirchen G, Bernier B, Varhol R, Delaney A, Thiessen N, Griffith OL, He A, Marra M, Snyder M, Jones S. Genome-wide profiles of STAT1 DNA association using chromatin immunoprecipitation and massively parallel sequencing. *Nature Methods* 4: 651-657 (2007).

[20] Schmid CD, Bucher P. ChIP-Seq Data reveal nucleosome architecture of human promoters. *Cell* 131: 831-832 (2007).

[21] Hackenberg M, Barturen G, Oliver JL. NGSmethDB: a database for next-generation sequencing single-cytosine-resolution DNA methylation data. *Nucleic Acids Res*. 2011; 39 (Database issue): D75–D79 (2011).

[22] Song L, Crawford GE. DNase-seq: A High-Resolution Technique for Mapping Active Gene Regulatory Elements across the Genome from Mammalian Cells. *Cold Spring Harb. Protoc*; doi:10.1101/pdb.prot5384 (2010).

[23] Wang Z, Gerstein M, Snyder M. RNA-Seq: a revolutionary tool for transcriptomics. *Nature Reviews Genetics*. 10 (1): 57–63 (2009).

[24] Green ED, Guyer MS. Charting a course for genomic medicine from base pairs to bedside. National Human Genome Research Institute. *Nature*. 10;470(7333):204-13 (2011).

[25] Lander ES. Initial impact of the sequencing of the human genome. *Nature*. 10;470(7333):187-97 (2011).http://en.wikipedia.org.

In: Bioinformatics
Editor: Renu Tuteja
ISBN 978-1-62100-913-9

Chapter VII

Phylogenomics

Monika Miglani and S. K. Gakhar
Centre for Biotechnology
Maharshi Dayanand University
Rohtak, India

1. Introduction

To deduce the evolution of the vast diversity of life is one of the vital objectives in biology. From the time of Darwin, the purpose of phylogenetic studies has been to decipher the evolutionary history of species by detecting the 'natural order', the one true relationship among them. He himself sketched species phylogenies in his celebrated notebooks, and presented one as the sole figure in The Origin of Species [1]. The consequent passion of biologists for the phylogenetic concept is revealed by the publication of Ernst Haeckel's famous "Trees" as early as 1866 [2]. Over the last half century, the ascent of genetics has given us many new tools to interrogate the evolution of species [3]. The essence of genomic sequence data and information regarding protein structure and function persists to grow. Genomic biology flourishes through a frequent reinterpretation of sequence data with the sustenance of new information streams and analysis methods [4]. To Charles Darwin apparently hardly anything in modern biology would have been more provoking than our new insights in the evolution of genomes.

At the core of phylogenetics is the supposition that groups of present day species are related by a species phylogeny – a tree in which present day species appear as leaf nodes and ancestral species as internal nodes (Box-1). *Phylogenetics – the reconstruction of evolutionary history* relies on using mathematical methods to infer the past from features of contemporary species, with only the fossil record to provide a glance on the evolutionary past of life on our planet. The boom in genome sequencing supplies a rich resource for reconstructing the evolutionary origins of gene families. This reconstruction involves the identification of homologous characters that are shared among different organisms, and the interpretation of phylogenetic trees from the correlation of these characters using reconstruction methods (Box-2). Tree and sequence alignment visualizations have a long history. Evolutionary tree diagrams can be found in even the earliest description of evolution, and their visualization still plays a key role in modern phylogenetics. However, the accuracy of the interpretation of evolutionary trees is profoundly accustomed upon the excellence of models for the evolution of such characters. Reconstructing the evolutionary history of life on Earth based solely on the information provided by living organisms seems out to be difficult as the underlying mechanisms are not yet well understood.

Until the 1970s, which brought the dawn of molecular techniques for sequencing proteins and DNA, phylogenetic reconstruction was primarily dependent on the analysis of morphological or ultra structural characters. The main group of animals and plants has been delineated fairly easily using these methods, thus proving the worth of comparative anatomy of fossils and extant species in some manner. However, this attempt is hindered by the limited number of reliable homologous characters available; these are almost non-existent in micro-organisms [5] and are rare even in complex organisms.

With the advent of molecular phylogenetics, the thought of the species phylogeny has become central in mathematical models of sequence evolution [6]. The use of molecular data has revolutionized the world of phylogenetics [7]. The idea that the DNA base sequence could be employed for evolutionary reconstruction seems to have been first expressed in print by Crick [8]. The actual principles and the first implementation of molecular evolutionary analysis were given a few years ago by several researchers in the late 19th century on the basis of microscopic study of plant cells that revealed conspicuous structural similarity between chloroplasts and cyanobacteria and than the striking discovery of organellar genomes further emphasized the issue [9]. The access to DNA sequences expanded the number of homologous characters that could be correlated from less than 100 to more than 1000, considerably refining the resolving power of phylogenetic inference [10]. The fundamental principles of molecular evolution were constituted, and many specific observations of major importance and impact on the fundamentals of neo-Darwinism were made in the pre- genomic era, the r-RNA based phylogeny being the premier case in point [11]. In particular, owing to its appreciable degree of conservation across all organisms, the gene encoding small subunit ribosomal RNA (SSUrRNA) was extensively used for the classification of micro organisms and accredited the recognition of the Archaea as a third discrete domain of the tree of life [12] (Box- 3).

However, with increasing number of genes being analyzed, topological conflicts between phylogenies based on individual genes were reported. Moreover, molecular phylogenies based on a single or few orthologous genes often yielded contradictory results [13]. Also, information from a single gene is often insufficient to obtain firm statistical support for particular nodes of a phylogeny. As a result, numerous parts of the tree of life remained poorly resolved simply because of sampling effects due to the limited availability of data [10]. After facing such topological conflicts, using multiple genes in a large concatenation was proposed to end these incongurences [13].

However, in an equivalent time taken to read these lines, thousand more base pairs of sequence information will have been generated by large scale genomic projects. This wealth of data, scarcely believable only a decade ago, has compelled the evolutionary biologists to think for a better solution. Thus, with increasing availability of whole genome data, *phylogenetic analysis is entering a new era – that of phylogenomics* - an intersection between the field of *evolution and* genomics [14]. Phylogenomics is the convergence of *genomic sciences* - the study of the function and structure of genes and genomes and *molecular phylogenetics* - the study of the hierarchial evolutionary relationships among organisms, their genes and genomes. In addition to furnishing approach to genome data, phylogenomics encloses adequate bioinformatic tools to manage and process the sequence information and to reproduce a proficient high throughput approach for genome wide phylogenetic analysis [15]. Indeed, existing models used in tree building algorithms only partially consider molecular evolutionary processes, and phylogenomics inference will benefit from an advanced understanding of these mechanisms. Fascinatingly, phylogenomics is also emphasizing on certain new issues such as Rare Genomic Changes (RGCs) [16,17] to provide an opportunity to use new "morphological – like" characters based on genome structure.

In this chapter, we describe current methods for phylogenomic inference and discuss their merits and pitfalls in light of their recent applications to distinct phylogenetic problems. Along with all other applications, the newly emerged concept of structural phylogenomics will also be discussed. The intimate link between protein structure and function is well known; the new concepts bring evolutionary modeling into this mix, to interpret and manipulate the connection between evolutionary events and innovations in protein function and structure. Inspite of the great achievements in the field of phylogenomics, challenges for the future have also been affirmed. Therefore, limitations of current phylogenetic reconstruction methods will also be elaborated, despite the continuous evolution of these methods. We have also discussed the important issue of statistical inconsistency and show how failing to fully capture the process of sequence evolution in the evolutionary models leads to tree reconstruction artefacts. Finally, we discussed the future potential of phylogenomics, and specifically address the issue of how to corroborate results from phylogenomic analyses.

2. Current Methods

The two important steps of classical phylogenetic inference – the identification of homologous characters and tree construction – are generally preserved in phylogenomics. The fundamental approach to access the certainty of phylogenetic tree reconstruction methods is the use of simulation studies [18]. While these studies are useful, current simulation studies do not sufficiently evaluate the issues that arise in protein superfamily evolution such as structural divergence, lack of positional homology, and site – lineage specific rate variation. The phylogenetic tree construction methods do not always agree in critical aspects of the predicted tree topologies given the same input despite the fact that the accuracy of a phylogenetic tree topology is critical to phylogenomic analysis [19]. Therefore, as for phylogenies based on morphological data or single genes, the authenticity of a phylogenomic tree confides on the quality of the characters and the accuracy of the reconstruction methods. Ideally reliable characters can be considered as those that have undergone only a few changes over time (preferably, a single change). Multiple changes create *homoplasy* (noise) in the form of convergence and reversal, masking genuine phylogenetic signal [20].

The three main types of standard phylogenetic reconstruction methods (distance, parsimony and likelihood methods) (Box-2) have been adapted for use in phylogenomics. Phylogenomic reconstruction methods can be categorized into *sequence based methods* and the methods based on *whole genome features*. These two approaches are fundamentally similar; the major distinction is of characters used. In both cases, it seems that probabilistic methods are more influential and more reliable. Firstly, they have a more robust theoretical justification in that they rely on an explicit account of their assumptions by using stochastic models describing the pattern of molecular evolution [21]. Secondly, not only do they allow estimates of the phylogeny and a confidence level, but they also provide general methods to evaluate the fit of the model used [22]. However, the sequence based methods are very similar to the classical tree reconstruction, and for which several reviews and textbooks are available, [21,23,24] while the methods based on whole genome features, that are becoming increasingly popular, but their relatively recent introduction limits their critical evaluation at this stage. As a result, methods that acquire an extensive methodological background and are based on multiple sequence alignment, currently remains the methods of choice. Both of these types of methods, as well as the study of rare genomic changes are discussed below.

2.1. Sequence Based Methods

Phylogenetic inference can be defined as the process of determining the estimated evolutionary history by analysis of a given data set. Increasingly, molecular data sets, such as DNA and protein sequences, are used to develop these phylogenies. Comparative sequence analysis is an important tool for evolutionary biologists. Minutes after obtaining a new sequence, BLAST searches can give researchers hints about the function and other properties of a gene. Any comparison of more than two related sequences implicitly assumes an underlying phylogeny.

2.1.1. Number of Characters versus Number of Species

Molecular characters, primarily DNA and derived protein sequences, provide tremendous new information that sheds light on many parts of the "tree of life". However, molecular phylogenies based on single genes generally direct to apparently conflicting results. Using multiple genes in a large concatenation was proposed to end these incongurences [25]. Primary sequences are solely responsible for sequence based phylogenomic methods, whereas phylogenomic trees are inferred from Multiple Sequence Alignments (MSA). During the starting of 20th century, the transition from single gene to multiple gene analysis, using sets of less than 20 genes, accelerated the growth of phylogenomics [26-30]. Later on, further progress was made on use of datasets containing more than 100 genes, [31-36] but at cost of considering fewer species than in single gene studies because of data availability and computational time constraints.

Whether the outstanding advancements in accuracy resulted from an increased number of characters (in this case gene) or species, [37,38] continued to be the persistent query in phylogenetics. However, evidences from computer simulations suggested the two as equivocal, [39,40] whereas empirical studies tend to affirm the caliber of extensive species sampling [37,41,42]. In practice, increasing the number of genes is straight forward for species for which complete genomes have been sequenced. However, the largest complete datasets (in which all genes are represented for all species) reinforced from sequence databases are disproportionate – including few species and many genes, or many species and few genes [43,44]. Thus, to generate more authentic results, [41] compiling phylogenomic datasets incorporating both species and genes is inevitable. Still, the results are invariably associated with missing data, because some genes are not depicted for all species [32, 35, 43] and the impact of these missing data are discussed below.

2.1.2. Super Matrices and Super Trees

The question of how to analyze multiple data sets has been the subject of extreme controversy [45,46]. After the complete convergence of multiple gene alignments from the chosen datasets as discussed above, two discrete approaches can be used for phylogenomic reconstruction; a) total evidence in which all datasets are combined together, [47] called hereafter the *supermatrix approach*; b) separate analysis, [48] in which the datasets are analyzed individually and resulting topologies are combined using consensus or supertree methods, called hereafter the *supertree approach*. The growth of phylogenomic studies that amass data sets to deduce phylogeny dependent on tens to thousands of markers has desperately increased character sampling, even though typically for a limited number of model taxa [34,49,50]. Supertree and supermatrix approaches, on the other hand, have featured increased taxon sampling especially with more modest gains in character sampling. [43,51,52]. The most popular strategy is to analyse the "supermatrix" constituted by the concatenation of individual genes using standard sequence based methods, [21] as indicated from the total evidence principle of using all the relevant available data. Supermatrices are capable to combine existing data directly and in principle furnish proficient syntheses of phylogenetic information that may disclose novel relationships [53]. In this approach, the sequences of genes are indicated as question marks that are not symbolized for some species. Certain amount of missing data can be even tolerated by tree reconstruction methods as is shown by several studies [26,30,32]. This proportion can be even surprisingly high without missing too much accuracy as exhibited by certain empirical [35,43,54] and simulation [35,55] studies. However, the consequences of missing data are limited because species are still represented by a large number of informative characters in phylogenomic studies [35] for which incomplete sequence information is available. Moreover, considering a species represented by only 10 genes out of 100 that are analysed in the whole study might generally be less problematic than not determining that species at all.

The robustness of the supermatrix approach to missing data makes it competent for phylogenetic reconstruction, as phylogenomic datasets can be gathered at low cost by mining existing databases [43,44] or by the sequencing of multiple PCR- targeted loci, [26,29,30] as well as cDNAs and ESTs [32,35]. Following this a large number of species can be incorporated, instead of being confined to model organisms for which complete genome sequences are attainable. Considering the existing applications of supermatrix approach, further progress have been continued to develop this approach as more effective. Phylogenomic trees based on sparse supermatrices are believed to produce important summaries of existing data and, at least occasionally, novel phylogenetic insights [53].

Another sequence based phylogenomic approach refers to *supertrees* - whose construction is a phylogenetic approach that comprises many overlapping source trees, instead of the character data used to derive those trees, and are incorporated to produce a single, large tree [56]. Since Sanderson *et al*'s review of supertrees in TREE, [57] there has been significant development in this field, as such supertrees are an even timelier, and more debatable topic. The prevalent manifestation of supertrees has improved upon the familiar method to enable conflicting estimates of phylogeny to be combined. Different methods for constructing supertrees have been proposed [58] but the breakthrough for supertrees came when the optimization based technique came into existence. Because of its intrinsic simplicity, the Matrix Representation using Parsimony (MRP) method [59,60,61] became the most popular [56]. MRP remains by far the best preferred supertree method, owing to a aggregation of historical precedence coupled with partial software implementation (for the parsimony analyses), universal applicability (it can combine all source trees), a methodology and optimization criterion that were transparent and conventional to biologists, and also results in a large, well resolved and apparently accurate supertrees. MRP has been the ultimate choice for constructing supertrees for many groups of mammals, [62-64] other vertebrates [65-67] and plants [68-70] It was visible only after MRP had captured general acceptance that the method does not show good accuracy in simulation, displaying almost similar with analyses of the primary character data at reconstructing a known model tree [71].

To provide an overview of the phylogeny of a given group by combining trees obtained from disparate sources of data (e.g. morphological and molecular data), supertree methods have been used. Keeping in view, the vast importance of supertrees in the field of phylogenomics, more advancement has been made regarding supertree construction methods. Recently a triplet based supertree approach to phylogenomic has been proposed i.e. *SuperTriplets.* Computer simulations and a biological case study indicate that supertriplets tend to propose more reliable but less resolved supertrees than MRP methods [72]. However, supertree reconstruction is currently an active area of research, and its use will probably expand in the near future.

The comparative merits of the two sequence – based approaches have not been perfectly explored. Empirical comparisons propose the predominance of the supermatrix approach over MRP in a study of crocodylians (crocodiles, caimans, alligators and gavials), [66] whereas the two approaches were found to be almost equivalent in terms of the results produced in case of grasses [73] and also during phylogenomic studies of bacteria [74,75]. *Supertree and Supermatrix approaches* have emphasized increased taxon sampling, usually with more modest gains in character sampling [43,51,52]. The different types of characters used in each method, made it difficult to compare the two approaches [76] However, more work is needed to address these issues more efficiently.

2.2. Whole Genome Based Methods

Because genomes are the results of evolution, virtually any features comparable between organisms can be used to infer phylogenies, as shown by vast amount of data published [77-84]. Phylogeny of organisms obtained using whole genome features cannot be equated to the phylogeny of single genes and also this classical phylogeny is sensitive to hidden paralogy, Horizontal Gene Transfer (HGT) or tree reconstruction artefacts [77-80]. The tree reconstruction artefacts are hard to detect and may affect any approach, while the new approaches – based on various character types might establish some important phylogenetic inference [48,85,86].

2.2.1. Gene Content and Gene Order

Methods of phylogenomic reconstruction based on the comparison of whole genome features beyond the sequence level – such as gene order and gene content – have been developed. The methods based on gene content and gene order do not rely on a multiple sequence alignment step, like sequence based approaches. However, they consider *homology* and *orthology* assessment. Many possibilities came into existence with changes in gene content and gene order within the genomes, as compared to only four states for nucleotide sequences. Thus, they may prove to be potentially good phylogenetic markers [20] being less prone to homoplasy by convergence or reversal. These methods still use the standard tree reconstruction algorithms [86] despite of the usage of different characters to those employed in sequence based approaches.

As phylogenetic trees of individual genes are inconsistent, the whole genome analysis – e.g., the *gene content (the presence/absence of gene families over genomes)* – is becoming a popular approach for interpreting the bulk phylogenetic signals [87]. A genuine and accessible way to explain and analyze complete genomes is by their gene repertoire [88]. Comparing genomes on the basis of the fraction of genes they share was one of the first comparative genomics activities to be developed with the availability of complete genome sequences [89]. The very first genome trees of complete organismal genomes that were published [77-79] were based on gene content approach. Gene content trees show reasonable correspondence to the known species tree. Phylogenomic analyses based on gene content mostly use orthologs, but some variants exist that use homologous instead of orthologous genes, [77] protein domain content, [84] or fold occurrence [80]. The unambiguous phylogenetic signals are not expected in case of gene content based phylogenies, these being more integrated than primary sequences. Moreover, gene content contains a strong phylogenetic signal [78,79] and has helped to clarify several taxonomic uncertainties [90].

Mainly the *Distance* [78-80,91-93] or *Parsimony* [77,80,81,93] methods (Box-2) are of choice while constructing phylogenetic trees from gene- content information. The invalid grouping of organisms with equivalent number of genes is the major task with gene content analysis [81,82,92,93]. This aspect, known as the 'big genome attraction' artefact, [94] is believed to occur from substantial convergent gene losses occurring in certain genomes, for example those of intracellular parasites [81,82]. Further progresses have been made with the development of probabilistic approaches for modeling the molecular details of genome evolution, with the hope that this should ultimately lead to more precise inferences based on gene content [87,94,95].

Because of the huge character space data, *gene order* constitutes the most promising genome feature - based method [96]. Earlier it was recognized as a valuable phylogenetic character [97]. However, the complex mathematical problem concerned, is to estimate the evolutionary distance from many rearrangements that are essential to transform one genome into another [98]. Even with the heuristic approach of breakpoint minimization, computational burden has notably confined the application of phylogenetic reconstruction based on gene order [99]. However, the evolution of gene order involves additional rearrangement mechanisms that are not even adequate for using these methods, like as translocation (transposition between chromosomes) and fusion or fission events of chromosomes [96].

Also, the methods based on *Maximum – Likelihood (ML) distances*, [100] or on *Bayesian inference* [101,102] has been proposed that consider multiple changes in gene order. An important level of saturation along with limited tree resolving power has been notified [102]. Further progress is still needed to realize the full potential of this approach. However, the genome trees of prokaryotes have been constructed through this approach using parsimony and distance methods on assuming that gene order can be described by the presence or absence of pairs of orthologous genes [92,93]. Still, this approach suffers because of the lack of evaluation regarding its accuracy.

2.2.2. Orthology Assessment

Orthologs and *paralogs* are two basically different types of homologous genes that evolved, respectively, by vertical descent from a single ancestral gene and by duplication [103]. However, an understanding of homology, orthology and paralogy has been challenged by other important evolutionary processes such as HGT and gene fusion or fission events, which are believed to be responsible for the formation of complex phylogenetic networks [104,105]. Various terms (e.g. in - paralogs, out – paralogs, super – orthologs or ultra – paralogs) have been coined to further elaborate the different evolutionary origins of sequence similarities. The term 'orthology' is often misunderstood to refer to functionally equivalent genes in different species; but, it is strictly an evolutionary concept, rather than a functional one [106].

The assessment of orthology is an important issue in phylogenomic studies. Orthology is a more fine grained definition of the sharing of a gene and therefore possibly yields better trees. This assessment is fundamentally based on sequence - similarity searches that can be misleading [107] because of differences in evolutionary rates and/or base composition between species, and the occurrence of HGT. Orthologs follow species divergence, [108,109] thus they have basically been used as evolutionary markers for inferring species phylogenies, and can also be used to link functionally equivalent genes across genomes, and facilitate to predict the function of an unknown protein using known orthologs in other species [103,110]. However, the primary issues of using orthologs in function annotation are domain shuffling, presence or absence of a domain, lineage – specific gene duplication and gene loss [111].

The individual genes must be precisely analysed for an efficient orthology assessment [112]. Inspite of the presence of automated procedures, [113] this step is usually neglected while reconstructing phylogenies based on gene content. However, analyses of homologous [81] or orthologous [93] genes using the gene content approach [91,114] have granted fairly congruent trees. Moreover, the noise introduced by HGT and parallel gene losses showed almost negligible effect on the results analysed using gene content approach. This recommends the potential of gene content approach in concern with the orthology assessment. The efficient modeling of HGT events and parallel gene losses will however, be essential to avoid the "big genome attraction" artefact [94].

2.2.3. DNA - String Approach

Early approaches were dependent upon paleontological and morphological data, but with the onset of molecular biology, molecular (amino acid and nucleotide) sequence data has gained all the attention. Entire genomes contain tremendous amount of sequence data that should definitely be useful in constructing phylogenetic trees [115]. Eventually, another striking approach derived from whole – genome features, is established on the distribution of oligonucleotides ("DNA Strings") in genomes [80,116-118]. This approach is not dependent on homology/orthology assessment, while it relies on the fact that each genome has an exclusive "signature" with regard to these strings; these refer to as the ratio between observed dinucleotide frequencies and those expected if neighbouring nucleotides were chosen at random [116]. Moreover, it is also possible to extract phylogenetic signal using this oligonucleotide 'word usage', [80,117-119] as reflected by the methods currently available.

This approach seems to evolve much faster than SSU rRNA, after comparing both in context of the prokaryotic phylogeny [118]. Thus, enhanced phylogenetic signal enclosed in oligonucleotides may overcome the use of such approaches for inferring ancient divergence events [10].

2.3. Rare Genomic Changes

The traditional methodology of comparative morphologists gave a new pathway to study genomes by looking for shared complex characters – known as Rare Genomic Changes (RGCs). The RGCs include intron positions, insertions and deletions, retroposon (SINE and LINE) integrations and gene fusion and gene fission events [17,20].

However, till now only a few characters of this type have been used to answer certain phylogenetic questions, such as the phylogeny of placental mammals [120,121] and jawed vertebrates, [122] or rooting the eukaryotic tree [123,124]. Only a few characters should not be solely considered for inferring the phylogenies, as homoplasy may also affect the RGCs [125-128]. The statistical study of large numbers of RGCs definitely presents an affirmed approach, with increasing number of completely sequenced genomes.

3. New Emerging Trends

The integration of evolution and genomics has been justified for many years as a fruitful and attainable feedback relationship [129-131]. One of the most challenging problems of central importance in the post – genomic era is the prediction of protein molecular function. Phylogenomics is a methodology for annotating the specific molecular function of a protein using the evolutionary history of that protein as conquered by a phylogenetic tree [129]. Many proteins encoded in a large number of recently sequenced genomes [132,133] and specific protein families have been assigned precise functional annotations using phylogenomics. Challenging problems in the field of computational biology such as this requires the fusion of an informed biological framework, effective bioinformatic tools, and high – quality experimental data [19]. Phylogenomic inference of protein function is a multistep process that involves selection of homologs, Multiple Sequence Alignment (MSA), and phylogenetic tree construction; overlaying annotations on the tree topology; differentiating between orthologs and paralogs and eventually inferring the function of a protein based on orthologs determined by this process and the annotations retrieved [134]. Many computational methods have been formulated to help in the prediction of gene function. These methods can be categorized into two main classes – the homology (or similarity) methods and the non – homology methods.

In the *homology methods,* functional predictions are based on identifying and characterizing similarity in sequence or structure of the gene or its encoded product to genes of known function. Whereas, in the *non homology methods*, properties of a gene other than its similarity to other gene/products are used to help in functional predictions. Both these functional prediction methods are comparative in nature. That is, they rely on comparing genes and genomes within and between species [130]. The Phylogenomic methodology is effective and robust for molecular function prediction on both, individual protein families and genome – wide datasets. However, evolution of novel functions and structures among protein families occurs through mutations at key positions, gene duplication, internal repeats, and gene fusion and fission events. Many proteins are constituted of multiple structural domains. The multiplicity of biological functions and structures that contribute to the diversity of life forms results from these evolutionary innovations.

Keeping these issues in mind, one of the newly emerged branches of phylogenomics came into existence. *Structural phylogenomics* [135] commits to the combined use of evolutionary and structural information in a bioinformatic analysis. This includes the construction and analysis of phylogenetic trees for amino acid data, and also encloses protein structure data and structure prediction to improve the accuracy of functional annotation. Instead of being an appropriate method, structural phylogenomics is more like a philosophical concept. The familiar link between protein structure and function is well known; to annotate and manipulate the correlation between evolutionary events and innovations in protein function and structure, the structural phylogenomics brings evolutionary modeling into this mix. During a phylogenomic inference of function structural information can be used at various levels. Like as, most annotation transfer protocols do not distinguish between homologs having only local similarity from those aligning along their entire length causing errors in annotation. To reduce the probability of such errors, incorporating domain architecture information (e.g. through the use of PFAM analysis), or structural phylogenomic clustering method such as Flower Power [136] can be used.

The phylogenetic tree topology accuracy can be affected by several factors [135]. First; dataset selection (how homologs are selected) can affect phylogenetic accuracy. Second, the input multiple sequence alignment is generally assumed correct by different phylogenetic reconstruction methods. The extreme sequence, structural and functional divergence recognized in most protein super families might not be effectively handled by phylogenetic tree reconstruction methods; appearing as the third challenge. For these reasons, errors in tree topology must be anticipated at the coarse branching order of a protein super family phylogeny. The gene duplication events and structural and functional changes are still not formulated in phylogenetic simulation protocols; however the modifications can be done. The corresponding fossil evidences for gene families are not yet available; however approximately equivalent of morphological data; assays for biochemical function; experimental data indicative of biological process and pathway association , cellular localization, enough 3D protein structures; protein – protein interaction and many more are available. The fundamental principle behind using all these experimental data to examine protein superfamily phylogenies is the conservation among function and structure and the phylogenetic tree that clusters functionally and structurally, the resultant similar proteins are committed to be more accurate.

Fortunately, there are several attributes that establish protein structure as an attractive basis for evaluating phylogenies. Firstly, the closely related proteins have higher structural similarity than more distantly related proteins, [137] thus emphasizing the concept that structural similarity correlates closely with evolutionary distance. Second, structural similarity is easy to quantify and also abundant software tool exist to superpose protein 3D structures and to compute the scores. Finally, many protein families have characteristic 3D structures for different subtypes. As indicated, the phylogenetic methods could be analyzed on the basis of their potential to cluster structurally similar proteins on the tree.

Even if structural phylogenomics may not be an absolute resolution to the problem of protein function prediction, it has been shown to furnish considerably eminent precision in *functional annotation* as compared to many other approaches to this task, and also provides a distinctive framework for interrogating changes in protein function and structure explored by evolution [135].

4. Future Challenges

Phylogenomics has caused tremendous amount of reduction in stochastic or sampling errors associated with the finite length of single genes in traditional phylogenetic analysis, because of the usage of many characters. Phylogenomics, the inference of phylogenetic trees using genome–scale data, is becoming the major criterion for resolving difficult parts of the tree of life. However, the systematic errors still occur which are dependent upon data quality and inference methods.

Moreover, incongurences can sometimes be well supported because of the following's [138] (i) hidden paralogy (primarily in protein markers, but also in rDNA), (ii) Horizontal Gene Transfer (HGT), (iii) incomplete lineage sorting and (iv) systematic errors (tree reconstruction artefacts). The classical sequence based methods (i.e. supermatrices) are also found to be affected by systematic errors. However, the other approaches can also get affected by systematic errors, as demonstrated by the occurrence of the "big genome attraction artefact" in gene content methods. This artefact is analogous to the problem of compositional bias in sequence data, and its influence can be diminished by the use of models of genome evolution that have been inferred from sequence based models [94].

4.1. Stochastic and Systematic Errors

The use of a large number of genes generates more primary information, which ultimately reduces sampling errors. Thus, increasing the statistical support as compared to the studies based on single or few genes, generally lead to trees with a hundred percent bootstrap support [31,36]. However, this remarkable level of resolving power [10] is miserably enlisted by an extremely enhanced sensitivity to systematic errors [96]. Generating a robustly supported tree does not necessarily mean that it is correct. Systematic errors correspond to cases in which tree reconstruction methods will recover an erroneous tree more and more consistently, with increased data being analysed [138]. All phylogenetic reconstruction methods draw hypothesis regarding the process of sequence evolution either implicitly (in the case of parsimony methods) or explicitly (in the case of distance and probabilistic methods). All the tree building methods are statistically consistent, as long as their assumptions are fulfilled; however, every method is known to be inconsistent under some conditions [21].

Model violations are persistent, because even a complex model is not capable to capture all the subtleties of the real world. A good example is the compositional heterogeneity of the sequences contained in an alignment, such as species with a high (or low) GC content. For e.g. in an organism, a preferential mutation of A and T occurs towards G and C under strong GC pressure. This biased mutation process will result in an increased fixation of G or C instead of A or T substitutions, especially at third codon positions. Also, if any other unrelated organism is under the same pressure, the probability of two organisms to acquire the same character state at many homologous positions rises. However, if the evolutionary processes are considered stationary, the model will conclude many fewer nucleotide positions with the same character state in these two unrelated species, than observed and the probabilistic methods will incorrectly conclude that the two GC rich taxa are closely related. When the model inadequately interprets multiple substitutions, the model violations get translated to tree reconstruction artefacts [139]. In phylogenomics, minor variations of amino acid compositions are even sufficient to establish a bias in the interpretation of multiple substitutions, thus making an intense spurious signal, leading to an incorrect topology [140].

4.2. Sources of Inconsistency

The independence of evolutionary changes at different site and the homogeneity of the nucleotide substitution are the major causes of model inadequacy. Moreover, the variations in the evolutionary rates among species are responsible for the widespread Long Branch Attraction (LBA) artefact, [141,142] because of the increased chances of convergence and reversal. These biases are considered among the best sources of inconsistency in sequence data.

5. Reducing the Perils of Inconsistency

During the pre-genomic era, the most apparent way of determining a fallacious phylogenetic result was the observation that incongruent trees are procured from different genes. For e.g., the misplacement of microsporidia at the base of the eukaryotic tree, instead of as a sister group of fungi, was affirmed by the comparison of various single gene phylogenies [143]. At the time of usage of whole genome information, it seems difficult to assign an erroneous result in a single tree produced, due to method inconsistency. In this way, the usage of the most accurate tree reconstruction methods available is of the extreme significance.

The three most effective approaches [138] to avoid the pitfalls of phylogenomics: (i) using probabilistic methods with complex models of sequence evolution that more accurately detect multiple substitutions, (ii) using a dense taxon sampling, (iii) removing the fastest evolving part of the data (e.g. species and positions). Reasonably, probability methods of phylogenomics inference, such as ML and Bayesian methods should be favored, as they explicitly consolidate the processes of sequence evolution in the models that they use [144]. However, in extreme cases [145] the probabilistic methods may also fail, as compared to their relative robustness in case of primary assumptions of a model [146-148]. Thus, more realistic models of sequence evolution are required. Therefore, the likelihood models that relax the assumptions of homogeneity and stationarity have also been designed to handle sequences with heterogeneous composition [149, 150]. A new concept of mixture models have also been developed in order to harmonize different models for distinct classes of site for accommodating various site heterogeneities in evolutionary dynamics [151, 152]. Eventually, it is supposed that evolutionary models should include all these improvements synchronously.

Only improving models is not sufficient to appropriately infer multiple substitutions as endless data is required by the complex models to evaluate the additional parameters. Moreover, increased taxa number, exclusively by breaking long branches, proves to be an outstanding way to precisely detect multiple substitutions [142, 153-155]. The effect of systematic error can be minimized by improving the taxon sampling, which can be achieved by enhancing the number of species, along with the omission of rogue taxa [156].

However, increasing taxon sampling is not possible forever, especially, due to extinction and thus it is not sure whether the models have correctly interpreted multiple substitutions or not. Another, substitutive strategy to avoid systematic error is to eliminate the data that evolves at a faster rate. For instance, one can easily remove the third codon position, which lead to aggregate the bias faster or even sequence can be recoded into a reduced alphabet that may be more homogeneous [157]. The removal of the fastest evolving positions has been accepted as the most prevalent approach, as it reduces the level of saturation and accordingly eliminates the positions that mostly violate a given model. Even, the genes can also be eradicated instead of positions, assuming that more information is accessible to determine; if they are fast evolving, thus the problematic species may be selected accurately [158]. However, the elimination of fast evolving data is dangerous instead of being so dynamic, as certain prior assumptions regarding the phylogeny and evolutionary model is required to determine the rate and these may bias the output of the data removal [140]. Thus, this idea may specifically be reversed for the time when it is not possible to break the long branches. Nevertheless, these types of approaches might be the only way to tackle the most complex cases, where the presence of various confounding factors misleads even the most reliable methods of phylogenetic inference.

6. Phylogenomics and Corroboration

The principal confirmation of evolutionary inferences is shown by the congruence of results generated from several datasets and / or various methods [48]. In order to corroborate results, classical, morphological and ultra structural studies being chosen to compare the single gene phylogenies, and the previously obtained molecular trees were selected to contrast with multiple phylogenies. Although, the ultimate origin of characters from which the evolutionary history of organisms can be reconstructed, proves to be whole genomes; thus, eventually it seems difficult to justify the phylogenomics results.

However, the first validation of the new large scale approaches comes out with the similarity between phylogenomic results and the SSU rRNA tree of prokaryotes. Thus, more sophistication and accuracy is required regarding methods based on whole genome features as that is case of sequence based methods.

With the advent of more sophisticated methods and the use of more reliable characters, inconsistency will be targeted more effectively. Ultimately the definite proof of corroboration of phylogenomics results will be their robustness towards species sampling. Corroboration in phylogenomics is an essential prerequisite for handling the large-scale resolution of the tree of life.

7. Perspectives towards Fully Resolved Tree of Life

Many collaborative efforts have been made towards realizing Darwin's dream of having "fairly true genealogical trees for each great kingdom of nature" in the form of concerted network initiatives for assembling the tree of life. Several phylogenomic research programs targeting distinct groups – such as eukaryotes, fungi, arthropods, nematodes, dipterans, birds – will soon lead to important improvements through the consideration of a dense taxon sampling for these groups.

Inspite of the great achievements in the field of phylogenomics, challenges for the future have also been affirmed. Infact, certain nodes have been still left unresolved as the resolution of phylogenetic tree ultimately depends on the evolutionary pattern by which organisms diversified. But with the increasing advancements in the field of phylogenomics, we might not have to accept the idea of a partially resolved tree of life. However, the present picture of the tree of life gave a full support to paraphrase *Theodosius Dobzhansky* [159] *"nothing in genomics makes sense except in the light of evolution"*.

Display items

Box 1. Phylogenetic Tree and Its Features

A phylogenetic tree or evolutionary tree is a branching diagram or "tree" showing the inferred evolutionary relationships among various biological species or other entities based upon similarities and differences in their physical and/or genetic characteristics

A tree is an acyclic connected graph that consists of a collection of nodes (internal and external) and branches connecting them so that every node can be reached by a unique path from every other branch

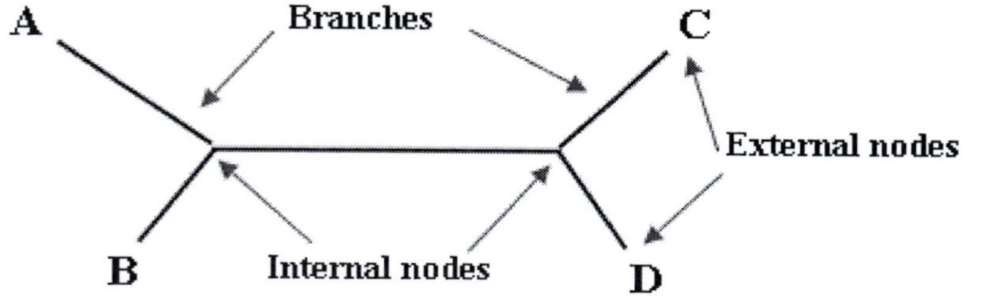

Figure 1. An unrooted phylogenetic tree joining 4 taxonomic units.

Features of a phylogenetic tree:

In the area of phylogenetic inference, trees are used as visual displays that represent hypothetical, reconstructed evolutionary events. The tree in this case consists of:

- Internal nodes which represent taxonomic units such as species or genes; the external nodes, those at the ends of the branches, represent living organisms.

- The lengths of the branches usually represent an elapsed time, measured in years, or the length of the branches may represent number of molecular changes (e.g. mutations) that have taken place between the two nodes. This is calculated is from the degree of differences when sequences are compared (refer to "alignments" later)
- Sometimes, the lengths are irrelevant and the tree represents only the order of evolution. [In a dendrogram, only the lengths of horizontal (or vertical, as the case may be) branches count].
- Finally the tree may be rooted or unrooted.

- *Unrooted trees*

An unrooted tree simply represents phylogenetic but does not provide an evolutionary path. In an unrooted tree, an external node represents a contemporary organism. Internal nodes represent common ancestors of some of the external nodes.

- *Rooted trees*

In case of a rooted tree, one of the internal nodes is used as an out-group, and, in essence, becomes the common ancestor of all the other external nodes. The out-group therefore enables the root of a tree to be located and the correct evolutionary pathway to be identified.

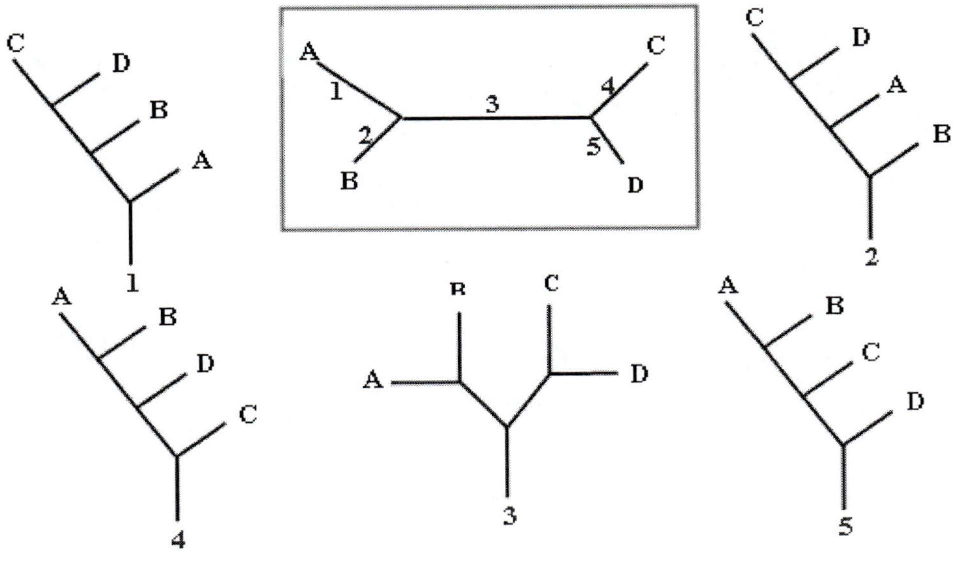

Figure 2. The five rooted trees that can be drawn from the unrooted tree (box).The positions of the roots are indicated by the number on the outline of the unrooted tree (box).

Box 2. Basic Principles and Methods of Phylogenetic Inference

Phylogenetic inference involves two crucial steps: first, homologous characters are identified among species; second, the evolutionary history of species is reconstructed from the comparison of these characters using tree – building methods for phylogenetic inference. Different types of characters (for example, morphological structures, ultra structural characteristics of cells, biochemical pathways, genes, amino- acids or nucleotides) can be used for inferring phylogenies, provided that they are homologous. After identifying the homologous characters, a character matrix is prepared, in which different character states are scored observed in each species.

Three main kinds of reconstruction methods can be used to infer phylogenetic trees from this character matrix as:

Distance methods: The optimal tree is generated by first calculating the estimated evolutionary distance between all pairs of sequences. Then these distances are used to generate a tree in which the branch patterns and length best represent the distance matrix.

Maximum parsimony: Possible trees are compared and each is given a score that is a reflection of the minimum number of character state changes (e.g. amino acid substitutions) that would be required over evolutionary time to fit the sequences into that tree. The optimal tree is considered to be the one requiring the fewest changes (the most parsimonious tree).

Likelihood methods: These are similar to parsimony methods in that possible trees are compared and given a score. The score is based on how likely the given sequences are to have evolved in a particular tree given a model of amino acid or nucleotide substitution probabilities. The optimal tree is considered to be the one that has the highest probability.

Box 3. Phylogenetic Tree of Life

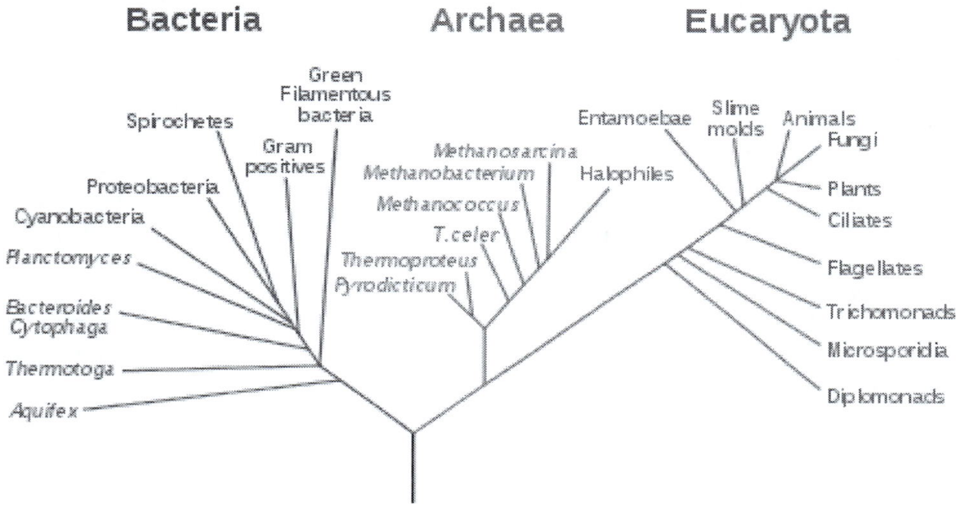

Figure 3. A phylogenetic tree based on rRNA data, showing the separation of bacteria, archaea, and eukaryotes.

References

[1] Darwin, C. *The origin of species by means of natural selection* (Murray, London, 1859).

[2] Haeckel, E. Generelle Morphologie der Organismen: Allgemeine Grundzüge der organischen Formen-Wissenschaft, mechanisch begründet durch die von Charles *Darwin reformirte Descendenz-Theorie* (Georg Reimer, Berlin, 1866).

[3] Rocchi, M. et al. Ancestral genomes reconstruction: An integrated,multi-disciplinary approach is needed. *Genome Res.* 16: 1441-1444 (2006).

[4] Tian, Y. and Dickerman, A.W. GeneTrees: a phylogenomics resource for Prokaryotes. *Nucleic Acids Research*. 35, D328–D331 (2007).

[5] Van Niel, C. B. in *Perspectives and Horizons in Microbiology* 3-12 (Rutgers University Press, New Brunswick, 1955).

[6] Siepel, A. Phylogenomics of primates and their ancestral populations. *Genome Res*. 19: 1929-1941 (2009).

[7] Zuckerkandl, E. and Pauling, L. Molecules as documents of evolutionary history. *J. Theor. Biol.* 8, 357-366 (1965).

[8] Crick, F.H. On protein synthesis. *Symp. Soc. Exp.* Biol. 12, 138–163 (1958).

[9] Sagan, L. On the origin of mitosing cells. *J. Theor. Biol*. 14, 255–274 (1967).

[10] Delsuc, F. et al. Phylogenomics and the reconstruction of the tree of life. *Nature Reviews Genetics*. 6, 361-75 (2005).

[11] Koonin, E.V. Darwinian evolution in the light of genomics. *Nucleic Acids Research.* 37, No. 4, 1011–1034 (2009).

[12] Woese, C. R. and Fox, G. E. Phylogenetic structure of the prokaryotic domain: The primary kingdoms. *Proc. Natl. Acad. Sci. USA* 74, 5088-5090 (1977).

[13] Jeffroy, O. et al. Phylogenomics: the beginning of incongruence? *Trends in Genetics.* 22 No.4, (2006).

[14] Eisen, J. A. and Fraser, C. M. Phylogenomics: intersection of evolution and genomics. *Science*. 300, 1706-1707 (2003).

[15] Reyes-Prieto, A. et al. Phylogenomics and its Growing Impact on Algal Phylogeny and Evolution. *Algae* 21(1): 1-10, (2006).

[16] Philippe, H. and Laurent, J. How good are deep phylogenetic trees? *Curr. Opin. Genet. Dev*. 8, 616-623 (1998).

[17] Rokas, A. and Holland, P. W. Rare genomic changes as a tool for phylogenetics. *Trends Ecol. Evol.* 15, 454-459 (2000).

[18] Huelsenbeck, J.P. Performance of phylogenetic methods in simulation. *Systems Biology*. 44(1):17–48 (1995).

[19] Engelhardt, B.E. *Predicting Protein Molecular Function*. (Electrical Engineering and Computer Sciences University of California at Berkeley, 2007).

[20] Gribaldo, S. and Philippe, H. Ancient phylogenetic relationships. *Theor. Popul. Biol.* 61, 391-408 (2002).

[21] Felsenstein J. Inferring Phylogenies. Sunderland, MA: Sinauer. 645 pp. (2004).

[22] Goldman, N. Statistical tests of models of DNA substitution. *J. Mol. Evol.* 36,182–98 (1993).

[23] Holder M, Lewis PO. Phylogeny estimation: traditional and Bayesian approaches. *Nat. Rev. Genet*. 4:275–84 (2003).

[24] Swofford, D.L. et al. Phylogenetic inference. In *Molecular Systematics*, ed. DM Hillis, C Moritz, BK Mable, pp. 407–514. Sunderland, MA: Sinauer (1996).

[25] Gee, H. Evolution: ending incongruence. *Nature*. 425, 782 (2003).

[26] Qiu, Y. L. et al. The earliest angiosperms: Evidence from mitochondrial, plastid and nuclear genomes. *Nature* 402, 404-407 (1999).

[27] Moreira, D. et al. The origin of red algae: Implications for the evolution of chloroplasts. *Nature*. 405, 69-72 (2000).

[28] Baldauf, S. L. et al. A kingdom-level phylogeny of eukaryotes based on combined protein data. *Science*. 290, 972-977 (2000).

[29] Madsen, O. et al. Parallel adaptive radiations in two major clades of placental mammals. *Nature*. 409, 610-614 (2001).

[30] Murphy, W. J. et al. Molecular phylogenetics and the origins of placental mammals. *Nature* 409, 614-618 (2001).

[31] Blair, J. E. et al. The evolutionary position of nematodes. *BMC Evol. Biol.* 2, 7 (2002).

[32] Bapteste, E. et al. The analysis of 100 genes supports the grouping of three highly divergent amoebae: Dictyostelium, Entamoeba, and Mastigamoeba. *Proc. Natl. Acad. Sci. USA* 99, 1414-1419 (2002).

[33] Lerat, E. et al. From gene trees to organismal phylogeny in Prokaryotes: The case of the -Proteobacteria. *PLoS Biol.* 1, e19 (2003).

[34] Rokas, A. et al. Genome-scale approaches to resolving incongruence in molecular phylogenies. *Nature* 425, 798-804 (2003).

[35] Philippe, H. et al. Phylogenomics of eukaryotes: Impact of missing data on large alignments. *Mol. Biol. Evol.* 21, 1740-1752 (2004).

[36] Wolf, Y. I. et al. Coelomata and not Ecdysozoa: Evidence from genome-wide phylogenetic analysis. *Genome Res.* 14, 29-36 (2004).

[37] Lecointre, G. et al. Species sampling has a major impact on phylogenetic inference. *Mol. Phylogenet. Evol.* 2, 205-224 (1993).

[38] Graybeal, A. Is it better to add taxa or characters to a difficult phylogenetic problem? *Syst. Biol.* 47, 9-17 (1998).

[39] Hillis, D. M. et al. Is sparse taxon sampling a problem for phylogenetic inference? *Syst. Biol.* 52, 124-126 (2003).

[40] Rosenberg, M. S. and Kumar, S. Taxon sampling, bioinformatics, and phylogenomics. *Syst. Biol.* 52, 119-124 (2003).

[41] Philippe, H. Rodent monophyly: Pitfalls of molecular phylogenies. *J. Mol. Evol.* 45, 712-715 (1997).

[42] Lin, Y.H. et al. Four new mitochondrial genomes and the increased stability of evolutionary trees of mammals from improved taxon sampling. *Mol. Biol. Evol.* 19, 2060-2070 (2002).

[43] Driskell, A. C. et al. Prospects for building the tree of life from large sequence databases. *Science* 306, 1172-1174 (2004).

[44] Sanderson, M. J. et al. Obtaining maximal concatenated phylogenetic data sets from large sequence databases. *Mol. Biol. Evol.* 20, 1036-1042 (2003).

[45] Bull, J.J. et al. Partitioning and combining characters in phylogenetic analysis. *Syst. Biol.* 42:384–97(1993).

[46] de Queiroz A. et al. Separate versus combined analysis of phylogenetic evidence. Annu. Rev. *Ecol. Syst*. 26: 657–81(1995).

[47] Kluge, A.G. A concern for evidence and a phylogenetic hypothesis of relationships among Epicrates (Boidae, Serpentes). *Syst. Zool*. 38:7–25 (1989).

[48] Miyamoto, M.M. and Fitch, W.M. Testing species phylogenies and phylogenetic methods with congruence. *Syst. Biol.* 44:64–76 (1995).

[49] Wildman, D.E. et al. Genomics, biogeography, and the diversification of placental mammals. *Proc. Natl. Acad. Sci. USA* 104:14395 (2007).

[50] Dunn, C.W. et al. Broad phylogenomic sampling improves resolution of the animal tree of life. *Nature*. 452:745–749 (2008).

[51] McMahon, M.M. and Sanderson, M.J. Phylogenetic supermatrix analysis of GenBank sequences from 2228 papilionoid legumes. *Syst. Biol.* 55:818–836 (2006).

[52] Bininda-Emonds, O.R.P. et al. The delayed rise of present-day mammals. *Nature*. 446: 507–512 (2007).

[53] Thomson, R.C. and Shaffer, H.B. Sparse Supermatrices for Phylogenetic Inference: Taxonomy, Alignment, Rogue Taxa, and the Phylogeny of Living Turtles. *Syst. Biol.* 59(1):42–58 (2010).

[54] Gatesy, J. et al. Resolution of a supertree/supermatrix paradox. *Syst. Biol.* 51, 652-664 (2002).

[55] Wiens, J. J. Missing data, incomplete taxa, and phylogenetic accuracy. *Syst. Biol.* 52, 528-538 (2003).

[56] Bininda-Emonds, O.R.P. The evolution of supertrees. *Trends in Ecology and Evolution*. 19 No.6 (2004).

[57] Sanderson, M.J. et al. Phylogenetic supertrees: assembling the trees of life. *Trends Ecol. Evol.* 13, 105–109 (1998).

[58] Bininda-Emonds, O. R. P. et al. The (Super) tree of life: Procedures, problems, and prospects. *Annu. Rev. Ecol. Syst.* 33, 265-289 (2002).

[59] Baum, B. Combining trees as a way of combining data sets for phylogenetic inference, and the desirability of combining gene trees. *Taxon* 41, 3-10 (1992).

[60] Ragan, M. A. Phylogenetic inference based on matrix representation of trees. *Mol. Phylogenet. Evol.* 1, 53-58 (1992).

[61] Doyle, J.J. Gene trees and species trees: molecular systematic as one-character taxonomy. *Syst. Bot.* 17, 144–163 (1992).

[62] Mahon, A.S. A molecular supertree of the Artiodactyla. In *Phylogenetic Supertrees: Combining Information to Reveal the Tree of Life* (Vol. 3) (Bininda-Emonds, O.R.P., ed.), pp. 411–437, Kluwer Academic (2004).

[63] Jones, K.E. et al. A phylogenetic supertree of the bats (Mammalia: Chiroptera). Biol. Rev. *Camb. Philos. Soc.* 77, 223–259 (2002).

[64] Grenyer, R. and Purvis, A. A composite species-level phylogeny of the 'Insectivora' (Mammalia: Order Lipotyphla Haeckel, 1866). *J. Zool.* 260, 245–257 (2003).

[65] Kelly, C.M. et al. Phylogenetics of advanced snakes (Caenophidia) based on four mitochondrial genes. *Syst. Biol.* 52, 439–459 (2003).

[66] Gatesy, J. et al. Inconsistencies in arguments for the supertree approach: Supermatrices versus supertrees of Crocodylia. *Syst. Biol.* 53, 342-355 (2004).

[67] Pisani, D. et al. A genus-level supertree of the Dinosauria. *Proc. R. Soc. Lond. Ser. B* 269, 915–921(2002).

[68] Plunkett, G.M. Relationship of the order Apiales to subclass Asteridae: a re-evaluation of morphological characters based on insights from molecular data. *Edinb. J. Bot.* 58, 183–200 (2001).

[69] Davies, T.J. et al. Darwin's abominable mystery: insights from a supertree of angiosperms. *Proc.Natl.Acad.Sci.USA*. 101, 1904–1909 (2004).

[70] Barker, N.P. et al. The paraphyly of Cortaderia (Danthonioideae; Poaceae): evidence from morphology and chloroplast and nuclear DNA sequence data. *Ann. Miss. Bot. Gard.* 90, 1–24 (2003).

[71] Bininda-Emonds, O.R.P. and Sanderson, M.J. Assessment of the accuracy of matrix representation with parsimony supertree construction. *Syst. Biol.* 50, 565–579 (2001).

[72] Ranwez, V. et al. SUPERTRIPLETS: a triplet-based supertree approach to Phylogenomics. *Bioinformatics* 26 ISMB, i115–i123 (2010).

[73] Salamin, N. et al. Building supertrees: An empirical assessment using the grass family (Poaceae). *Syst. Biol.* 51, 136-150 (2002).

[74] Brochier, C. et al. Eubacterial phylogeny based on translational apparatus proteins. *Trends Genet.* 18, 1-5 (2002).

[75] Daubin, V. et al. A phylogenomic approach to bacterial phylogeny: Evidence of a core of genes sharing a common history. *Genome Res.* 12, 1080-1090 (2002).

[76] Bininda-Emonds, O. R. P. Trees versus characters and the supertree/supermatrix "paradox". *Syst. Biol.* 53, 356-359 (2004).

[77] Fitz-Gibbon, S.T. and House, C.H. Whole genome-based phylogenetic analysis of free living microorganisms. *Nucleic Acids Res.* 27:4218–22 (1999).

[78] Snel, B. et al. Genome phylogeny based on gene content. *Nat. Genet.* 21:108–10 (1999).

[79] Tekaia, F. et al. The genomic tree as revealed from whole proteome comparisons. *Genome Res.* 9:550–57 (1999).

[80] Lin, J. and Gerstein, M. Whole-genome trees based on the occurrence of folds and orthologs: implications for comparing genomes on different levels. *Genome Res.* 10: 808–18 (2000).

[81] House, C.H. and Fitz-Gibbon, S.T. Using homolog groups to create a whole-genomic tree of free-living organisms: an update. *J. Mol. Evol.* 54:539–47(2002).

[82] House, C.H. et al. Geobiological analysis using whole genome based tree building applied to the Bacteria, Archaea and Eukarya. *Geobiology* 1:15–26 (2003).

[83] Henz, S.R. et al. Whole-genome prokaryotic phylogeny. *Bioinformatics* 21: 2329–35 (2005).

[84] Yang, S. et al. Phylogeny determined by protein domain content. *Proc. Natl. Acad. Sci. USA* 102:373–78 (2005).

[85] Swofford, D.L. When are phylogeny estimates from molecular and morphological data incongruent? In *Phylogenetic Analysis of DNA Sequences*, ed. MM Miyamoto, J Cracraft, pp. 295–333. New York: Oxford Univ. Press (1991).

[86] Wolf, Y.I. et al. Genome trees and the tree of life. *Trends Genet.* 18:472–79 (2002).

[87] Xun, Gu. and Zhang, H. Genome Phylogenetic Analysis Based on Extended Gene Contents *Mol. Biol. Evol.* 21(7):1401–1408 (2004).

[88] Huynen, M.A. and Bork, P. Measuring genome evolution. *Proc. Natl. Acad. Sci. USA* 95: 5849–56 (1998).

[89] Koonin, E.V. and Mushegian, A.R. Complete genome sequences of cellular life forms: glimpses of theoretical evolutionary genomics. *Curr. Opin. Genet. Dev.* 6:757–62 (1996).

[90] Snel, B. et al. Genome trees and the nature of genome evolution. *Annu. Rev. Microbiol.* 59:191–209 (2005).

[91] Clarke, G. D. et al. Inferring genome trees by using a filter to eliminate phylogenetically discordant sequences and a distance matrix based on mean normalized BLASTP scores. *J. Bacteriol.* 184, 2072-2080 (2002).

[92] Korbel, J. O. et al. SHOT: A web server for the construction of genome phylogenies. *Trends Genet.* 18, 158-162 (2002).

[93] Wolf, Y. I. et al. Genome trees constructed using five different approaches suggest new major bacterial clades. *BMC Evol. Biol.* 1, 8 (2001).

[94] Lake, J. A. and Rivera, M. C. Deriving the genomic tree of life in the presence of horizontal gene transfer: Conditioned reconstruction. *Mol. Biol. Evol.* 21, 681-690 (2004).

[95] Huson, D. H. and Steel, M. Phylogenetic trees based on gene content. *Bioinformatics* 20, 2044-2049 (2004).

[96] Philippe, H. et al. Phylogenomics Annu. *Rev. Ecol. Evol. Syst*. 36:541–62 (2005).

[97] Sankoff, D. et al. Gene order comparisons for phylogenetic inference: Evolution of the mitochondrial genome. *Proc. Natl. Acad. Sci. USA* 89, 6575-6579 (1992).

[98] Hannenhalli, S. and Pevzner, P. A. Transforming cabbage into turnip: Polynomial algorithm for sorting signed permutations by reversals. *J. ACM* 46, 1-27 (1999).

[99] Blanchette, M. et al. Gene order breakpoint evidence in animal mitochondrial phylogeny. *J. Mol. Evol.* 49, 193-203 (1999).

[100] Wang L-S, W.T. Distance-based genome rearrangement phylogeny. *See Gascuel* 2005, pp. 353–83 (2005).

[101] Larget, B. et al. ABayesian analysis of metazoan mitochondrial genome arrangements. *Mol. Biol. Evol*. 22:486–95 (2005).

[102] York, T.L. et al. Bayesian estimation of the number of inversions in the history of two chromosomes. *J. Comput. Biol.* 9:805–18 (2002).

[103] Koonin, E.V. Orthologs, Paralogs, and Evolutionary Genomics. *Annu. Rev. Genet.* 39:309–38 (2005).

[104] Rivera, M.C. and Lake, J.A. The ring of life provides evidence for a genome fusion origin of eukaryotes. *Nature* 431, 152–155 (2004).

[105] Doolittle, W.F. and Bapteste, E. Pattern pluralism and the Tree of Life hypothesis. *Proc. Natl. Acad. Sci. USA*. 104, 2043–2049 (2007).

[106] Fitch, W.M. Homology a personal view on some of the problems. *Trends Genet.* 16, 227–231(2000).

[107] Koski, L. B. and Golding, G. B. The closest BLAST hit is often not the nearest neighbor. *J. Mol. Evol.* 52, 540-542 (2001).

[108] Blair, J.E. and Hedges, S.B. Molecular phylogeny and divergence times of deuterostome animals. *Mol. Biol. Evol*. 22, 2275–2284(2005).

[109] Ciccarelli, F.D. et al. Toward automatic reconstruction of a highly resolved Tree of Life. *Science*. 311, 1283–1287 (2006).

[110] Tatusov, R.L. et al. A genomic perspective on protein families. *Science* 278, 631–637 (1997).

[111] Sjolander, K. Phylogenomic inference of protein molecular function: advances and challenges. *Bioinformatics* 20, 170–179 (2004).

[112] Fitch, W. M. Distinguishing homologous from analogous proteins. *Syst. Zool.* 19, 99-113 (1970).

[113] Sicheritz-Ponten, T. and Andersson, S. G. A phylogenomic approach to microbial evolution. *Nucleic Acids Res*. 29, 545-552 (2001).

[114] Dutilh, B. E. et al. The consistent phylogenetic signal in genome trees revealed by reducing the impact of noise. *J. Mol. Evol*. 58,527-539 (2004).

[115] Ulitsky, I. et al. The Average Common Substring Approach to Phylogenomic Reconstruction. *J.Comput. Biol*. 13(2):336-50 (2006).

[116] Campbell, A. et al. Genome signature comparisons among prokaryote, plasmid, and mitochondrial DNA. *Proc. Natl. Acad. Sci. USA* 96, 9184- 9189 (1999).
[117] Edwards, S. V. et al. A genomic schism in birds revealed by phylogenetic analysis of DNA strings. *Syst. Biol.* 51, 599-613 (2002).
[118] Pride, D. T. et al. Evolutionary implications of microbial genome tetranucleotide frequency biases. *Genome Res.* 13, 145-158 (2003).
[119] Qi, J., et al. Whole proteome prokaryote phylogeny without sequence alignment: A K-string composition approach. *J. Mol. Evol.* 58, 1-11 (2004).
[120] Nikaido, M. et al. Phylogenetic relationships among cetartiodactyls based on insertions of short and long interspersed elements: hippopotamuses are the closest extant relatives of whales. *Proc. Natl. Acad. Sci. USA* 96, 10261-10266 (1999).
[121] van Dijk, M. A. et al. Protein sequence signatures support the African clade of mammals. *Proc. Natl. Acad. Sci. USA* 98, 188-193 (2001).
[122] Venkatesh, B. et al. Molecular synapomorphies resolve evolutionary relationships of extant jawed vertebrates. *Proc. Natl. Acad. Sci. USA* 98, 11382-11387 (2001).
[123] Philippe, H. et al. Early-branching or fast-evolving eukaryotes? An answer based on slowly evolving positions. *Proc. R. Soc. Lond. B Biol. Sci.* 267, 1213-1221 (2000).
[124] Stechmann, A. and Cavalier-Smith, T. Rooting the eukaryote tree by using a derived gene fusion. *Science* 297, 89-91 (2002).
[125] Snel, B. et al. Genome evolution. Gene fusion versus gene fission. *Trends Genet.* 16, 9-11 (2000).
[126] Bapteste, E. and Philippe, H. The potential value of indels as phylogenetic markers: Position of trichomonads as a case study. *Mol. Biol. Evol.* 19, 972-977 (2002).
[127] Krzywinski, J. and Besansky, N. J. Frequent intron loss in the white gene: A cautionary tale for phylogeneticists. *Mol. Biol. Evol.* 19, 362-366 (2002).
[128] Pecon-Slattery, J. et al. Phylogenetic assessment of introns and SINEs within the Y chromosome using the cat family Felidae as a species tree. *Mol. Biol. Evol.* 21, 2299-2309 (2004).
[129] Eisen, J.A. Phylogenomics: improving functional predictions for uncharacterized genes by evolutionary analysis. *Genome Res.* 8:163–167 (1998).
[130] Eisen JA, Wu M. Phylogenetic analysis and gene functional predictions: phylogenomics in action. *Theor Popul Biol.* 61: 481–488 (2002).
[131] Mitchell-Olds, T. and Clauss, M.J. Plant evolutionary genomics. *Curr. Opin. Plant Biol.* 5:74–79 (2002).
[132] Nierman, W.C. et al. Complete genome sequence of Caulobacter crescentus. *Proc. Natl. Acad. Sci. USA* 98: 4136–4141(2001).
[133] Heidelberg, J.F. et al. The genome sequence of Vibrio cholerae and the etiologic agent of cholera. *Nature* 406: 477–483 (2000).
[134] Brown, D. and Sjolander, K. Functional Classification Using Phylogenomic Inference. *Computational Biology*. 2, Issue 6: e77 (2006).
[135] Sjolander, K. Getting Started in Structural Phylogenomics. *Computational Biology*. 6, Issue 1: e1000621 (2010).
[136] Krishnamurthy, N. et al. FlowerPower: Clustering proteins into domain architecture classes for phylogenomic inference of protein function. *BMC Evol Biol.* 7 Suppl 1: S12 (2007).

[137] Baker, D. and Sali, A. Protein structure prediction and structural genomics. *Science* 294: 93–96 (2001).

[138] Brinkmann, H. and Philippe, H. Animal phylogeny and large-scale sequencing: progress and pitfalls. *Journal of Systematics and Evolution*. 46 (3): 274–286 (2008).

[139] Lartillot, N. et al. Suppression of long-branch attraction artefacts in the animal phylogeny using a site-heterogeneous model. *BMC Evolutionary Biology*. 7 Suppl 1: S4 (2007).

[140] Rodriguez-Ezpeleta, N. et al. Detecting and overcoming systematic errors in genome-scale phylogenies. *Systematic Biology* 56: 389–399 (2007).

[141] Felsenstein, J. Cases in which parsimony or compatibility methods will be positively misleading. *Syst. Zool.* 27, 401-410 (1978).

[142] Hendy, M. and Penny, D. A framework for the quantitative study of evolutionary trees. *Syst. Zool.* 38, 297-309 (1989).

[143] Keeling, P. J. and Fast, N. M. Microsporidia: Biology and evolution of highly reduced intracellular parasites. *Annu. Rev. Microbiol.* 56, 93-116 (2002).

[144] Whelan, S. et al. Molecular phylogenetics: state-of-the-art methods for looking into the past. *Trends Genet.* 17, 262-272 (2001).

[145] Kolaczkowski, B. and Thornton, J. W. Performance of maximum parsimony and likelihood phylogenetics when evolution is heterogeneous. *Nature* 431, 980-984 (2004).

[146] Sullivan, J. and Swofford, D. L. Should we use model-based methods for phylogenetic inference when we know that assumptions about among-site rate variation and nucleotide substitution pattern are violated? *Syst. Biol.* 50, 723-729 (2001).

[147] Huelsenbeck, J. P. The robustness of two phylogenetic methods: four-taxon simulations reveal a slight superiority of maximum likelihood over neighbor joining. *Mol. Biol. Evol.* 12, 843-849 (1995).

[148] Gaut, B. S. and Lewis, P. O. Success of maximum likelihood phylogeny inference in the four-taxon case. *Mol. Biol. Evol.* 12, 152-162 (1995).

[149] Galtier, N. and Gouy, M. Inferring pattern and process: maximum-likelihood implementation of a nonhomogeneous model of DNA sequence evolution for phylogenetic analysis. *Mol. Biol. Evol.* 15, 871-879 (1998).

[150] Foster, P. G. Modeling compositional heterogeneity. *Syst. Biol.* 53, 485-495 (2004).

[151] Lartillot, N. and Philippe, H. A Bayesian mixture model for across-site heterogeneities in the amino-acid replacement process. *Mol. Biol. Evol.* 21, 1095-1109 (2004).

[152] Pagel, M. and Meade, A. A phylogenetic mixture model for detecting pattern heterogeneity in gene sequence or character-state data. *Syst. Biol.* 53, 571-581(2004).

[153] Hillis, D.M. Inferring complex phylogenies. *Nature* 383: 130–131(1996).

[154] Philippe, H. and Douzery, E. The pitfalls of molecular phylogeny based on four species, as illustrated by the Cetacea/Artiodactyla relationships. *Journal of Mammalian Evolution* 2: 133–152 (1994).

[155] Zwickl, D.J. and Hillis, D.M. Increased taxon sampling greatly reduces phylogenetic error. *Systematic Biology* 51: 588–598 (2002).

[156] Sanderson, M.J. and Shaffer, H.B. Troubleshooting molecular phylogenetic analyses. *Annual Review of Ecology and Systematics* 33: 49–72 (2002).

[157] Woese, C.R. et al. Archaeal phylogeny: reexamination of the phylogenetic position of Archaeoglobus fulgidus in light of certain composition-induced artifacts. *Systematic and Applied Microbiology* 14: 364–371 (1991).

[158] Brinkmann, H. et al. An empirical assessment of long-branch attraction artefacts in deep eukaryotic phylogenomics. *Systematic Biology* 54: 743–757 (2005).

[159] Dobzhansky, T. Nothing in biology makes sense except in the light of evolution. *Am. Biol. Teacher* 35, 125-129 (1973).

In: Bioinformatics
Editor: Renu Tuteja

ISBN 978-1-62100-913-9

Chapter VIII

Transposons – The Alchemists in 'Junk' Disguise

Vipin Singh[*]
Asst. Professor, Amity Institute of Biotechnology, Amity University
Noida, Uttar Pradesh, India

Abstract

One of the recurrent themes of most sequencing projects has been the high content of repetitive sequences, more specifically transposons, in eukaryotic genomes. Barring a few exceptions, transposon or transposon derived sequences account for the highest fraction of most genomes e.g. 45% in human, 37.5% in mouse, 52% in Opossum, about 80% in barley , 60 % in maize and 22% in *Drosophila*, and they continue to proliferate in virtually all known eukaryotic species. With numbers close to millions and still not assigned a well-defined function, the transposons remain an unsolved puzzle. Long relegated as selfish or junk DNA, they are turning out to be a treasure trove of genomic novelties as their impact on host genome evolution is beginning to be understood. Bioinformatics has played a key role in this recent resurrection of transposons. In this chapter, I briefly introduce the diversity among transposons, followed by citation of recent literature signifying the impact of transposons on the host genomes and finally emphasize on various bioinformatics resources that are geared towards identifying and annotating transposons and other repeats.

1.1. Background

The term 'repetitive sequences' (repeats, DNA repeats, repetitive DNA) refers to homologous DNA fragments present in multiple copies in the genome. Repetitive DNA was originally discovered based on reassociation kinetics and classified into 'highly' and 'middle' repetitive sequences, roughly corresponding to tandem and interspersed repeats, respectively.

[*] Email – vipin@ccmb.res.in or ashvip@gmail.com.

Tandem repeats represent arrays of DNA fragments immediately adjacent to each other in head-to-tail orientation. In contrast, interspersed repeats are DNA fragments with an upper size limit of 25-30 kb, inserted more or less at random into host DNA. Interspersed repeats are mostly inactive and often incomplete copies of transposable elements (TEs) [1].

Ever since the Nobel prize winning pioneering work of Barbara Mc. Clintock on the 'controlling elements' at the 'dissociation locus' in maize, TEs (or mobile elements or transposons) continue to be an enigma in the genome [2]. In the post-sequencing era it is being increasingly realized that a major portion of almost all eukaryotic genomes consists of transposons and transposon derived sequences, and that they continue to proliferate in virtually all known eukaryotic species. Almost 45% of the human genome is accounted for by transposons and their derivatives [3]. In mouse 37.5% of the genome consists of TEs [4]. 52% of the genome in short tailed opossum is attributed to TEs [5]. This number is greater than 80% for some grasses (> 90% in wheat), 60 % for maize and 35% for rice. In *Drosophila*, the TEs account for 3.1% of the genome while in *C.elegans* this number stands at 6.5%. As is evident from the above data, the transposons are ubiquitous. The only known exceptions are *Plasmodium falciparum* and probably several closely related species, in which no transposable element has been identified [6]. Thus, with numbers close to millions and still not assigned a well-defined function, the transposons continue to be an enigma.

In this chapter, I first introduce the diversity among transposons, followed by various mechanisms of transposition, citation of recent literature signifying the impact of transposons on the host genomes, a brief review of various bioinformatics resources geared towards identifying and annotating transposons, followed by a brief appraisal of my own work in the field. I finally end with a quick perusal of the diametrically opposite school of thoughts on the evolutionary significance of transposons and conclude with the emerging consensus on the same in the light of current knowledge.

1.2. The Transposons

Transposons represent one of the most diverse kinds of repeats in the genome. This diversity is represented not only in the structure but also in the transposition mechanism employed. Classically, and more recently, the transposons have been divided into two major classes [7-10].

1. *Class 1 or Retrotransposons* – These transpose from one location in the genome to another via an RNA intermediate. The mechanism of transposition is commonly referred to as "copy-paste" since the original transposon remains intact and a copy is integrated elsewhere in the genome (Figure 1.1).
2. *Class 2 transposons or DNA transposons* – no RNA intermediate is formed during transposition. The transposon is excised from its original location and integrates elsewhere in the genome. This mechanism is commonly referred to as "cut – paste" mechanism of transposition (Figure 1.1).

The broad outlines of this classification are as follows.

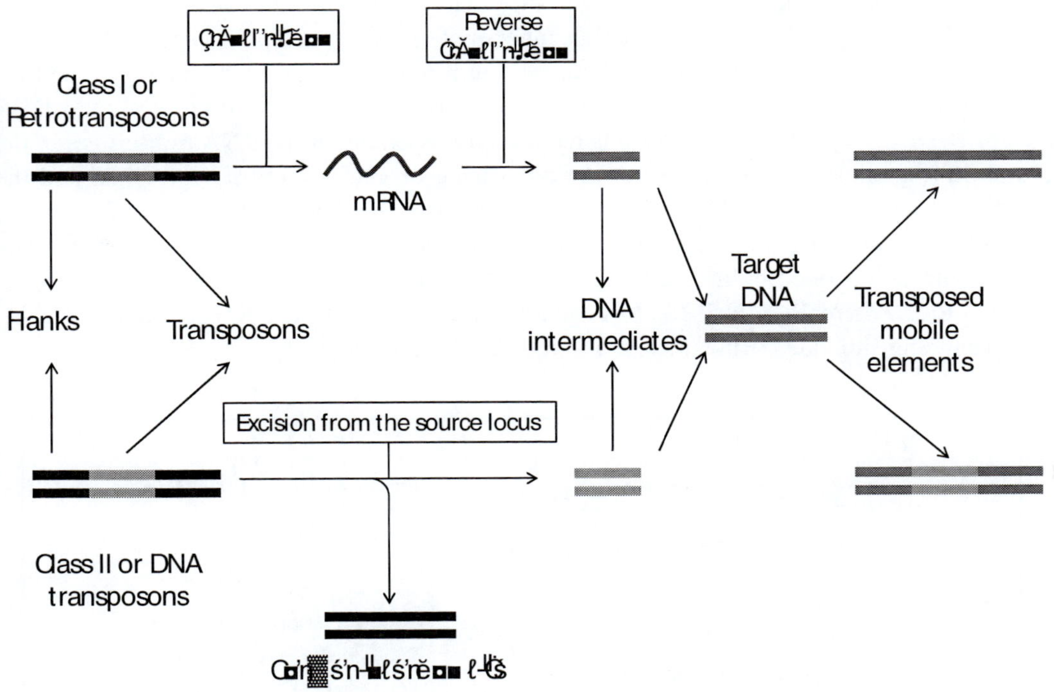

Figure 1.1. Class I and II Transposons – Based on the mechanism of transposition, eukaryotic transposons are divided into two major classes – a) Class I or Retrotransposons (red) – transpose via an RNA intermediate, the source locus remains intact and a copy of the transposon is integrated into the target locus (copy – paste mechanism). b) Class II or the DNA transposons (blue), on the other hand transpose by excision from the source locus and integration into the target locus (cut – paste mechanism).

1.2.1. Class I or Retrotransposons

Class 1 TEs transpose via RNA intermediate. The RNA intermediate is transcribed from the genomic copy, reverse transcribed, and the resulting cDNA integrated into a new target site in the genome. Since the source locus remains intact and a new copy is integrated elsewhere, retrotransposons are the predominant fraction of the repeat content of many genomes and are mojor contributors in genome expansion. Retrotransposons can be divided into four subclasses on the basis of their mechanistic features, organization and reverse transcriptase phylogeny. These are listed below.

1.2.1.1. Long Terminal Repeat (LTR) Retrotransposons

Characterized by the presence of long terminal repeats (LTRs) at either end, the LTR retrotransposons are less abundant in animals but the predominant order in plants. Size ranges from a few hundred base pairs to 25 kb. The long terminal repeats themselves vary considerably in size from a few hundred base pairs to more than 5 kb and start with 5'-TG-3' and end with 5'-CA-3'. The LTRs contain the promoters and terminators associated with transcription. Upon integration the LTR retrotransposons produce 4-6 bp target site duplications (TSDs - small direct repeats produced upon transposition because of the repair of the staggered cut introduced at the target site). The LTR retrotransposons are similar to retroviruses except that they lack a functional *env* (envelope) gene. Thus LTR retrotransposons contain ORFs for GAG (a structural protein for virus like particles) and for POL (encodes aspartic protease - AP, reverse transcriptase -RT, RNase H (RH), and DDE integrase -DDE INT). The Gag protein is required for assembly of the RNA transcript into the cytoplasmic particles; the protease for the processing of gag and pol polyproteins; and the reverse transcriptase (RT), RNase H (RH), and integrase (IN) domains are responsible for the formation of the DNA intermediate and its integration into the genome. These genes and proteins are all encoded by a single mRNA which consists of 5'-R-U5-PBS-coding region-PPT-U3-R-3' where R, U5, PBS, PPT and U3 stand for repeated RNA, unique 5'RNA, primer binding site, polypurine tract, and unique 3'RNA, respectively [8,9] (Figure 1.2).

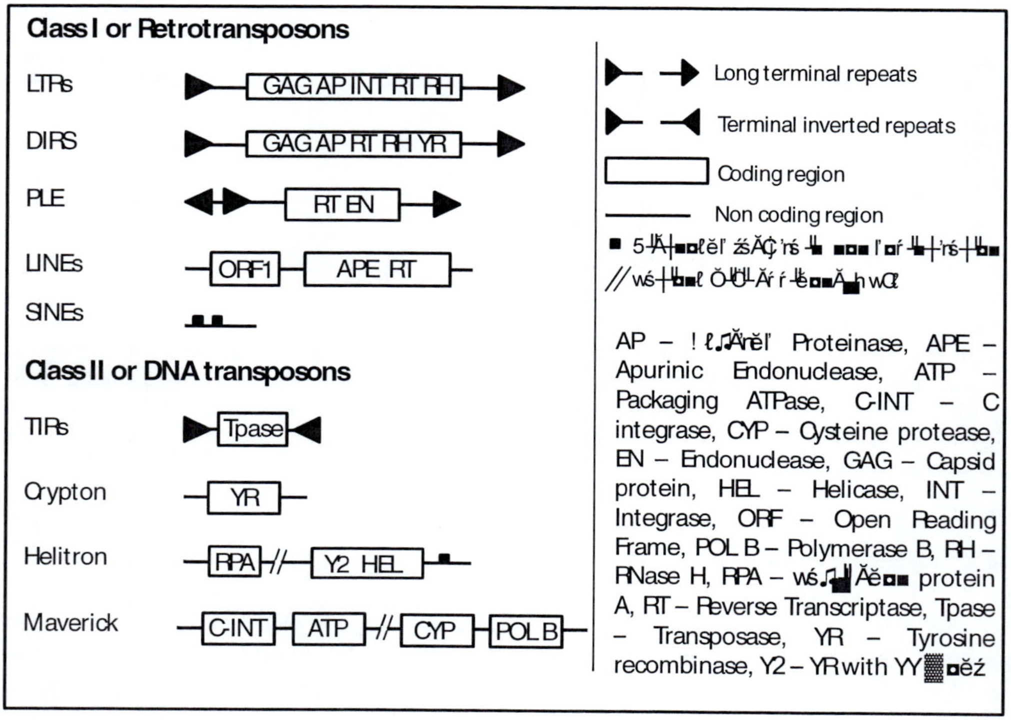

Figure 1.2. Transposon classification - Transposons are divided into two classes – Retrotransposons and DNA transposons. The Retrotransposons are further subdivided into orders LTR, DIRs, PLE, LINE and SINE. The DNA transposons are further divided into orders TIR, Crypton, Helitron and Maverick.

1.2.1.2. Non – LTR Retrotransposons

As the name indicates these TEs lack long terminal repeats. Their 3' end is generally made of poly(A) tail. They are further classified into LINEs, SINEs and SVAs. Of these LINES are autonomous i.e. they encode the entire machinery required for retrotransposition while SINEs and SVAs are non autonomous and are retroposed in trans by LINEs. LINES represent autonomous non-LTR retrotransposons and contain one or two open reading frames and an endogenous promoter at the 5' end. They are found in all eukaryotic kingdoms. Autonomous LINEs encode at least a RT and a nuclease in their pol ORF for transposition

[9,11] (Figure 1.2). *SINEs or Small Interspersed Elements* represent nonautonomous non-LTR retrotransposons. Typically they are mosaic structures derived from transfer RNA, or 7SL RNA or 5s ribosomal RNA. They have an internal Pol III promoter and do not possess any reading frames. SINES are small (80 – 500 bp) and generate target site duplications of 5 – 15 base pairs upon insertion. The 3' ends of SINEs are either derived from LINE elements or contain poly (A) tail recognizable by L1 elements [9] (Figure 1.2).

SVA or SINE-VNTR-Alu are composite non-autonomous retrotransposons named after their composite elements, SINE, VNTR (Variable Number Tandem Repeats) and Alu. There are ~ 3000 SVA elements in the human genome. The SVA elements are most likely transcribed by Polymerase II promoter [12].

1.2.1.3. DIRS or Dictyostelium Intermediate Repeat Sequences (DIRS)

DIRs were first reported in slime mould *Dycteostelium discoideum*. Other DIRS elements include PAT from nematode *Panagrellus redivivus* and Prt1 from zygomycetes fungus *Phymycoces blakesleeanus*. Members of the DIRS order contain a tyrosine recombinase (YR) gene instead of an integrase, and therefore do not form TSDs. They have unusual termini resembling either split direct repeats (SDRs) or nonidentical inverted repeats. Members of this order have a wide host range and have been reported from green algae to animals (nematodes, insects, sea urchins, fish, amphibians and remnants in human genome) and fungi. DIRS most likely evolved from a Gypsy like ancestral LTR retrotransposon after recruiting a Tyrosine recombinase [13] (Figure 1.2).

1.2.1.4. PLE or Penelope Like Elements

First reported in *Drosophila virilis*, PLE has been reported in more than 50 species, including unicellular animals (amoeba, cnidarians, rotifers, flatworms, roundworms, fish, amphibians and reptilians), fungi and plants.Its structure and the amino acid sequences of its conserved RT domains differ greatly from those of LTR and non-LTR retrotransposons, yet it shares structural features with both classes. PLE encode a single ORF composed of the RT and EN domains. The PLE reverse transcriptase is more closely related to telomerase than to reverse transcriptase from LTR or non-LTR retrotransposons. The endonuclease is similar to GIY-YIG intron encoded endonucleases (named after the conserved amino acid motif Gly-Ile-Tyr-Xn-Tyr-Ile-Gly) and the bacterial DNA repair protein Urvc.

As in non-LTR retrotransposons, Penelope has frequent 5' truncations and its TSDs, when present, are of variable length. As with LTR retrotransposons, Penelope may have LTR like structures, which may be in either direct or inverse orientation. The 5' untranslated region (5'UTR) contains a functional intron of 75 bp bounded by GT/AG donor/acceptor splice sites. Penelope is responsible for hybrid dysgenesis in *Drosophila virilis* characterized by simultaneous mobilization of several unrelated TE families in the family of dysgenic crosses [14,15] (Figure 1.2).

1.2.2. Class II Elements or DNA Transposons

Class II elements, like Class I, are ancient and found in almost all eukaryotes. They are also found in prokaryotes in simple forms called Insertion Sequences or as part more complex structures. As mentioned earlier, the class II or DNA transposons most commonly move by the non replicative "cut and paste" mechanism. However, there are a few notable exceptions that move by a "copy-paste" mechanism but without an RNA intermediate [16]. Hence, eukaryotic DNA transposons are currently divided into three major subclasses.

i. Classic cut and paste DNA transposons
ii. Rolling circle transposons or Helitrons
iii. Those whose mechanism of transposition is not well understood.

1.2.2.1. Classical Cut and Paste DNA transposons or Terminal Inverted Repeats

10 superfamilies of the cut and paste DNA transposons are recognized, namely – Tc1/mariner, hAT, P element, MuDR/Foldback, CACTA, PiggyBac, PIF/Harbinger, Merlin, Transib and Banshee. These DNA transposons are characterized by a transposase flanked by terminal inverted repeats (TIRs) of variable lengths. Members of this subclass hop from one locus to another by excising as double stranded DNA from the original location and reinserting elsewhere [16]. Most cut and paste DNA transposons encode DDE transposase.

1.2.2.2. Rolling Circle Transposons or Helitrons

The rolling circle eukaryotic transposons are represented by Helitrons. They were initially discovered in plants (*Arabidopsis thaliana* and *Oriza sativa*) and in the nematode *Caenorhabditis elegans* and are now known to occur in diverse range of species from protists to mammals. Helitrons represent a unique class of transposons fundamentally different from the classical transposons both in terms of structure and mechanism of transposition. All autonomous helitrons encode a protein called `RepHel' (1000-3000 amino acids), a Y2 type Tyrosine recombinase, which consists of a replication initiator (Rep) domain and a DNA helicase (Hel) domain. Two additional domains - the apurinic endonuclease (EN) domain and cysteine protease (CP) are found in animals. Helitron ends are defined by 5' TC and 3' CTRR motif (R= purine) and a short (15 to 20 bp) hairpin structure, a few nucleotides (10-12 bp) from the 3' end. Helitrons transpose between 5'A-T3' sites in the host and do not form target site duplications. The conserved Rep domain and absence of target site duplications strongly suggest that Helitrons use Rolling Circle model of transposition. Functionally, the 5' end of Helitron is equivalent to origin of replication (ori) and the 3' end to terminus (ter) [17,18]. Recently, helitrons have been reported to be amplifying massively in the lineage of the bat *Myotis lucifugus*. This is the first instance of helitrons being reported in mammals [19] (Figure 1.2).

1.2.2.3. Mavericks or Polintons

They represent the most complex eukaryotic transposons known till date, 15 - 20 kb long, with 6 bp TSDs, 100 to 1000 bp TIRs at both ends and 5'AC and TC 3' termini. Polintons have a wide host range and are found in protists, fungi and animals from *Entamoeba* to chicken. Polintons appear to have evolved from a linear plasmid that acquired a retroviral integrase, at least a billion year ago. They encode up to 11 proteins including DNA polymerase B (POLB), a retrovirus like INT (c-int type), an A transposase, and an adenoviral-like Cysteine protease. However, they do not encode a RT, suggesting that they undergo replicative transposition without RNA intermediates [20] (Figure 1.2).

1.3. The Transposition Mechanism

The mechanism of transposition depends on the type of transposase recruited by the element. Till date five different protein families have been described to mediate transposition [21]. These include:

a) DDE transposases - Most DDE transposons excise from the flanking DNA to generate an excised linear transposon, which is the substrate for integration into the target. Retrotransposons copy out by reverse transcribing (RT) a full length copy of their RNA. Long terminal repeat retrotransposons make a full length cDNA copy from their RNA and integrate this into a target using a DDE transposase.
b) Y2 or rolling circle transposases - encode a serine transposase which mediates excision of the transposon to form a circular intermediate. A reversal of the catalytic effects results in transposon insertion. Y2 transposons paste one strand of the transposon into the target and use it as a template for DNA replication.
c) Tyrosine (Y) transposases - retrotransposons are thought to generate a circular cDNA intermediate by reverse transcription. A Y transposase integrates the element into the target.
d) Serine (S) transposases - encode a serine transposase, which mediates excision of the transposon to form a circular intermediate.
e) RT/En transposases - retrotransposons use reverse transcriptase (RT) to copy their RNA directly into a target that has been nicked by a transposon-encoded nuclease (En). Since the first strand synthesis is facilitated by the priming of poly Ts woth the poly A tail at the excision site, the process is referred to as Target Primed Reverse Transcription.

These proteins employ different catalytic mechanisms to facilitate various DNA breaks and joins imperative to transpose to a new target, effecting cut-paste and/or copy-paste reactions.

Given the wide range of diversity in structure and as also the mechanism of transposition and the fact that genome-sequencing projects have added momentum to discovery of novel transposons, development of a unified classification system for eukaryotic transposable elements is an arduous task. However, Wicker *et al* have proposed a unified classification system for eukaryotic transposons [9] and a similar classification has been implemented in Repbase [10].

1.4. The Impact of TEs on Host Genome

TEs represent miniature genomes with a versatile repertoire of cis regulatory elements and/or trans acting factors. Long relegated as selfish DNA, they are turning out to be treasure trove of genomic novelties as their impact on host genome evolution is beginning to be understood. Besides serving as inexhaustible source of novel genes and exons, gene functions and regulatory motifs and signals, the insertion of a transposon at a locus may change its properties drastically with local and/or long range or global consequences. Discussed below are a few instances to underscore the importance of TEs in host genome evolution.

1.4.1. Genes Evolved from TEs

All transposon genes – Transposase, Integrase, Reverse transcriptase, Gag and Envelope - have been co-opted by host genomes for cellular functions. A few examples of genes owing their origin to transposons or transposon derived sequences are discussed below.

Two indispensable components of V(D)J recombination, a gene assembly mechanism responsible for the adaptive immune system in vertebrates, are – (1) the RAG1 and RAG2 proteins, which interact to form the recombinase, and (2) the recombination signal sequences (RSS) flanking the V, D, and J segments, which define the specific sequences bound , cleaved and joined by the RAG1/2 protein complex [22,23]. The shared ancestry of V(D)J recombinases and transposases was established with the discovery that RAG1, which provides the catalytic core for V(D)J recombination, is closely related in sequence to transposases encoded by Transib elements, a group of DNA transposons recently identified in the genomes of diverse invertebrates [24].

The fact that Transib elements form 5 bp target site duplications upon transposition as do most cut and paste reactions mediated by RAG1 in vitro, as also the conservation of spatial and sequence characteristics between TIRs of Transib and RSS unequivocally indicate that V(D)J recombination is a product of fortuitous event of DNA transposon domestication.

Other examples of genes originating from transposons include *Daysleeper* in *Arabidopsis thaliana* [25], *Gary* in several grasses [26], *Tram (Zbed1)* [27] , *Buster1-3, Zbed4* [28] *and P52Ripk* [29] *in* mammals, *Dref* in *Drosophila* and *Lin-15B* in *Caenorhabditis elegans* [8] (all derived from Transposase gene of hAT DNA transposons). Genes possibly derived and domesticated from Pogo DNA transposons include *Cenp-B* in eukaryotes [30], *Pdc2/Rag3* in Yeast [31], *Jerky* [32], *Jerky like gene* [33]and *KIAA0461* [28] in mammals. Five *Pgbd* genes [34], derived from PiggyBac like transposons, have been detected in the human genome. *Mustang*, a family of genes derived from transposase gene of Mutator transposon, has been found in diverse angiosperms [35]. Domesticated integrases include *Gin1* in eutherian mammals (derived from Ty3/Gypsy LTR retrotransposons) [36] and *C- integrase* (possibly derived from Mavericks) [37,38]. *Mart*, a family of ten gag-derived genes, is found in eutherian mammals [39]. Ma is another family of Gag derived genes, which encode neuronal proteins that are target of the auto immune response associated with paraneoplastic neurological disorders. Ma genes are derived either from LTR retrotransposons or retroviruses [40]. Fv1 gene is another gag derived sequence which restricts murine leukemia virus replication [41,42].

Domesticated envelope genes include Syncytin-1 [43,44] and Syncytin-2 [45,46] in human encoded by defective proviruses, HERV-W and HERV-FRD, respectively, and Syncytin-A and Syncytin-B in mouse [47]. Syncytins are specifically expressed in placenta and play a role in formation of syncytiotrophoblast [48]. Another intriguing potential function of Env proteins is the possible role of the protection of the fetus by naturally existing retroviral immunosuppressive function [49].

Volff enlists 40 domesticated genes derived from transposons [50]. Feschotte and Pritham, despite applying stringent criteria, enlist 54 domesticated genes derived from DNA transposons alone [16]. Some of the domesticated genes may have diverged so much from their ancestral sequences that they escape detection. Thus, the exact quantum of domesticated genes may be much higher than currently estimated.

1.4.2. Novel Gene functions

Not only have the TEs given rise to a large number of new genes as mentioned above, in a few instances, they also ensure that the gene function is not compromised in absence of the gene. The HetA and TART retrotransposons accounting for the telomerase function in Drosophila is a case in point. Telomeres are nucleoprotein structures that cap chromosomal ends. Generally telomeres consist of 6 to 10 bp, GT rich, tandem repeats. Maintenance and regulation of telomere length is imperative for normal cellular functions as they prevent chromosomes from shortening by under replication and thus prevent degradation of coding sequences.

They also distinguish true chromosomal ends from double strand breaks and thus prevent end fusion, recombination, or activation of cell cycle checkpoints. A specialized multi-subunit ribonucleoprotein, telomerase, a reverse transcriptase that adds telomeric repeats de novo after each cell division thus counteracting the end replication problem, is recruited by most eukaryotes [51]. However, telomerase is not encoded in all the *Drosophila* species studied till date.

The absence of telomerase is compensated for by two non-LTR retrotransposons – HetA and TART, which have been co-opted for telomere maintenance. Telomeres in drosophila are composed of tandem arrays of repeated sequences formed by successive transpositions of HetA and TART. Thus the Drosophila telomere repeats are 6 to 14 kb long. HetA and TART specifically retropose to chromosomal ends forming a polar array with the 3' end of all elements oriented towards the centromere [52-54].

TEs can also be recruited as non-coding host gene, e.g. BC200 (Brain cytoplasmic RNA, 200 nt), a small neural non-messenger RNA formed from a monomeric Alu element, is a component of ribonucleoprotein particle that is thought to act as a regulator of decentralized translation of dendrites in anthropoid primates, thought to contribute to the maintenance of long-term synaptic plasticity [55,56]. BC 200 levels are reduced by > 60% during normal ageing, but are significantly upregulated in Alzheimer's brains. There is a linear relationship between BC200 upregulation and disease progression. However, the cause –effect relationship between two is unclear [57,58].

1.4.3. Gene Duplication, Exon Shuffling and Evolution of New Genes

TEs also play an important role in two of the most important mechanisms of evolution of new genes and biological functions, namely, 'exon shuffling' and 'gene duplication'. The 'Exon shuffling' hypothesis was put forth by Walter Gilbert (1978) and entails that exons code for functional units of a protein and that evolution of new genes has proceeded by recombination or exclusion of exons. Gene duplication, as the name indicates, refers to duplication of regions of DNA containing genes. Since the second copy of the gene is often free from selective pressure, it mutates faster than the functional gene copy and may eventually become redundant (Pseudogene) or may evolve into a new gene with subtle or striking differences in structure and function from the original gene. The gene duplication hypothesis has been very strongly advocated by Susomo Ohno in his book 'Evolution by gene duplication' (1970).

TEs can potentially shuffle the genome by carrying the 5' and 3' flanks to new locations in the genome with (retrotransposons) or without (DNA transposons) duplication. This process is called 'flank transduction' and is particularly well studied and understood in retrotransposons. 3' flank transduction was first reported in LINES in the human genome and has since been reported for other retro-elements as well. It has been suggested that 3' flank transduction occurs because the endogenous poly (A) signal of the retro-element is weak, resulting in a read through transcript until the next poly (A) signal. This transcript is then integrated elsewhere in the genome resulting in duplication of the flank [12,59-64]. In case a retrotransposon inserts downstream of a strong promoter, the transcription may start from the upstream promoter instead of the endogenous promoter, resulting in transduction of 5' flank.

Exon shuffling by L1 retrotransposns mediated flank transduction, has been proposed to be an important mechanism for evolution of novel genes [64-66]. Examples of exon shuffling by retrotransposon include exon 9 of CFTR (Cystic fibrosis transmembrane conductance regulator) [67], and ATM (AT, mutated) exon 30 and flanking intron [68]. Till date, only one example of gene duplication by retrotransposition mediated flank transduction has been elucidated. The AMAC (Acyl-malonyl condensing enzyme) gene has been duplicated thrice in the human genome through SVA mediated 3' flank transduction events. The three transduced copies of the AMAC gene contain intact ORFs in the human genome, and at least two are actively transcribed in different human tissues [69].

DNA transposons can also copy gene sequences from one region in the genome to another. This is particularly true of Pack- MULEs (Mutator Like Transposable Elements). They are non-autonomous MULEs but differ from other non-autonomous elements in not being deletion derivatives of autonomous MULEs. Instead, these are chimaeric and generally contain a variety of sequences between their terminal inverted repeats (TIRS), including fragments from host genes. This phenomenon was first reported for Maize Mu1 element, which contains part of a gene of unknown function called MRS-A. Arabidopsis and rice MULEs are also known to harbor gene sequences. Over 3000 Pack-MULEs in rice containing fragments from more than 1000 cellular genes have been reported. Pack-MULEs frequently juxtapose fragments from multiple chromosomal loci to form novel open reading frames, some of which are expressed as chimaeric transcripts. Comparison of captured gene fragments and their source genes indicate that fragments of genomic DNA have been captured, rearranged and amplified over millions of years. Given that MULEs are ubiquitous in all characterized plant genomes, gene fragment capture by Pack-MULES might represent an important mechanism for the origin of new genes in higher plants [70].

Thus, TEs have the potential to shuffle coding sequences in the genome, juxtaposing gene fragments derived from diverse regions and functions, effecting new permutations and combinations and placing them in new genomic contexts, thereby serving as efficient handles to tinker with the genome and contribute to evolution of novel genes, new functions and possibly to origin of new species.

1.4.4. Gene Regulation by Transposons

The evolutionary significance of cis regulatory mutations is no less important than mutations within genes. Active TEs have a rich repertoire of regulatory sequences, which include basal promoters for RNA polymerase II or III, enhancers, insulators, splice sites and polyadenylation signals and a number of transcription factor binding sites. Hence TEs can directly influence the expression of nearby genes at transcriptional and post-transcriptional levels.

Insertion of a TE 5' of a gene may

a) *Provide alternative promoter for host genes - e.g.* LTR mediated transcription of Agouti [71], PTN [72], apoC-1, EDNRB [73], CYP19 [74], βGAL-T5 [75], and SPAM1 [76] genes. 49 L1 antisense promoter driven chimeric mRNAs corresponding to sense and anti sense strands of human genes and have been reported. It has also been shown that the L1 antisense promoter is capable of functioning as tissue specific alternative promoter expanding the transcriptional flexibility of several human genes [77-79].
b) *Disrupt existing cis regulatory elements - e.g.* Gypsy elements in Drosophila exert mutagenic effects by functioning as insulators, blocking the activation of the gene by distal enhancers. The insulator effect is mediated by Suppressor of Hairy-wing which binds to 12 octomeric binding sites in the 5' UTR of gypsy [80].
c) *Introduce new cis elements - e.g.* Insertion of a Ty element in the 5' UTR of CYC-7 locus results in 20 fold increase in the expression of the gene as well as the Ty element, in haploid cells. The LTR of the Ty element contains SV40 enhancer like sequences and mating-type responsive elements, which explains haploid cell specific overexpression [81]. Transposons may also import tissue specific enhancers as happens in case of *tom* retrotransposons in *Drosophila annanasae* resulting in mutant eye phenotypes [82]. Orientation specific insertion effects have also been documented. Thus, *Tam3* insertion into the intron of *plena*, a homologue of Arabidopsis floral homeotic gene *Agamous*, in the same orientation as the gene results in loss of function 'ple' phenotype in which the sterile floral organs replace the sex organs. Tam3 insertion at the same site in opposite orientation, however, results in gain of function 'ovulata' phenotype, in which *ple* is ectopically expressed and sterile organs are replaced by sex organs [83].
d) *Premature termination signals* – Retrotransposon sequences can provide polyadenylation signals that induce termination of gene transcripts. Of the 1.1 million Alus in the human genome, about 10,000 are found in 3' UTRs of protein coding genes. 1% of these provide active polyadenylation sites for their respective genes. Known mammalian polyadenylation signals comprise the major hexamer AATAAA plus at least 12 single nucleotide variants. Alu sequences contain 3 sites that potentially harbor such hexamers either directly or through a single base mutation. [84].
e) *Alteration of splice sites* - A TE inserted within the gene may alter the normal splice sites resulting in novel chimaeric transcripts through exon skipping, intron retention and retention of TE derived sequences in the transcript. This can result in translation of new protein isoforms, or in the destabilization or degradation of the mRNA by the

nonsense mediated decay pathway. For instance, the human Leptin receptor gene is expressed in two different isoforms OBRa and OBRb due to alternative splicing within the LTR of the endogenous retrovirus HERV-K, resulting in a unique peptide sequence of 67 amino-acids in the C terminus of the ORBa protein [85].

Besides the above mentioned local and context specific changes, TEs may also bring about global changes in gene regulation. The dispersal of expanding TE families throughout the genome potentially allows the same regulatory motif or motifs to be recruited at many chromosomal locations, pulling multiple genes into a single regulatory network. This model revisits the Britten and Davidson model [86] and has recently received experimental support.

Johnson *et al* show that transposon mediated duplication and insertion of repressor element 1 (RE1 or Neuron restrictive silencing element, NRSE), a long (21 bp) conserved transcription factor binding site, has led to the acquisition of novel target genes by REST (Neuron restrictive silencing factor), a transcription repressor specific to RE1.

REST binds to numerous neuron specific target genes and plays important role in multiple biological processes and disease states [87]. More than 1500 near perfect binding sites for p53 have been dispersed in the human genome by two families of endogenous retrovirus long terminal repeats.

These sites account for 30% of all p53-binding sites. Furthermore, in all five individual cases that were studied in details, the p53 binding site within the LTR could be directly associated with p-53 dependent transcriptional activation of the closest adjacent gene. Since the p53 binding site containing LTRs are primate specific, it seems plausible that the dispersal of p53 binding site by these LTRs brought several new genes into the p53 network promoting the assembly of a primate-specific transcriptional network of p53 regulated genes [88].

1.4.5. Transposons and the Epigenetic Regulation of the Genome

Most of the epigenetic regulation concerned with transposons is directed towards silencing their expression and mobility [89]. Silencing mechanisms that suppress TEs include post transcriptional silencing of TEs directly by RNAi or RNAi mediated chromatin modifications, and RNAi independent chromatin modifications. Moreover, the RNAi silencing mechanisms of TEs differ in germline and somatic cells.

a) Post Transcriptional Silencing of TEs Directly by RNAi

TEs give rise to numerous siRNAs in most species including humans [90-93]. TE siRNA levels are correlated with element activity, and mutation in both argonaute and dicer family proteins lead to increased TE activity in most eukaryotes. In *Caenorhabditis elegans,* Tc1 transposons in the germ line are silenced via RNAi. Small interfering RNAs of Tc1 were detected from the precursor double stranded RNAs which in turn are derived from snapback of the readthrough transcripts, spanning the entire Tc1 element, in the region of terminal inverted repeats [90]. The presence of both sense and antisense promoters in the L1 5'UTR and the resulting bidirectional transcripts account for RNA interference and suppression of L1s in humans [92]. Transposon insertions in regions that are bidirectionally transcribed may also seed siRNA silencing of the TEs. The Aubergine dependent germline silencing of homologous TEs elsewhere in the genome upon insertion of TEs into heterochromatic loci in Drosophila is a case in point [94].

As opposed to plants, in animals the germ line is sequestered early in embryogenesis. In Drosophila as also in other animals including mammals, the germ line and somatic cell silencing of TEs is mediated by distinct pathways involving Argonaute2 (siRNA) [93,95] and Piwi (piRNA) [96-98] respectively. The mechanism of piRNA biogenesis and silencing are not well understood. Recently, HP1 homolog, Rhino has been shown to indispensible for nuage organization, transposon silencing and ping pong amplification of piRNAs in Drosophila [99]. Interestingly, the piwi subclade is not found in plants, where the germ cells differentiate from somatic cells in adult. It has been recently shown that in the pollen of Arabidopsis, TEs are unexpectedly reactivated and transpose in the pollen vegetative nucleus owing to the down regulation of DDM1 (Decrease in DNA methylation 1) and many of the TE siRNAs. However, 21 nucleotide siRNA generated from Athila retrotransposon in the vegetative nucleus translocate into the sperm nucleus and suppress TE activity in sperm. The authors propose a conserved role for reprogramming in germ-line companion cells, such as nurse cells in insects and vegetative nuclei in plants, to reveal intact TEs in the genome and regulate their activity in gametes [100].

b) Silencing of TEs by Chromatin Modifications

Several types of chromatin modifications including histone tail modifications, DNA methylation and alterations in chromatin packaging and condensation suppress TE expression. Nucleosomes associated with TEs are enriched for methylation of histone H3 at lysine 9 [101,102], a signal for repression and inactivation. DNA methylation on cytosine residues is another potent signal for silencing TEs [103,104]. In plants chromatin remodeling proteins like SWI/SNF play a role in TE silencing e.g. in A. thaliana, the SWI2/SNF2 like protein DDM1 is specifically required for TE silencing and chromatin condensation [105].

c) RNAi mediated Chromatin Modifications

RNAi may also mediate chromatin modifications leading to suppression of TE activity. For example, in A. thaliana DCL3 (Dicer like 3) gene generates 24-26 bp si RNAs from TEs which bind to AGO4 (Argonaute4). The si-RNA loaded AGO4 complex silences different TEs by targeting asymmetrical DNA methylation [106].

Thus, elaborate epigenetic mechanisms have evolved both in plants and animals to check the unbridled expansion of TEs.

1.4.6. Transposons and Genome Expansion and Contraction

Genomes are dynamic entities and undergo expansion or contraction during evolution. Forces contributing to genome expansion include polyploidy (duplication of the whole genome – common in plants), segmental duplication, and transposon amplification (mainly retrotransposons) [107]. While polidy leads to doubling of the genome size in a very short time (within one lifespan) the genome expansion accounted for by transposons is cumulative, takes place over evolutionary time scales and is quantitatively smaller.

However, there are instances of doubling of the genome size without ployploidization. In *Oryza australiensis*, a wild relative of the Asian cultivated rice, *Oryza sativa*, a recent amplification of 3 LTR retrotransposon families has resulted in the accumulation of more than 90,000 copies, leading to a rapid two-fold increase in its size in the last 3 million years [108]. A genome-wide comparative analysis of related organisms – *Arabidopsis thaliana* (125 Mb) and *Brassica oleracea* (600 Mb) reveals that several families of DNA transposons have amplified preferentially in *B. oleracea* significantly contributing to the genome expansion in this species since their divergence from a common ancestor. Genome compaction by large-scale removal of TEs in *A. thaliana* is ruled out since the two genus share virtually all TE lineages [109]. Intra-species differences in genome size may also arise because of differentials in transposon activity. Genome sizes of 9 subspecies of the *Drosophila melanogaster* group were estimated and shown to positively correlate with the amount of reverse transcriptase related sequences in all species except one [110].

A recent study elucidates a unique mechanism of genome expansion via polyploidy and the consequent burst of retrotransposon activity. In this study the authors link the expansion of the CRM1 retrotransposon subfamily in corn (allotetrapliod) with repeated formation of novel recombinant elements derived from two parental retrotransposon genotypes which may have been brought together during the hybridization of two sympatric species that make up the present day corn genome [111].

Forces contributing to genome contraction are less understood and one of the possible mechanisms includes loss of intervening DNA by intra-strand recombination between related transposons via unequal homologous recombination. For such recombination events, the candidate transposon should have multiple copies in the genome, which are not too far separated or too diverged from each other. Alu retrotransposons, with greater than one million copies in the human genome, one copy every 3000 base pairs, and remarkable sequence similarity (70% to 100%) among subfamilies of different ages, specifically predispose the human genome to recombination mediated deletions.

492 human specific Alu recombination mediated deletions accounting for loss of 400kb of DNA with respect to chimpanzee were identified. The fact that 60% of these deletions mapped with known or predicted genes (including 3 that deleted functional exons, as compared with orthologous chimpanzee genes) underlines the importance of such events in introducing subtle, yet significant genomic differences between human and chimpanzee since their divergence from a common ancestor [112]. Conversely, 663 lineage specific deletions (~771 kb of DNA, including 13 exons in 6 genes either demonstrably or putatively functional in the human genome) were found in chimpanzee, which balance out the genomic expansion caused by chimpanzee specific Alu inserts [113]. Similarly, L1 recombination mediated deletion accounts for 320 kb of the human genome with a single event (of the 56 human specific L1 recombination mediated deletion) responsible for the loss of 64 kb of DNA, the largest known mobile element recombination mediated deletion reported till date [114].

Another potential mechanism of reduction in genome size is the deletions that occur during retrotransposition. It is estimated that Alu retrotransposition may be responsible for 3000 deletion events eliminating 900 kb of DNA concurrently from the human genome [115]. Further insights were gained with investigations into L1 insertion mediated deletions. 50 such deletion events in the human and chimpanzee genomes resulting in loss of 18 and 15 kb of DNA in human and chimpanzee genome, respectively, were identified. These deletions were distinctly in two different size ranges suggesting two different molecular mechanisms effecting these deletions [116].

Loss of DNA may also occur by recombination in homologous regions within the repeat element. This is particularly conspicuous for LTR retrotransposons where the two LTRs of the same element recombine resulting in loss of intervening retrotransposon sequence and formation of solo LTRs. However, this still leaves the genome richer by a single LTR [107]. Formation of solo-LTRs through unequal homologous recombination counterbalances amplification of LTR retrotransposons in rice [117]. Recent and rapid DNA loss of 192 Mb of LTR retrotransposon sequence in the last 8 million years in rice is attributed to unequal homologous recombination and illegitimate recombination in LTR retrotransposons [118]. Lastly, illegitimate recombination may itself play an important role in containing genome expansion as has been shown in *Arabidopsis* [119].

Thus, polyploidy and expansion of repetitive DNA offer a plausible explanation for the C-value (the amount of DNA in a haploid genome) paradox [120] , the apparent lack of correlation between the genome size and genomic and morphological complexity. The dynamics of recombination within transposons and between transposons (intra and inter element unequal homologous recombination and illegitimate recombination) also clearly show that genomes do not possess a "one way ticket to genomic obesity" [121].

1.4.7. TEs and the Developmental Genome Rearrangements in Ciliates

A special type of chromatin diminution is observed in ciliates wherein extensive rearrangements of the germ line genome occur during the development of a new somatic macronucleus from the germ line micronucleus.The elimination of germ line sequences is directed by non coding RNAs and is initiated by double strand breaks [reviewed in122]. The enzyme responsible for the double strand breaks, however, remained elusive. It has been recently shown that PiggyMac, a domesticated piggy Bac transposase, is required for these rearrangements in *Paramecium tetraurelia* [123, reviewed in124].

1.4.8. TEs and Microsatellites

Several non-LTR retrotransposons carry homopolymeric tracts. Their abundance in the genome allows seeding of microsatellites at random locations in the genome. In particular, this has been studied for Alu elements which harbor two potential seeds for microsatellites: the linker region in the middle of the element and 3' polyA tail. In addition, individual Alu elements are generally flanked by A+T rich target site duplications which may themselves server as nucleation centers for microsatellites. These homopolymeric or heteropolymeric repeats can be subject to mutations, such as nucleotide substitutions and replication slippage, which can produce microsatellites of varying length and complexity.

Thus, it is not surprising that 20% of all microsatellites (including ~50% of mononucleotide microsatellites) shared by the human and chimpanzee genomes lie within Alu elements. In some cases, this association might result from a random integration of Alu elements near existing microsatellites. Alternatively, and more commonly, Alu elements themselves are the source of homopolymeric simple sequences that may give rise to microsatellite sequence motifs, following additional mutational events [125]. As per estimates, Alu elements provide at least 2.2 million potential sites for generating microsatellite repeats [126]. Triplet repeat expansion in the middle A rich region of an intronic Alu in the frataxin gene is responsible for Freidreich ataxia [127].

Given such wide ranging and far reaching impact of TEs on the host genomes, they have come to occupy the centre stage in genome research. The advent of whole genome sequencing and high end computing has contributed immensely to the development of transposon bioinformatics. I now briefly discuss the advances made in transposon bioinformatics ending in a brief prelude to our own research work.

1.5. Transposons Bioinformatics

Given their ubiquitous nature, high copy number, high divergence between members of the same class, widespread truncation (loss of sequence due to complexities in transposition or post insertion recombination), fragmentation (disruption by insertion of other sequences) and nesting (the tendency of some transposons to preferentially insert into other transposons), as also the problems they pose in sequencing and assembly of genomes, transposons present unique challenges to bioinformaticians.

Transposon bioinformatics is a relatively young, but rapidly expanding discipline, propelled by the recent strides in whole genome sequencing technologies and the tremendous expansion in storage space and computational processing. Most of transposon bioinformatics effort is geared towards TE discovery and annotation. A 'dense forest' of programs (more than 50 as enlisted by Lerat) for repeat detection and annotation has mushroomed over the last ten years.

These have been extensively reviewed in recent papers [128-132]. Major effort has gone into developing strategies and algorithms to dramatically reduce the demands of time and space for repeat detection and annotation of large eukaryotic genomes, without compromising on specificity and sensitivity. I will now briefly describe the strategies for TE discovery and annotation followed by a prelude to my comparative genomics based repeat induced sequence changes detection pipeline. The readers may refer to the aforementioned reviews or the original papers for details.

1.5.1. TE Discovery

TE discovery refers to identification of new TE classes and families in a genome. There are four broad approaches to TE discovery; viz. De novo repeat discovery, homology, structure and comparative genomics based methods. These are briefly discussed below.

1.5.1.1. De Novo Repeat Discovery

As the name indicates prior information about TE structure or homology to known TE sequences is not used. The inherent repetitive nature of transposons and other types of repeats such as tandem repeats is employed for identification and classification. De novo repeat discovery basically involves i) identification of repeats in the sequence and ii) defining repeat families.

i) Identification of Repeats in the Sequence

Several strategies including dot-plot, k-mer, spaced–seed, self comparison and periodicity approaches have evolved to identify repeats in a given sequence ab initio.

a) Dot Matrix Approach

Dot matrix or dot plot is the earliest and simplest de novo repeat finding technique wherein a sequence is plotted against itself, the common nucleotides on the x and y axis represented by dots. The lines parallel to the diagonal represent repeats in a dot plot. A major problem with the initial dot plots was the high level of background noise. Improvised dot plots, where a dot represents a threshold length of identical or nearly identical nucleotides helps improve the signal. An extension of the dot pot, the Auto Dot PLOT (Adplot) computes the diagonal band region of corresponding dot plot [133].

b) k-mer Approach

k-mer search algorithm, which treat a repeat as an exact substring of size k occurring multiple times in the string of length n, in conjunction with the suffix tree data structure (a compact representation that stores all possible suffixes of a given sequence) provides a fast and viable alternative to identify repeats. Intuitively, a small k-mer may lead to a large number of false positives, while a longer k-mer may miss out on actual repeat sequences. The length of k-mer is determined by the formula $k >= \log_4(n)$ where n is the length of the sequence. Suffix trees can be used to search for strings in linear space and time with a complexity of $O(n+z)$ where z is the number of maximal repeats.

The k-mer approach with suffix tree data structure is implemented in REPuter [134,135] and Repeat-match [136]. REPuter facilitates the detection of all exact repeats contained in complete genomes and also has a visualization component. Both RepeatFinder [137] and RepeatGluer [138] use the suffix tree data structure to generate the initial set of maximal repeats (longest exact match). In contrast to other k-mer based approaches as also most other de novo repeat finding algorithms which use assembled sequence as starting point, ReAS – Recovery of ancestral repeats [139], takes assembly reads from whole genome shotgun as input. This overcomes the errors introduced during assembly.

A major drawback of the k-mer approach is the large memory requirement, particularly for larger sequences, since all possible suffixes of a given sequence are computed and stored. The use of suffix arrays, an array of integers giving the starting position of suffixes of a string in lexicographical order, redresses the large space requirement for suffix trees. Suffix array data structure has been implemented in Vmatch [140] which subsumes REPuter.

c) Spaced Seed Approach

'Spaced seed' approach allows a predefined level of tolerance for variation in sequence identity and/or length of the k-mer. It was first used for repeat detection in PatternHunter [141] and extended to 'Multiple spaced seeds' in PatternHunter II [142]. RAP or Repeat Analysis Program [143] extends the k-mer approach by using gapped words. This produces better signal than single ungapped word resulting in higher specificity. The program works on direct indexing of gapped words, allowing identification of exact and inexact patterns.

d) Self Comparison Approaches

Self comparison or pairwise similarity searches using BLAST as implemented in Repeat Pattern Toolkit [144], BLASTER suite and RECON [145] or PALS (pairwise alignment of long sequences) as in PILER [146] also provide fast and efficient algorithms to identify repeats ab initio.

e) Other Approaches

Besides, periodicity approaches such as Fourier transform have also been used for de novo repeat finding, as in Spectral Repeat Finder [147]. The specific regions of a sequence that contribute to a given periodicity are located through a sliding window analysis. FORRepeats [140] uses a novel space and time efficient, and alphabet independent data structure called 'factor oracle' for repeat detection and works well for high copy number and not too diverged repeats.

II) Defining Repeat Families

Clustering refers to the process of organizing objects into groups whose members are similar in some way(s), so as to determine the intrinsic grouping in a set of unlabelled data. Clustering is the most crucial step of de novo repeat discovery. Nested repeats as well as fragmentation arising out of insertions or deletions lead to complexities and necessitate filtering steps.

Two general approaches are used for clustering and filtering out false positives (non TE clusters). In the first approach, sub clusters are generated when several sequences end at the same or similar position in the multiple sequence alignment of all repeats in a cluster. Sequences are split according to these boundaries and reclustered resulting in detection and separation of nested repeats. This approach has been implemented in RECON [145]. RECON uses heuristics and extends single linkage clustering so as to utilize multiple alignment information to define the boundaries of individual copies of the repeat, as also to segregate homologous but distinct repeat families. In a second approach, the longest sequence in a cluster is arrived at by merger of overlapping or nearby hits and accounting for insertions-deletions in an attempt to identify potential complete and functional copies of the transposon. The remaining copies are then clustered using single link clustering. GROUPER program from the BLASTER suite [148] and PILER de novo repeat discovery system [146] follow this approach.

Repeat Pattern Tool kit [144] builds a repeat graph using only perfect local alignments from the clustering step. The vertices V represent repetitive sequences. Weighted edges E represent relationships among similar elements. Minimum spanning trees are constructed from the connected components. Each minimum spanning tree represents a repeat family. A consensus sequence is deduced by reducing each tree to a single vertex.

RepeatGluer [138] represents all repeats in the genome as a mosaic of subrepeats and employs modified De Bruijn graphs. Every k-mer in the genome sequence is represented as a node in this graph. Two nodes are then connected by a directed edge if they overlap in the genome. RepeatFinder [137] on the other hand, merges overlapping or nearby exact repeats (derived either from REPuter or Repeat-match) to generate a set of merged repeats. Two merged repeats are kept in the same category if they contain at least one exact repeat. In the final round, BLAST is used to compare each category against all other categories. Finally, an element each, representing the resulting categories is selected as reference sequence.

The last step in de novo repeat discovery involves the construction of a multiple sequence alignment so as to generate a consensus sequence. The consensus sequence is expected to carry all diagnostic features of the family and can be used as a reference sequence for repeat annotation. While it allows identification of new TE families, one of the major drawbacks of the de novo repeat discovery approach is that it is biased towards the more abundant repeats. The low copy number repeats escape detection. Also, if the repeats are too diverged, the classification may be faulty.

1.5.1.2. Homology Based Methods

Based on detecting homology to known TE protein coding sequences, it is the most common approach to detect TE families. It offers several advantages over *de novo* TE discovery being more accurate and in identifying TEs with very low copy numbers. However, it is biased towards detection of previously identified TEs and to TEs not too diverged from the known ones. Another disadvantage is that it is not applicable to TEs that consist entirely of non-coding sequences such as SINES (Small interspersed nuclear elements eg Alu in humans) or MITES (miniature inverted repeat transposable elements). Homology based methods are more commonly used for detection as in RepeatMasker and Censor (discussed later).

1.5.1.3. Structure Based Methods

Structure based methods rely on identifying structural hallmarks of a particular class of TEs e.g. LTRs for LTR-retrotransposons. These have high specificity and are not biased against TEs with low copy numbers. However, these are limited by the fact that there is great diversity in TEs structure and some TEs do not have well defined architectural features so as to merit structure based identification.

LTR retrotransposons are particularly suited to this approach since full length elements possess several distinct architectural features such as LTRs, target site duplications (TSDs), primer binding sites (PBSs), polypurine tracts (PPTs) and ORFs for gag and pol, which facilitate structure based identification. LTR_STRUC [149] and LTR_FINDER [150] represent structure based algorithms for LTR retrotransposons. Both scan a query sequence for pairs of similar LTRs not separated from each other by a distance greater than average LTR-retrotransposon length. These are then further analyzed for LTR-retrotransposn signatures such as PBS, PPT and TSD. LTR_FINDER was the first web server for de novo LTR detection. LTRharvest [151] is an open source de novo LTR detection algorithm and can efficiently handle large datasets from finished and unfinished genomes.

RTanalyzer utilizes the hallmarks of LINE retroposition such as formation of target site duplications, endonuclease cleavage site and poly (A) to identify LINE mediated transposition events. A retrotransposition score indicates the likelihood of a given sequence arising from LINE mediated retrotransposition. TSDfinder improves the RepeatMasker LINE annotations by unambiguously identifying the TSDs using Blast2seq, thus demarcating the exact boundaries of transposition.

Similarly, inverted repeat structure (a hallmark of most DNA transposons) is used to identify miniature inverted-repeat transposable elements (MITEs - nonautonomous DNA transposons with terminal inverted repeats without transposase gene) by FINDMITE [152],

TRANSPO [153], MUST [154] and MITE-Hunter [155] programs. SMaRTFinder [156], a generic structure based algorithm uses Kurtzv suffix tree based algorithm to conduct efficient searches in DNA for an ordered set of motifs and a list of intervals specifying the distance between motifs. TE-HMM [157], a Hidden Markov Model based structural method, takes advantage of the fact that the nucleotide composition of TE ORFs often differs from that of the host genes in the same genome. For a given species, TE-HMM builds three different models for class I and class II TEs, and genes respectively using respective training sets. The input sequence is then segregated into the above three classes based on the compositional bias.

1.5.1.4. Comparative Genomics Methods

Capsi and Patcher [158] have very innovatively utilized the fact that transposition creates large insertions that can be detected in multiple sequence alignments to design a comparative genomics based method to detect new TE families. They start with whole genome multiple alignments and scan for insertion regions (IRs) where multiple alignments of orthologous genome sequences are disrupted by a large (> 200 bp) insertion in one or more species. These are then filtered for simple and tandem repeats. The IRs are then locally aligned with all other IRs to identify repeat insertion regions (RIRs). This process is constrained to cluster only IRs that are inferred to have occurred on the same branch of phylogenetic tree. This method can identify new TE families and instances, as well as date TEs to branches of a phylogeny. However, this method is heavily dependent on the quality of whole genome alignments which are usually poor in TE rich regions.

1.5.2. TE Detection

TE detection refers to annotation of a given sequence with respect to position and nature of all TEs it harbors. TE detection entails assembling a reference set of TE sequences produced by methods described above, followed by determination of consensus sequence and classification of TEs. Repbase [159,160] is the definitive repository of such reference sets for eukaryotes. The most commonly used repeat detection alogorithms are RepeatMasker [161] and Censor [162].

RepeatMasker uses a library of predefined repeat reference sequences to annotate repeats in input sequence. The CrossMatch, WU-BLAST or BLAST search engine is used for identifying the homology. The divergence of the query sequences is compared to a curated repeat library using a set of statistically optimized scoring matrices for a range of background GC levels. Large insertions or deletions that occur in TEs after they integrate in the genome can cause detection methods to identify two matches to a query sequence, instead of one match with a long gap. This occurs through both TE nesting and small scale insertion and deletion processes. Therefore, a post processing step, called 'defragmentation' is often required to join fragments of a single TE insertion event into a biologically meaningful annotation. For example, the ProcessRepeats script distributed with RepeatMasker joins LTRs to body of their corresponding LTR retrotransposon and links polyA sequences to the tail of non-LTR retrotransposons. Maskeraid, a software enhancement to RepeatMasker facilitating transparent replacement of CrossMatch by WU-BLAST increases the speed of masking more than 30-fold at the most sensitive settings [163].

Besides other files the RepeatMasker output consists of a masked file, an annotation file and a summary table. RepeatMasking is often the first essential step for annotation of newly sequenced genomes as it demarcates the unique regions of the genome and gives a general estimate of the repeat content of the genome. RepeatMasker annotation files are generally available alongside the genome sequence and in the PreMasked tab of RepeatMasker web site as well.

RepeatMasker takes only DNA sequence as input. In contrast, Censor accepts either DNA or protein sequence as input and uses the appropriate WU-BLAST program to identify repeats. Censor also provides an interactive positional map of the annotated repeats on the query sequence. PLOTREP, principally a defragmentation algorithm, post-processes censor output by merging diagonals that are within a threshold distance of each other and belong to the same family [164].

Greedier [165] is another library based iterative transposon detection algorithm which has an inbuilt mechanism for defragmentation. Identified local similarities between the repeat library and the target genome are transformed into a graph, the vertex representing a similar subsequence pair. The edges denote pairs of subsequences that can be connected to form higher similarities. Greedier scans these graph greedily to find matches to individual repeat units in the library. Based on similarity scores, matches with scores greater than a threshold are removed and the rest of the genome is stitched together. The threshold is reduced with successive iterations until no hits are returned from the comparison.

It may be noted that the quality of repeat annotation obtained from library based methods largely depends on the quality of reference sequences in the library as unequivocally established in a recent study [166] and also on the quality of the sequence assembly. Given the complexity of transposon biology, a combined evidence approach is now being taken so as to reaffirm and enhance the results by a particular algorithm by one or more alternate algorithm(s) facilitating precise identification and detection as well as avoiding false positives [148]. One such pipeline is DAWGPAWS [167], a Perl based workflow for plant genomes. The sequence in DAWGPAWS is fed to both gene annotation programs and to several repeat annotation programs like LTR_STRUC, LTR_FINDER, LTR_seq, FIND_LTR, FIND_ MITE, RepSeek, RepeatMasker, HMMER, TE nest etc. All programs are run in batch mode and the output from each program is converted into the GFF format.

I will now briefly describe our recently published pipeline, RISCI, which combines homology search, structure based approach and comparative genomics to identify differential repeat induced sequence changes in closely related genomes [168] which may then be analyzed for downstream functions. It also provides an improvement over RepeatMasker annotations, particularly for non-LTR retrotransposons, in cases where the transposon locus is differentially occupied in the two genomes under comparison.

RISCI – Repeat Induced Sequence Changes Identifier

Transposons continue to be an enigma in the genome, with numbers close to millions and still not assigned a well defined function. Long relegated as selfish or junk DNA, they are turning out to be a treasure trove of genomic novelties. Besides serving as an inexhaustible source of novel genes, exons, gene function and regulatory motifs and signals (as discussed above), the insertion of a transposon at a locus may change its properties drastically with local and/or long range or global consequences [77,169-172]. These changes are more palpable when a transposon insertion results in gene disruption and is manifested as a disease condition [173-175]. Such insertions may be subject to negative selection and lost in due course [176].

Most transposon insertions that persist are, therefore, either silent or result in subtle and/or adaptive changes. The cumulative impact of these subtle changes may account for the observed phenotypic, physiological and behavioral differences between closely related genomes that share a high degree of sequence similarity [177]. Notable examples include human-specific inactivation of the CMP-N-acetylneuraminic acid hydroxylase gene via Alu-mediated replacement resulting in widespread biochemical difference between human and non human primates [178] and the loss of exon 34 of tropleolastin gene in human via an Alu recombination-mediated deletion [179].

The challenge, then, is to selectively identify these differential insertions and the consequent alteration of the target locus. To this end, we have designed RISCI – "Repeat Induced Sequence Changes Identifier", a comprehensive comparative genomics based *in silico* subtractive hybridization pipeline to identify such changes, if exclusive to one of the two genomes being compared. It is modeled on LINEs or Long Interspersed Nuclear Elements (non Long Terminal Repeat retrotransposons) [reviewed in 180], since they display a wide array of sequence changes upon insertion, such as target site duplication, 3'and 5' flank transduction, deletion of target locus upon insertion, inversion and truncation of repeat sequence during transposition besides post insertion modifications like disruption and recombination [181].

RISCI is a comparative genomics-based pipeline which sequentially picks the transposon loci in one genome ('Reference' or 'Main' genome), using one of the three repeat mining options (see materials and methods), and precisely zooms into the corresponding orthologous loci in other genome(s) ('Comparative genome(s)') using user defined length of flanks (default 5000 bases) extending 50 bases into the transposon (repeat overhangs) and Blastn [182]. It then infers the nature of alteration either at the transposon locus in the reference genome or the ortholog in the comparative genome(s), based on event specific-alignment signatures (discussed below). The genomic context (intergenic or genic, if genic – exonic or intronic) of the transposon locus in the reference genome and the ortholog in the comparative genome(s) is also integrated by parsing the annotation files, if available. For each transposon locus in the reference genome, RISCI sequentially assesses whether the orthologous locus in the comparative genome is occupied (indicating shared ancestry), has undergone post insertion changes, or is empty. If empty, RISCI infers insertion-associated sequence changes based on the location of target site duplication (TSD - discussed later). If TSD is not found, the orthologous locus is checked for insertion-mediated deletion or parallel independent insertions or insertion deletion at the orthologous locus (Figure 1.3).

RISCI offers both whole genome as well specific region analyses. It runs on contig as well as on assembled chromosome sequence, allows multiple genome comparisons, offers three repeat mining utilities (RISCI_RM, RISCI_NON_RM and RISCI_BLAST, and two filters 'length' and 'gene' (see materials and methods). Wherever possible, the upstream query sequence is tagged with a user defined length of non repeat sequence (default- 500 bp) to avoid spurious hits (see materials and methods). In most cases this non repeat tag forms a part of the upstream blast hit used in RISCI annotation. RISCI also uses improvised soft masking (see materials and methods) to arrive at the orthologous locus in the comparative genome.

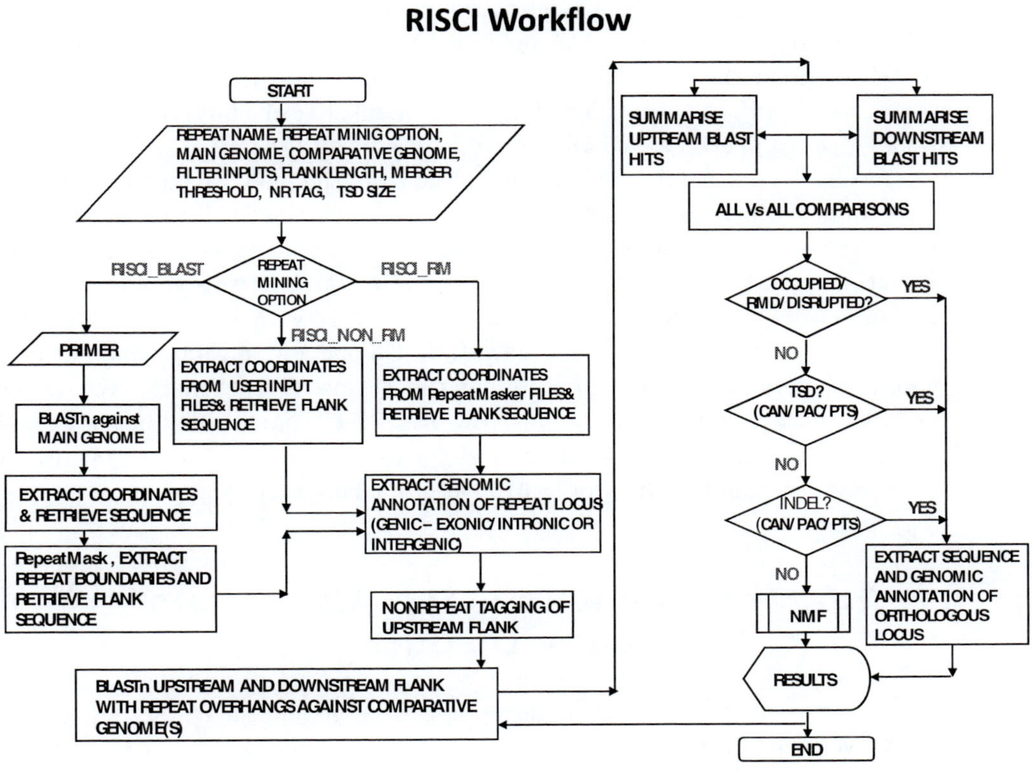

Figure 1.3. RISCI flow chart depicting basic steps of the pipeline. Major inputs include the repeat name (as identified by Repeat Masker), repeat mining option, filter inputs, flank length (default 5000 bases), merger threshold (default 50 bases), size of non-repeat tag (default 500 bases), maximum target site duplication (TSD) size (default 50 bases) and speed options. The genomic annotation of the repeat locus is parsed from the gen bank file if made available. The upstream sequence is tagged with user defined length of non repeat sequence wherever possible (default 500 bases). The upstream and downstream flanks, each carrying 50 base overhang into the repeat, are BLASTed separately against the comparative genome(s) and the BLAST alignment files summarized. For each repeat locus in the main genome, all upstream blast hits are compared against downstream blast hits in the same orientation and sequentially checked for shared ancestry (occupied), post insertion changes (recombination-mediated deletions, disruptions etc.), target site duplication (orthologous locus empty) and classified as CAN, PAC or PTS. If no matches for TSD are found both on the corresponding chromosomal homologue as well as on other chromosomes, the locus is checked for miscellaneous events (INDELs) like insertion-mediated deletion or parallel insertion or insertion deletion on corresponding chromosomal homolog, else the locus is reported as no match found (NMF). In the final results file, RISCI annotation for each locus, genomic annotation of the repeat loci in the main or reference genome and of the orthologous locus in the comparative genome, percentage repeat content of the flanks, Blastn coordinates (query and subject) for the flanks, size of TSD or INDEL or RMD are reported. In case of TSD, sequence of 5' and 3' TSD in the reference genome and of the lone copy of TSD in the comparative genome are also reported.

The blast databases of the genomes are made with the – o option set to T to enable use of fastacmd so as to speedily retrieve flank sequence from the reference genome and the ortholog sequence from the comparative genome. A merger option is also provided so as to merge BLAST hits in the comparative genome if the gap between two similarly oriented Blast HSPs is not greater than the user defined length (default 50 bp) both in terms of the query and subject coordinates. A scoring scheme has also been implemented to assign confidence scores in cases where multiple orthologous loci are predicted (see materials and methods). As mentioned above, specialized modules to take care of complications involved in truncated repeat analysis are inbuilt. Confirmation module for flank transduction is also inbuilt in RISCI. Besides, 3 speed options are inbuilt.

Though modeled on LINES, the pipeline is generic and may be applied on other transposons as well. We showcase the utility of RISCI by comparing full length and truncated L1HS and AluYa5 retrotransposons in the reference human genome with the chimpanzee genome and the alternate human assemblies (Celera and HuRef). Comparison of the reference human genome with alternate human assemblies using RISCI predicts 14 novel polymorphisms in full length L1HS, 24 in truncated L1HS and 140 novel polymorphisms in AluYa5 insertions, besides several insertion and post insertion changes. We present comparison with two previous studies to show that RISCI predictions are broadly in agreement with earlier reports. We also demonstrate its versatility by comparing various strains of *Mycobacterium tuberculosis* for IS 6100 insertion polymorphism.

1.6. Transposons – The Alchemists in Junk Disguise

Ever since their discovery, the transposons have continued to be an enigma. Paradoxically, two diametrically opposite views exist about the contribution of TEs to host genome function and evolution. The neo-Darwinians invoke the 'Phenotype paradigm' in their quest for the selective advantage (function) provided by the transposable elements to the host genome. As per the phenotypic paradigm, "Genes ensure their survival and representation in subsequent generations by providing a selective advantage for their host organism". Since the transposons have survived and continue to proliferate, the phenotypic paradigm logically explains the quest for transposon function. In consonance with the phenotypic paradigm, Britten and Davidson proposed in their theory for 'Gene regulation for higher cells', that 'activator' RNAs encoded by 'integrator' genes seek out and bind to 'receptor' sites upstream from structural genes and bring about expression, with the integrators and receptors constituting the majority of interspersed repetitive DNA [86].

However, two seminal papers published simultaneously in Nature brought about a paradigm shift in this quest [183-185]. In these papers, the authors very strongly condemned the application of Phenotypic Paradigm to transposable elements.

> "When a given DNA, or class of DNA, of unproven phenotypic function can be shown to have encoded a strategy (such as transposition) which ensures its genomic survival, then no other explanation for its existence is necessary. The search for other explanations may prove if not intellectually sterile, ultimately futile". (Orgel and Crick, 1980)

They concluded that the emergence and spread of transposable elements could be explained solely by their ability to replicate themselves in the genome. Further, they argued that, "the replicative capacity of the transposable elements provides them with a biased transmission relative to host genes" and "because they can out replicate their hosts, their evolutionary success is largely irrelevant to any selective advantage provided to their host genome". This theory found further support when it was theoretically shown that transposable elements could spread and flourish in natural populations even if they were selectively disadvantageous to their host organisms [186]. These findings formed the apostles of the "Selfish DNA" theory, which unequivocally states that transposable elements are merely genomic parasites and that their selfish nature prevents them from playing an important role in genome evolution. It discouraged evolutionary studies into ways in which elements could influence their host genomes.

The truth, as always, is somewhere in between.

Jacob remarks, "Evolution is a tinkerer. As opposed to creating new forms de novo, it will more often rearrange whatever materials are at hand to create novelty". While the transposable elements are by and large parasitic, molecular evidence of their co-option by the host genomes or "molecular domestication" or "exaptation" is more than a sporadic episode in evolution [187]. Hence, the relationship between transposable elements and their hosts can be best appreciated in a parasite – mutualist continuum.

TEs add a new dimension to natural selection and struggle for existence. Classically, natural selection is known to act at the organism level. However, the presence of semi-autonomous, self-replicating transposable elements in the host genome results in an intragenomic conflict. In this conflict, the TEs are interested in multiplying and proliferating to the maximum so as to ensure transmission to the next generation while the host genome is constrained to maintain its own integrity owing to the mutant nature of TEs.In this struggle for existence, both transposons and the host genome produce variations, which are then subjected to natural selection. The evolution of horizontal transfer and transposition only in the germline by transposons, and the evolution of host defense mechanisms (methylation, hetrochromatinization, imprinting, rna-i etc.) against transposons are cases in point. Thus, TEs may have played a pivotal role in the two main macroevolutionary transitions in the history of life, that is, chromatin formation for the prokaryotic/eukaryotic transition and methylation for the invertebrate/vertebrate transition [187].

Another dimension to this interaction is added when some of the TEs get co-opted and specialized to perform specific cellular functions. In effect, the presence of transposons in the genome creates additional evolutionary pressures enforcing a greater dynamism on the genome than possible otherwise.

References

[1] Britten RJ, Kohne DE: Repeated Sequences in DNA. *Science* 1968, 161: 529-540.

[2] McClintock B. The significance of responses of the genome to challenge. 1983. Ref Type: Hearing.

[3] Lander ES, Linton LM, Birren B, Nusbaum C, Zody MC, Baldwin J *et al.*: Initial sequencing and analysis of the human genome. *Nature* 2001, 409: 860-921.

[4] Waterston RH, Lindblad-Toh K, Birney E, Rogers J, Abril JF, Agarwal P *et al.*: Initial sequencing and comparative analysis of the mouse genome. *Nature* 2002, 420: 520-562.

[5] Mikkelsen TS, Wakefield MJ, Aken B, Amemiya CT, Chang JL, Duke S *et al.*: Genome of the marsupial Monodelphis domestica reveals innovation in non-coding sequences. *Nature* 2007, 447: 167-177.

[6] Gardner MJ, Hall N, Fung E, White O, Berriman M, Hyman RW *et al.*: Genome sequence of the human malaria parasite Plasmodium falciparum. *Nature* 2002, 419: 498-511.

[7] Finnegan DJ: Eukaryotic transposable elements and genome evolution. *Trends Genet* 1989, 5: 103-107.

[8] Craig NL, Craige R, Gellert M, Lambowitz AM: *Mobile DNA II*. 2002.

[9] Wicker T, Sabot F, Hua-Van A, Bennetzen JL, Capy P, Chalhoub B *et al.*: A unified classification system for eukaryotic transposable elements. *Nat. Rev. Genet.* 2007, 8: 973-982.

[10] Kapitonov VV, Jurka J: A universal classification of eukaryotic transposable elements implemented in Repbase. *Nat. Rev. Genet* 2008, 9: 411-412.

[11] Ostertag EM, Kazazian HH, Jr.: Biology of mammalian L1 retrotransposons. *Annu Rev Genet* 2001, 35: 501-538.

[12] Ostertag EM, Goodier JL, Zhang Y, Kazazian HH, Jr.: SVA elements are nonautonomous retrotransposons that cause disease in humans. *Am. J. Hum. Genet* 2003, 73: 1444-1451.

[13] Poulter RT, Goodwin TJ: DIRS-1 and the other tyrosine recombinase retrotransposons. *Cytogenet Genome Res* 2005, 110: 575-588.

[14] Arkhipova IR: Distribution and phylogeny of Penelope-like elements in eukaryotes. *Syst. Biol.* 2006, 55: 875-885.

[15] Evgen'ev MB, Arkhipova IR: Penelope-like elements--a new class of retroelements: distribution, function and possible evolutionary significance. *Cytogenet Genome Res.* 2005, 110: 510-521.

[16] Feschotte C, Pritham EJ: DNA transposons and the evolution of eukaryotic genomes. *Annu. Rev. Genet.* 2007, 41: 331-368.

[17] Kapitonov VV, Jurka J: Rolling-circle transposons in eukaryotes. *Proc. Natl. Acad. Sci. USA* 2001, 98: 8714-8719.

[18] Kapitonov VV, Jurka J: Helitrons on a roll: eukaryotic rolling-circle transposons. *Trends Genet.* 2007, 23: 521-529.

[19] Pritham EJ, Feschotte C: Massive amplification of rolling-circle transposons in the lineage of the bat Myotis lucifugus. *Proc. Natl. Acad. Sci. USA* 2007, 104: 1895-1900.

[20] Kapitonov VV, Jurka J: Self-synthesizing DNA transposons in eukaryotes. *Proc. Natl. Acad. Sci. USA* 2006, 103: 4540-4545.

[21] Curcio MJ, Derbyshire KM: The outs and ins of transposition: from Mu To Kangaroo. *Nat Rev* 2003, 4: 1-13.

[22] Cooper MD, Alder MN: The evolution of adaptive immune systems. *Cell* 2006, 124: 815-822.

[23] Gellert M: V(D)J recombination: RAG proteins, repair factors, and regulation. *Annu. Rev. Biochem.* 2002, 71: 101-132.

[24] Kapitonov VV, Jurka J: RAG1 core and V(D)J recombination signal sequences were derived from Transib transposons. *PLoS Biol.* 2005, 3: e181.

[25] Bundock P, Hooykaas P: An Arabidopsis hAT-like transposase is essential for plant development. *Nature* 2005, 436: 282-284.

[26] Muehlbauer GJ, Bhau BS, Syed NH, Heinen S, Cho S, Marshall D *et al.*: A hAT superfamily transposase recruited by the cereal grass genome. *Mol. Genet Genomics* 2006, 275: 553-563.

[27] Esposito T, Gianfrancesco F, Ciccodicola A, Montanini L, Mumm S, D'Urso M *et al.*: A novel pseudoautosomal human gene encodes a putative protein similar to Ac-like transposases. *Hum. Mol. Genet* 1999, 8: 61-67.

[28] Smit AF: Interspersed repeats and other mementos of transposable elements in mammalian genomes. *Curr. Opin. Genet Dev* 1999, 9: 657-663.

[29] Gale M, Jr., Blakely CM, Hopkins DA, Melville MW, Wambach M, Romano PR *et al.*: Regulation of interferon-induced protein kinase PKR: modulation of P58IPK inhibitory function by a novel protein, P52rIPK. *Mol. Cell Biol* 1998, 18: 859-871.

[30] Smit AF, Riggs AD: Tiggers and DNA transposon fossils in the human genome. *Proc. Natl. Acad. Sci. USA* 1996, 93: 1443-1448.

[31] Hohmann S: Characterisation of PDC2, a gene necessary for high level expression of pyruvate decarboxylase structural genes in Saccharomyces cerevisiae. *Mol. Gen. Genet* 1993, 241: 657-666.

[32] Toth M, Grimsby J, Buzsaki G, Donovan GP: Epileptic seizures caused by inactivation of a novel gene, jerky, related to centromere binding protein-B in transgenic mice. *Nat. Genet.* 1995, 11: 71-75.

[33] Zeng Z, Kyaw H, Gakenheimer KR, Augustus M, Fan P, Zhang X *et al.*: Cloning, mapping, and tissue distribution of a human homologue of the mouse jerky gene product. *Biochem. Biophys. Res. Commun.* 1997, 236: 389-395.

[34] Sarkar A, Sim C, Hong YS, Hogan JR, Fraser MJ, Robertson HM *et al.*: Molecular evolutionary analysis of the widespread piggyBac transposon family and related "domesticated" sequences. *Mol. Genet Genomics* 2003, 270: 173-180.

[35] Cowan RK, Hoen DR, Schoen DJ, Bureau TE: MUSTANG is a novel family of domesticated transposase genes found in diverse angiosperms. *Mol. Biol. Evol.* 2005, 22: 2084-2089.

[36] Llorens C, Marin I: A mammalian gene evolved from the integrase domain of an LTR retrotransposon. *Mol. Biol. Evol.* 2001, 18: 1597-1600.

[37] Gao X, Voytas DF: A eukaryotic gene family related to retroelement integrases. *Trends Genet.* 2005, 21: 133-137.

[38] Feschotte C, Pritham EJ: Non-mammalian c-integrases are encoded by giant transposable elements. *Trends Genet* 2005, 21: 551-552.

[39] Brandt J, Schrauth S, Veith AM, Froschauer A, Haneke T, Schultheis C *et al.*: Transposable elements as a source of genetic innovation: expression and evolution of a family of retrotransposon-derived neogenes in mammals. *Gene* 2005, 345: 101-111.

[40] Wills NM, Moore B, Hammer A, Gesteland RF, Atkins JF: A functional -1 ribosomal frameshift signal in the human paraneoplastic Ma3 gene. *J. Biol. Chem.* 2006, 281: 7082-7088.

[41] Best S, Le TP, Towers G, Stoye JP: Positional cloning of the mouse retrovirus restriction gene Fv1. *Nature* 1996, 382: 826-829.

[42] Benit L, de PN, Casella JF, Callebaut I, Cordonnier A, Heidmann T: Cloning of a new murine endogenous retrovirus, MuERV-L, with strong similarity to the human HERV-L element and with a gag coding sequence closely related to the Fv1 restriction gene. *J. Virol.* 1997, 71: 5652-5657.

[43] Mi S, Lee X, Li X, Veldman GM, Finnerty H, Racie L *et al.*: Syncytin is a captive retroviral envelope protein involved in human placental morphogenesis. *Nature* 2000, 403: 785-789.

[44] Chang C, Chen PT, Chang GD, Huang CJ, Chen H: Functional characterization of the placental fusogenic membrane protein syncytin. *Biol. Reprod.* 2004, 71: 1956-1962.

[45] Blaise S, de PN, Benit L, Heidmann T: Genomewide screening for fusogenic human endogenous retrovirus envelopes identifies syncytin 2, a gene conserved on primate evolution. *Proc. Natl. Acad. Sci. USA* 2003, 100: 13013-13018.

[46] Chen CP, Chen LF, Yang SR, Chen CY, Ko CC, Chang GD *et al.*: Functional Characterization of the Human Placental Fusogenic Membrane Protein Syncytin 2. *Biol. Reprod.* 2008.

[47] Dupressoir A, Marceau G, Vernochet C, Benit L, Kanellopoulos C, Sapin V *et al.*: Syncytin-A and syncytin-B, two fusogenic placenta-specific murine envelope genes of retroviral origin conserved in Muridae. *Proc. Natl. Acad. Sci. USA* 2005, 102: 725-730.

[48] Potgens AJ, Drewlo S, Kokozidou M, Kaufmann P: Syncytin: the major regulator of trophoblast fusion? Recent developments and hypotheses on its action. *Hum. Reprod. Update* 2004, 10: 487-496.

[49] Kidwell MG, Lisch DR: Perspective: transposable elements, parasitic DNA, and genome evolution. *Evolution* 2001, 55: 1-24.

[50] Volff JN: Turning junk into gold: domestication of transposable elements and the creation of new genes in eukaryotes. *Bioessays* 2006, 28: 913-922.

[51] Pardue ML, DeBaryshe PG: Telomeres and telomerase: more than the end of the line. *Chromosoma* 1999, 108: 73-82.

[52] Pardue ML, DeBaryshe PG: Drosophila telomeres: two transposable elements with important roles in chromosomes. *Genetica* 1999, 107: 189-196.

[53] Pardue ML, DeBaryshe PG: Retrotransposons provide an evolutionarily robust non-telomerase mechanism to maintain telomeres. *Annu. Rev. Genet.* 2003, 37: 485-511.

[54] Casacuberta E, Pardue ML: HeT-A and TART, two Drosophila retrotransposons with a bona fide role in chromosome structure for more than 60 million years. *Cytogenet Genome Res.* 2005, 110: 152-159.

[55] Martignetti JA, Brosius J: BC200 RNA: a neural RNA polymerase III product encoded by a monomeric Alu element. *Proc. Natl. Acad. Sci. USA* 1993, 90: 11563-11567.

[56] Tiedge H, Chen W, Brosius J: Primary structure, neural-specific expression, and dendritic location of human BC200 RNA. *J. Neurosci.* 1993, 13: 2382-2390.

[57] Basile V, Vicente A, Martignetti JA, Skryabin BV, Brosius J, Kennedy JL: Assignment of the human BC200 RNA gene (BCYRN1) to chromosome 2p16 by radiation hybrid mapping. *Cytogenet Cell Genet.* 1998, 82: 271-272.

[58] Mus E, Hof PR, Tiedge H: Dendritic BC200 RNA in aging and in Alzheimer's disease. *Proc. Natl. Acad. Sci. USA* 2007, 104: 10679-10684.

[59] Goodier JL, Ostertag EM, Kazazian HH, Jr.: Transduction of 3'-flanking sequences is common in L1 retrotransposition. *Hum. Mol. Genet* 2000, 9: 653-657.

[60] Pickeral OK, Makalowski W, Boguski MS, Boeke JD: Frequent human genomic DNA transduction driven by LINE-1 retrotransposition. *Genome Res.* 2000, 10: 411-415.

[61] Szak ST, Pickeral OK, Makalowski W, Boguski MS, Landsman D, Boeke JD: Molecular archeology of L1 insertions in the human genome. *Genome Biol.* 2002, 3: research0052.

[62] Szak ST, Pickeral OK, Landsman D, Boeke JD: Identifying related L1 retrotransposons by analyzing 3' transduced sequences. *Genome Biol.* 2003, 4: R30.

[63] Moran JV, Holmes SE, Naas TP, DeBerardinis RJ, Boeke JD, Kazazian HH, Jr.: High frequency retrotransposition in cultured mammalian cells. *Cell* 1996, 87: 917-927.

[64] Moran JV, DeBerardinis RJ, Kazazian HH, Jr.: Exon shuffling by L1 retrotransposition. *Science* 1999, 283: 1530-1534.

[65] Eickbush T: Exon shuffling in retrospect. *Science* 1999, 283: 1465.

[66] Boeke JD, Pickeral OK: Retroshuffling the genomic deck. *Nature* 1999, 398: 108-9, 111.

[67] Rozmahel R, Heng HH, Duncan AM, Shi XM, Rommens JM, Tsui LC: Amplification of CFTR exon 9 sequences to multiple locations in the human genome. *Genomics* 1997, 45: 554-561.

[68] Ejima Y, Yang L: Trans mobilization of genomic DNA as a mechanism for retrotransposon-mediated exon shuffling. *Hum. Mol. Genet* 2003, 12: 1321-1328.

[69] Xing J, Wang H, Belancio VP, Cordaux R, Deininger PL, Batzer MA: Emergence of primate genes by retrotransposon-mediated sequence transduction. *Proc. Natl. Acad. Sci. USA* 2006, 103: 17608-17613.

[70] Jiang N, Bao Z, Zhang X, Eddy SR, Wessler SR: Pack-MULE transposable elements mediate gene evolution in plants. *Nature* 2004, 431: 569-573.

[71] Duhl DM, Vrieling H, Miller KA, Wolff GL, Barsh GS: Neomorphic agouti mutations in obese yellow mice. *Nat. Genet* 1994, 8: 59-65.

[72] Schulte AM, Lai S, Kurtz A, Czubayko F, Riegel AT, Wellstein A: Human trophoblast and choriocarcinoma expression of the growth factor pleiotrophin attributable to germ-line insertion of an endogenous retrovirus. *Proc. Natl. Acad. Sci. USA* 1996, 93: 14759-14764.

[73] Medstrand P, Landry JR, Mager DL: Long terminal repeats are used as alternative promoters for the endothelin B receptor and apolipoprotein C-I genes in humans. *J. Biol. Chem.* 2001, 276: 1896-1903.

[74] Landry JR, Rouhi A, Medstrand P, Mager DL: The Opitz syndrome gene Mid1 is transcribed from a human endogenous retroviral promoter. *Mol. Biol. Evol.* 2002, 19: 1934-1942.

[75] Dunn CA, Medstrand P, Mager DL: An endogenous retroviral long terminal repeat is the dominant promoter for human beta1,3-galactosyltransferase 5 in the colon. *Proc. Natl. Acad. Sci. USA* 2003, 100: 12841-12846.

[76] Dunn CA, Mager DL: Transcription of the human and rodent SPAM1 / PH-20 genes initiates within an ancient endogenous retrovirus. *BMC Genomics* 2005, 6: 47.

[77] Speek M: Antisense promoter of human L1 retrotransposon drives transcription of adjacent cellular genes. *Mol. Cell Biol.* 2001, 21: 1973-1985.

[78] Nigumann P, Redik K, Matlik K, Speek M: Many human genes are transcribed from the antisense promoter of L1 retrotransposon. *Genomics* 2002, 79: 628-634.

[79] Matlik K, Redik K, Speek M: L1 antisense promoter drives tissue-specific transcription of human genes. *J. Biomed. Biotechnol.* 2006, 2006: 71753.

[80] Ramos E, Ghosh D, Baxter E, Corces VG: Genomic organization of gypsy chromatin insulators in Drosophila melanogaster. *Genetics* 2006, 172: 2337-2349.

[81] Roelants F, Potier S, Souciet JL, de MJ: Delta sequence of Ty1 transposon can initiate transcription of the distal part of the URA2 gene complex in Saccharomyces cerevisiae. *FEMS Microbiol. Lett.* 1997, 148: 69-74.

[82] Tanda S, Corces VG: Retrotransposon-induced overexpression of a homeobox gene causes defects in eye morphogenesis in Drosophila. *EMBO J.* 1991, 10: 407-417.

[83] Bradley D, Carpenter R, Sommer H, Hartley N, Coen E: Complementary floral homeotic phenotypes result from opposite orientations of a transposon at the plena locus of Antirrhinum. *Cell* 1993, 72: 85-95.

[84] Chen C, Ara T, Gautheret D: Using Alu elements as polyadenylation sites: A case of retroposon exaptation. *Mol. Biol. Evol.* 2009, 26: 327-334.

[85] Kapitonov VV, Jurka J: The long terminal repeat of an endogenous retrovirus induces alternative splicing and encodes an additional carboxy-terminal sequence in the human leptin receptor. *J. Mol. Evol.* 1999, 48: 248-251.

[86] Britten RJ, Davidson EH: Gene regulation for higher cells: a theory. *Science* 1969, 165: 349-357.

[87] Johnson R, Gamblin RJ, Ooi L, Bruce AW, Donaldson IJ, Westhead DR *et al.*: Identification of the REST regulon reveals extensive transposable element-mediated binding site duplication. *Nucleic Acids Res.* 2006, 34: 3862-3877.

[88] Wang T, Zeng J, Lowe CB, Sellers RG, Salama SR, Yang M *et al.*: Species-specific endogenous retroviruses shape the transcriptional network of the human tumor suppressor protein p53. *Proc. Natl. Acad. Sci. USA* 2007, 104: 18613-18618.

[89] Slotkin RK, Martienssen R: Transposable elements and the epigenetic regulation of the genome. *Nat. Rev. Genet* 2007, 8: 272-285.

[90] Sijen T, Plasterk RH: Transposon silencing in the Caenorhabditis elegans germ line by natural RNAi. *Nature* 2003, 426: 310-314.

[91] Watanabe T, Takeda A, Tsukiyama T, Mise K, Okuno T, Sasaki H *et al.*: Identification and characterization of two novel classes of small RNAs in the mouse germline: retrotransposon-derived siRNAs in oocytes and germline small RNAs in testes. *Genes Dev* 2006, 20: 1732-1743.

[92] Yang N, Kazazian HH, Jr.: L1 retrotransposition is suppressed by endogenously encoded small interfering RNAs in human cultured cells. *Nat.Struct Mol .Biol.* 2006, 13: 763-771.

[93] Ghildiyal M, Seitz H, Horwich MD, Li C, Du T, Lee S *et al.*: Endogenous siRNAs derived from transposons and mRNAs in Drosophila somatic cells. *Science* 2008, 320: 1077-1081.

[94] Ronsseray S, Josse T, Boivin A, Anxolabehere D: Telomeric transgenes and trans-silencing in Drosophila. *Genetica* 2003, 117: 327-335.

[95] Kawamura Y, Saito K, Kin T, Ono Y, Asai K, Sunohara T *et al.*: Drosophila endogenous small RNAs bind to Argonaute 2 in somatic cells. *Nature* 2008, 453: 793-797.

[96] Kalmykova AI, Klenov MS, Gvozdev VA: Argonaute protein PIWI controls mobilization of retrotransposons in the Drosophila male germline. *Nucleic Acids Res.* 2005, 33: 2052-2059.

[97] Girard A, Sachidanandam R, Hannon GJ, Carmell MA: A germline-specific class of small RNAs binds mammalian Piwi proteins. *Nature* 2006, 442: 199-202.

[98] Aravin AA, Sachidanandam R, Girard A, Fejes-Toth K, Hannon GJ: Developmentally regulated piRNA clusters implicate MILI in transposon control. *Science* 2007, 316: 744-747.

[99] Klattenhoff C, Xi H, Li C, Lee S, Xu J, Khurana JS *et al.*: The Drosophila HP1 homolog Rhino is required for transposon silencing and piRNA production by dual-strand clusters. *Cell* 2009, 138: 1137-1149.

[100] Slotkin RK, Vaughn M, Borges F, Tanurdzic M, Becker JD, Feijo JA *et al.*: Epigenetic reprogramming and small RNA silencing of transposable elements in pollen. *Cell* 2009, 136: 461-472.

[101] Gendrel AV, Lippman Z, Yordan C, Colot V, Martienssen RA: Dependence of heterochromatic histone H3 methylation patterns on the Arabidopsis gene DDM1. *Science* 2002, 297: 1871-1873.

[102] Martens JH, O'Sullivan RJ, Braunschweig U, Opravil S, Radolf M, Steinlein P *et al.*: The profile of repeat-associated histone lysine methylation states in the mouse epigenome. *EMBO J.* 2005, 24: 800-812.

[103] Walsh CP, Chaillet JR, Bestor TH: Transcription of IAP endogenous retroviruses is constrained by cytosine methylation. *Nat. Genet* 1998, 20: 116-117.

[104] Woodcock DM, Lawler CB, Linsenmeyer ME, Doherty JP, Warren WD: Asymmetric methylation in the hypermethylated CpG promoter region of the human L1 retrotransposon. *J. Biol. Chem.* 1997, 272: 7810-7816.

[105] Lippman Z, Gendrel AV, Black M, Vaughn MW, Dedhia N, McCombie WR *et al.*: Role of transposable elements in heterochromatin and epigenetic control. *Nature* 2004, 430: 471-476.

[106] Qi Y, He X, Wang XJ, Kohany O, Jurka J, Hannon GJ: Distinct catalytic and non-catalytic roles of ARGONAUTE4 in RNA-directed DNA methylation. *Nature* 2006, 443: 1008-1012.

[107] Bennetzen JL: Mechanisms and rates of genome expansion and contraction in flowering plants. *Genetica* 2002, 115: 29-36.

[108] Piegu B, Guyot R, Picault N, Roulin A, Saniyal A, Kim H *et al.*: Doubling genome size without polyploidization: dynamics of retrotransposition-driven genomic expansions in Oryza australiensis, a wild relative of rice. *Genome Res.* 2006, 16: 1262-1269.

[109] Zhang X, Wessler SR: Genome-wide comparative analysis of the transposable elements in the related species Arabidopsis thaliana and Brassica oleracea. *Proc. Natl. Acad. Sci. USA* 2004, 101: 5589-5594.

[110] Boulesteix M, Weiss M, Biemont C: Differences in genome size between closely related species: the Drosophila melanogaster species subgroup. *Mol. Biol. Evol.* 2006, 23: 162-167.

[111] Sharma A, Schneider KL, Presting GG: Sustained retrotransposition is mediated by nucleotide deletions and interelement recombinations. *Proc. Natl. Acad. Sci. USA* 2008, 105: 15470-15474.

[112] Sen SK, Han K, Wang J, Lee J, Wang H, Callinan PA *et al.*: Human genomic deletions mediated by recombination between Alu elements. *Am. J. Hum. Genet* 2006, 79: 41-53.

[113] Han K, Lee J, Meyer TJ, Wang J, Sen SK, Srikanta D *et al.*: Alu recombination-mediated structural deletions in the chimpanzee genome. *PLoS Genet* 2007, 3: 1939-1949.

[114] Han K, Lee J, Meyer TJ, Remedios P, Goodwin L, Batzer MA: L1 recombination-associated deletions generate human genomic variation. *Proc. Natl. Acad. Sci. USA* 2008, 105: 19366-19371.

[115] Callinan PA, Wang J, Herke SW, Garber RK, Liang P, Batzer MA: Alu retrotransposition-mediated deletion. *J. Mol. Biol.* 2005, 348: 791-800.

[116] Han K, Sen SK, Wang J, Callinan PA, Lee J, Cordaux R *et al.*: Genomic rearrangements by LINE-1 insertion-mediated deletion in the human and chimpanzee lineages. *Nucleic Acids Res.* 2005, 33: 4040-4052.

[117] Vitte C, Panaud O: Formation of solo-LTRs through unequal homologous recombination counterbalances amplifications of LTR retrotransposons in rice Oryza sativa L. *Mol. Biol. Evol.* 2003, 20: 528-540.

[118] Ma J, Devos KM, Bennetzen JL: Analyses of LTR-retrotransposon structures reveal recent and rapid genomic DNA loss in rice. *Genome Res* 2004, 14: 860-869.

[119] Devos KM, Brown JK, Bennetzen JL: Genome size reduction through illegitimate recombination counteracts genome expansion in Arabidopsis. *Genome Res.* 2002, 12: 1075-1079.

[120] Thomas CA, Jr.: The genetic organization of chromosomes. *Annu. Rev. Genet* 1971, 5: 237-256.

[121] Bennetzen JL, Kellogg EA: Do Plants Have a One-Way Ticket to Genomic Obesity? *Plant Cell* 1997, 9: 1509-1514.

[122] Duharcourt S, Lepere G, Meyer E: Developmental genome rearrangements in ciliates: a natural genomic subtraction mediated by non-coding transcripts. *Trends Genet.* 2009, 25: 344-350.

[123] Baudry C, Malinsky S, Restituito M, Kapusta A, Rosa S, Meyer E *et al.*: PiggyMac, a domesticated piggyBac transposase involved in programmed genome rearrangements in the ciliate Paramecium tetraurelia. *Genes Dev.* 2009, 23: 2478-2483.

[124] Motl JA, Chalker DL: Subtraction by addition: domesticated transposases in programmed DNA elimination. *Genes Dev.* 2009, 23: 2455-2460.

[125] Arcot SS, Wang Z, Weber JL, Deininger PL, Batzer MA: Alu repeats: a source for the genesis of primate microsatellites. *Genomics* 1995, 29: 136-144.

[126] Batzer MA, Deininger PL: Alu repeats and human genomic diversity. *Nat. Rev. Genet* 2002, 3: 370-379.

[127] Campuzano V, Montermini L, Molto MD, Pianese L, Cossee M, Cavalcanti F *et al.*: Friedreich's ataxia: autosomal recessive disease caused by an intronic GAA triplet repeat expansion. *Science* 1996, 271: 1423-1427.

[128] Bergman CM, Quesneville H: Discovering and detecting transposable elements in genome sequences. *Brief Bioinform.* 2007, 8: 382-392.

[129] Saha S, Bridges S, Magbanua ZV, Peterson DG. Computational Approaches and Tools Used in Identification of Dispersed Repetitive DNA Sequences. *Tropical Plant Biol.* 1, 85-96. 2008. Ref Type: Generic.

[130] Saha S, Bridges S, Magbanua ZV, Peterson DG: Empirical comparison of ab initio repeat finding programs. *Nucleic Acids Res.* 2008, 36: 2284-2294.

[131] Lerat E: Identifying repeats and transposable elements in sequenced genomes: how to find your way through the dense forest of programs. *Heredity* 2010, 104: 520-533.

[132] Flutre T, Duprat E, Feuillet C, Quesneville H: Considering transposable element diversification in de novo annotation approaches. *PLoS One* 2011, 6: e16526.

[133] Taneda A: Adplot: detection and visualization of repetitive patterns in complete genomes. *Bioinformatics* 2004, 20: 701-708.

[134] Kurtz S, Schleiermacher C: REPuter: fast computation of maximal repeats in complete genomes. *Bioinformatics* 1999, 15: 426-427.

[135] Kurtz S, Choudhuri JV, Ohlebusch E, Schleiermacher C, Stoye J, Giegerich R: REPuter: the manifold applications of repeat analysis on a genomic scale. *Nucleic Acids Res.* 2001, 29: 4633-4642.

[136] Delcher AL, Kasif S, Fleischmann RD, Peterson J, White O, Salzberg SL: Alignment of whole genomes. *Nucleic Acids Res.* 1999, 27: 2369-2376.

[137] Volfovsky N, Haas BJ, Salzberg SL: A clustering method for repeat analysis in DNA sequences. *Genome Biol.* 2001, 2: RESEARCH0027.

[138] Pevzner PA, Tang H, Tesler G: De novo repeat classification and fragment assembly. *Genome Res.* 2004, 14: 1786-1796.

[139] Li R, Ye J, Li S, Wang J, Han Y, Ye C *et al.*: ReAS: Recovery of ancestral sequences for transposable elements from the unassembled reads of a whole genome shotgun. *PLoS Comput Biol* 2005, 1: e43.

[140] Abouelhoda MI, Kurtz S, Ohlebusch E. Replacing suffix trees with enhanced suffix arrays. *J. Discrete Algorithm* 2, 53-86. 2004. Ref Type: Generic

[141] Ma B, Tromp J, Li M: PatternHunter: faster and more sensitive homology search. *Bioinformatics* 2002, 18: 440-445.

[142] Li M, Ma B, Kisman D, Tromp J: PatternHunter II: highly sensitive and fast homology search. *Genome Inform* 2003, 14: 164-175.

[143] Campagna D, Romualdi C, Vitulo N, Del FM, Lexa M, Cannata N *et al.*: RAP: a new computer program for de novo identification of repeated sequences in whole genomes. *Bioinformatics* 2005, 21: 582-588.

[144] Agarwal P, States DJ: The Repeat Pattern Toolkit (RPT): analyzing the structure and evolution of the C. elegans genome. *Proc. Int. Conf Intell Syst Mol. Biol.* 1994, 2: 1-9.

[145] Bao Z, Eddy SR: Automated de novo identification of repeat sequence families in sequenced genomes. *Genome Res.* 2002, 12: 1269-1276.

[146] Edgar RC, Myers EW: PILER: identification and classification of genomic repeats. *Bioinformatics* 2005, 21 Suppl 1: i152-i158.

[147] Sharma D, Issac B, Raghava GP, Ramaswamy R: Spectral Repeat Finder (SRF): identification of repetitive sequences using Fourier transformation. *Bioinformatics* 2004, 20: 1405-1412.

[148] Quesneville H, Bergman CM, Andrieu O, Autard D, Nouaud D, Ashburner M *et al.*: Combined evidence annotation of transposable elements in genome sequences. *PLoS Comput. Biol.* 2005, 1: 166-175.

[149] McCarthy EM, McDonald JF: LTR_STRUC: a novel search and identification program for LTR retrotransposons. *Bioinformatics* 2003, 19: 362-367.

[150] Xu Z, Wang H: LTR_FINDER: an efficient tool for the prediction of full-length LTR retrotransposons. *Nucleic Acids Res.* 2007, 35: W265-W268.

[151] Ellinghaus D, Kurtz S, Willhoeft U: LTRharvest, an efficient and flexible software for de novo detection of LTR retrotransposons. *BMC Bioinformatics* 2008, 9: 18.

[152] Tu Z: Eight novel families of miniature inverted repeat transposable elements in the African malaria mosquito, Anopheles gambiae. *Proc. Natl. Acad. Sci. USA* 2001, 98: 1699-1704.

[153] Santiago N, Herraiz C, Goni JR, Messeguer X, Casacuberta JM: Genome-wide analysis of the Emigrant family of MITEs of Arabidopsis thaliana. *Mol. Biol. Evol.* 2002, 19: 2285-2293.

[154] Chen Y, Zhou F, Li G, Xu Y: MUST: a system for identification of miniature inverted-repeat transposable elements and applications to Anabaena variabilis and Haloquadratum walsbyi. *Gene* 2009, 436: 1-7.

[155] Han Y, Wessler SR: MITE-Hunter: a program for discovering miniature inverted-repeat transposable elements from genomic sequences. *Nucleic Acids Res.* 2010, 38: e199.

[156] Morgante M, Policriti A, Vitacolonna N, Zuccolo A: Structured motifs search. *J. Comput Biol.* 2005, 12: 1065-1082.

[157] Andrieu O, Fiston AS, Anxolabehere D, Quesneville H: Detection of transposable elements by their compositional bias. *BMC Bioinformatics* 2004, 5: 94.

[158] Caspi A, Pachter L: Identification of transposable elements using multiple alignments of related genomes. *Genome Res.* 2006, 16: 260-270.

[159] Jurka J, Kapitonov VV, Pavlicek A, Klonowski P, Kohany O, Walichiewicz J: Repbase Update, a database of eukaryotic repetitive elements. *Cytogenet Genome Res.* 2005, 110: 462-467.

[160] Kohany O, Gentles AJ, Hankus L, Jurka J: Annotation, submission and screening of repetitive elements in Repbase: RepbaseSubmitter and Censor. *BMC Bioinformatics* 2006, 7: 474.

[161] Tarailo-Graovac M, Chen N: Using RepeatMasker to identify repetitive elements in genomic sequences. *Curr. Protoc Bioinformatics* 2009, Chapter 4: Unit.

[162] Jurka J, Klonowski P, Dagman V, Pelton P: CENSOR--a program for identification and elimination of repetitive elements from DNA sequences. *Comput Chem* 1996, 20: 119-121.

[163] Bedell JA, Korf I, Gish W: MaskerAid: a performance enhancement to RepeatMasker. *Bioinformatics* 2000, 16: 1040-1041.

[164] Toth G, Deak G, Barta E, Kiss GB: PLOTREP: a web tool for defragmentation and visual analysis of dispersed genomic repeats. *Nucleic Acids Res.* 2006, 34: W708-W713.

[165] Li X, Kahveci T, Settles AM: A novel genome-scale repeat finder geared towards transposons. *Bioinformatics* 2008, 24: 468-476.

[166] Buisine N, Quesneville H, Colot V: Improved detection and annotation of transposable elements in sequenced genomes using multiple reference sequence sets. *Genomics* 2008, 91: 467-475.

[167] Estill JC, Bennetzen JL: The DAWGPAWS pipeline for the annotation of genes and transposable elements in plant genomes. *Plant Methods* 2009, 5: 8.

[168] Singh V, Mishra RK: RISCI--Repeat Induced Sequence Changes Identifier: a comprehensive, comparative genomics-based, in silico subtractive hybridization pipeline to identify repeat induced sequence changes in closely related genomes. *BMC Bioinformatics* 2010, 11: 609.

[169] Medstrand P, van de Lagemaat LN, Dunn CA, Landry JR, Svenback D, Mager DL: Impact of transposable elements on the evolution of mammalian gene regulation. *Cytogenet Genome Res.* 2005, 110: 342-352.

[170] von SR, Shapiro JA: How repeated retroelements format genome function. *Cytogenet Genome Res.* 2005, 110: 108-116.

[171] Feschotte C: Transposable elements and the evolution of regulatory networks. *Nat. Rev. Genet* 2008, 9: 397-405.

[172] Tomilin NV: Regulation of mammalian gene expression by retroelements and non-coding tandem repeats. *Bioessays* 2008, 30: 338-348.

[173] Kazazian HH, Jr.: Mobile elements and disease. *Curr. Opin. Genet Dev.* 1998, 8: 343-350.

[174] Callinan PA, Batzer MA: Retrotransposable elements and human disease. *Genome Dyn.* 2006, 1: 104-115.

[175] Belancio VP, Hedges DJ, Deininger P: Mammalian non-LTR retrotransposons: for better or worse, in sickness and in health. *Genome Res.* 2008, 18: 343-358.

[176] Boissinot S, Entezam A, Furano AV: Selection against deleterious LINE-1-containing loci in the human lineage. *Mol. Biol. Evol.* 2001, 18: 926-935.

[177] De S, Teichmann SA, Babu MM: The impact of genomic neighborhood on the evolution of human and chimpanzee transcriptome. *Genome Res.* 2009, 19: 785-794.

[178] Hayakawa T, Satta Y, Gagneux P, Varki A, Takahata N: Alu-mediated inactivation of the human CMP- N-acetylneuraminic acid hydroxylase gene. *Proc. Natl. Acad. Sci. USA* 2001, 98: 11399-11404.

[179] Szabo Z, Levi-Minzi SA, Christiano AM, Struminger C, Stoneking M, Batzer MA *et al.*: Sequential loss of two neighboring exons of the tropoelastin gene during primate evolution. *J. Mol. Evol.* 1999, 49: 664-671.

[180] Ostertag EM, Kazazian HH, Jr.: Biology of mammalian L1 retrotransposons. *Annu Rev. Genet* 2001, 35: 501-538.

[181] Cordaux R, Batzer MA: The impact of retrotransposons on human genome evolution. *Nat. Rev. Genet* 2009, 10: 691-703.

[182] Altschul SF, Gish W, Miller W, Myers EW, Lipman DJ: Basic local alignment search tool. *J. Mol. Biol.* 1990, 215: 403-410.

[183] Doolittle WF, Sapienza C: Selfish genes, the phenotype paradigm and genome evolution. *Nature* 1980, 284: 601-603.

[184] Orgel LE, Crick FH: Selfish DNA: the ultimate parasite. *Nature* 1980, 284: 604-607.

[185] Orgel LE, Crick FH, Sapienza C: Selfish DNA. *Nature* 1980, 288: 645-646.

[186] Hickey DA: Selfish DNA: a sexually-transmitted nuclear parasite. *Genetics* 1982, 101: 519-531.

[187] Miller WJ, McDonald JF, Nouaud D, Anxolabehere D: Molecular domestication--more than a sporadic episode in evolution. 2. *Genetica* 1999, 107: 197-207.

In: Bioinformatics
Editor: Renu Tuteja
ISBN 978-1-62100-913-9

Chapter IX

Computational Insights into the Binding Mechanisms of Withaferin A, A Herbal Anticancer Drug

Abhinav Grover, Vibhuti Agrawal, Ashutosh Shandilya, Virendra S. Bisaria and Durai Sundar*
Department of Biochemical Engineering and Biotechnology,
Indian Institute of Technology (IIT) Delhi,
Hauz Khas, New Delhi, India

Abstract

Cancer is a disease marked by genetic instability and thus specific inhibition of individual proteins or signalling pathways holds a great potential for subversion of this genetic plasticity of cancers. Many active compounds from traditional medicinal sources could serve as good scaffolds for rational drug design. Combinatorial chemistry approaches based on natural product scaffolds are being used to create screening libraries that closely resemble drug-like compounds. Most of these compounds are part of routinely used traditional medicines and hence their tolerance and safety are relatively better known than any other chemical entities that are new for human use. Thus traditional medicine-based bio-prospecting offers unmatched structural variety as promising new leads. Recent surge towards usage of high-specificity drugs having reduced side effect profile is urging explorations with naturally occurring herbal drug candidates. Computational tools can be used to elucidate the interactions between the drug and its target molecules and to identify the stability of such interactions. The understanding obtained from these studies will help in developing future approaches towards cure of these nefarious cancerous diseases.

In this chapter, we discuss the progress in computational approaches for drug discovery and its potential application in biomedicine.

* Corresponding author: Tel: +91-11-26591901. Fax: +91-11-26582282. Email: sundar@dbeb.iitd.ac.in.

Cancer is one of the most dreaded diseases of the 20th century and is spreading further continuously with increasing incidence in 21st century. It is a major public health burden in both developed and developing countries. A study conducted in 2011 projected around 1.59 million new cases and 0.59 million deaths due to cancer in the US alone [1]. Cancer is the second leading cause of death in the United States [2] where one in every four deaths is due to cancer. Multidisciplinary scientific investigations are being carried out in different parts of the world with the aim to combat this disease but a sure-shot perfect cure is still eluding the world of medicine. Recently, a greater emphasis is being laid on finding alternative and complementary fields of medicine to provide a better strategy to deal with cancer. 'Ayurveda' as it means in Sanskrit (the ancient language of India) – 'ayus' (life) and 'ved' (knowledge), often translated as science of life, is one such alternative field of medicine. It is a 5000 years old system of Indian medicine which uses natural herbal extracts to prevent diseases, rejuvenate body systems and extend lifespan. It has been successful in preventing or suppressing various tumors since very early times. Today, several studies are being conducted on these herbs under a multitude of ethno-botanical grounds.

Natural products have long been attractive tools for cell biologists. These molecules are often synthesized by organisms in response to stress and can exhibit a plethora of biological activities including anti-cancer, anti-viral and anti-microbial activities. Due to these interesting properties, these natural products, mostly derived from plants have consistently been subjected to various pharmacological, clinical and chemical studies as early as the beginning of 19th century.

Many molecules such as aspirin, digitoxin, morphine, quinine, streptomycin, chloramphenicol, erythromycin and vancomycin, which have been described decades or centuries ago, are still being used as drugs. It is estimated that over hundred new natural product-based leads are in clinical development [3]. It would be cheaper and perhaps more productive to re-examine plant remedies described in ancient texts [4]. Many active compounds from traditional medicinal sources could serve as good scaffolds for rational drug design. Combinatorial chemistry approaches based on natural product scaffolds are being used to create screening libraries that closely resemble drug-like compounds [5]. Most of these compounds are part of routinely used traditional medicines and hence their tolerance and safety are relatively better known than any other chemical entities that are new for human use [6]. Thus traditional medicine based bio-prospecting offers unmatched structural variety in promising new leads [7].

Cancer and Natural Products

Currently, more than 30 compounds of natural origin are in different phases of clinical study for the treatment of different types of cancer [8].

Additionally, many new natural compounds of diverse structures isolated from plant sources have been considered leads or prototypes and their structural modifications have afforded compounds with significant pharmacological activities. Several plant derived compounds are currently successfully employed in cancer treatment, few of which are mentioned below and others in Table 1.

Table 1. Anticancer drugs from plants

Anticancer metabolite	Source
Agrimonolide	*Agrimonia pilosa*
Bruceantin	*Brucea antidysenterica*
Caesalpins	*Bruguiera paviflora*
Chrysophanol, isochrysophanol	*Cassia absus*
Dikitonepongamol, glabrin and karanjin	*Pongamia pinnata*
Flavopiridol	*Dysoxylum binectariferum*
Geranin, tannin and citric acid	*Geranium robertianum*
Glycyrrhizin, glabranin	*Glycerrhiza glabra*
Hydrocyanic acid, delphinidin and cyaniding	*Cassia absus*
Polyphenols, epigallo-catechin-3-gallate	*Camellia sinensis*
Quercetin	*Cajanus cajan*
Selenium, ayanin	*Physalis angulata*
Taxol	*Taxodium distichum*
Thalicarpin	*Thalictrum revolutum*
Zizphoisides	*Alphitonia zizphoides*

Vincristine and Vinblastine

One of the most significant examples of plant derived anti-cancer drug is the group of alkaloids belonging to vinca alkaloid family. These have been isolated from the periwinkle *Catharanthus roseus* which is found in the rain forests of Madagascar. Vincristine and vinblastine are the predominant vinca alkaloids present in this plant. These secondary metabolites have similar structures but show differences in their activity spectrum and toxicity [9,10]. Vinca alkaloids impede cancerous growth by inhibiting mitosis cell division process. They prevent microtubules from assembling in a mitotic spindle, thus providing hindrance to cell division cycle [11].

Etoposide and Teniposide

Epipodophyllotoxin is an isomer of podophyllotoxin which was isolated as the active antitumor agent from the roots of *Podophyllum* species, *Podophyllum peltatum* Linnaeus and *Podophyllum emodi* Wallich (Berberidaceae). The semi-synthetic derivatives of epipodophyllotoxin namely etoposide, teniposide and etopophos are widely used in the treatment of small-cell lung, bronchial and testicular cancers, lymphoma and leukemia [3,12,13]. Etoposide-like compounds are potent irreversible inhibitors of DNA topoisomerase II and their action is based on the formation of a nucleic acid-drug-enzyme complex which prevents the re-annealing of single and double-stranded breaks in DNA associated with DNA-topoisomerase binding. Accumulation of such breaks leads to cell death [14,15]. However, these compounds suffer from several limitations such as myelosuppression, the development of drug resistance and cytotoxicity against normal cells. In order to overcome the limitations of these compounds and to develop new compounds with better anti-tumour activity, many structural modifications have been performed on the cyclolignan skeleton [16,17].

Paclitaxel and Docetaxel

The discovery of paclitaxel (Taxol®) from the bark of the Pacific Yew, *Taxus brevifolia* Nutt. (Taxaceae) is another evidence of success of a natural product as a drug molecule. Various parts of *Taxus brevifolia* and other *Taxus* species (e.g., *Taxus Canadensis* Marshall, *Taxus baccata* L.) have been used by several native American tribes for the treatment of some non-cancerous cases [13] while *Taxus baccata* was reported to be used in Indian Ayurvedic medicine for the treatment of cancer. Paclitaxel was approved by the US FDA in 1992 and was then clinically introduced to the US market [18]. Paclitaxel is significantly active against ovarian cancer, advanced breast cancer, small and non-small cell lung cancer [18-21].

The mechanism of action of these taxanes is unique [22,23]. They act by disrupting the normal microtubule function and hence, can be described as mitotic inhibitors. But, unlike vinca alkaloids who destabilise the microtubule assembly, these act as microtubule stabilising agents. Vinca alkaloids affect the rates of tubulin polymerisation while taxanes induce the inhibition of microtubule depolymerisation. Despite of differences in their mode of action, both – vinca alkaloids and taxanes's mechanism is dose-dependent [24,25]. Docetaxel is a semi-synthetic analogue of paclitaxel obtained from 10-deacetyl-baccatin III. It shows potent anticancer activity, better than that of taxol and has better pharmacokinetic properties such as improved water solubility. It is also a stabiliser of microtubules and is cytotoxic against several human cells.

Camptothecin

Camptothecin is a quinoline alkaloid isolated from the Chinese ornamental tree *Camptotheca acuminita* Decne (Nyssaceae) whose antitumor activity was discovered long ago in 1958 [26]. It exhibits the property of inhibiting DNA topoisomerase I, a cellular enzyme that maintains the topographic structure of DNA during translation, transcription and mitosis [27,28]. Camptothecin has been modified chemically to reduce its toxic effects and improve its solubility in water. Two of its derivatives which have been approved for cancer chemotherapy and have been commercialised are topotecan and irinotecan [29,30]. Topotecan is applied to metastatic ovarian cancer and has efficacy in small-cell lung cancer [31]. Irinotecan is a pro-drug of the active metabolite and it is useful in metastatic colorectal cancer in combination with fluorouracil and oxaliplatin [32,33].

Withanolides

The plant *Withania somnifera* is held in high repute in traditional Indian medicine, largely due to the presence of withanolides such as withaferin A (WA), withanolide A (WLA), withanone (WN) and many more [34]. Structures of these withanolides are shown in

Figure 1. The withanolides, mainly localized in leaves in the concentration range of 0.001 to 0.5% dry weight (DW) [35], are a group of naturally occurring C-28 steroidal lactones [36] . They are built on an intact or rearranged ergostane framework in which C-22 and C-6 are appropriately oxidized to form a six-membered lactone ring. Apart from withanolides, several groups of other chemical constituents such as steroidal lactones, alkaloids, flavonoids, tannin etc. have been identified, extracted, and isolated [37]. At present, more than 12 alkaloids, 40 withanolides and several sitoindosides (withanolides containing a glucose molecule at carbon 27) have been isolated and reported from aerial parts, roots and berries of *Withania* species. Withanolides have diverse pharmacological properties ranging from anti-tumor, anti-inflammatory, anti-stress, anti-oxidant, immunomodulatory, hemopoetic, cardio-protective and neuro-protective [38-41]. *Withania* has been compared to ginseng and thus is also referred to as Indian ginseng.

Though the therapeutic effects of these withanolides are well documented, the detailed specific action mechanisms on the respective disease associated targets are still unknown. Towards understanding this, we have elucidated the mode of action of withanolides on various crucial targets like NF-kB, mammalian proteasomes including human, Hsp90/Cdc37 and many others using *in silico* experiments.

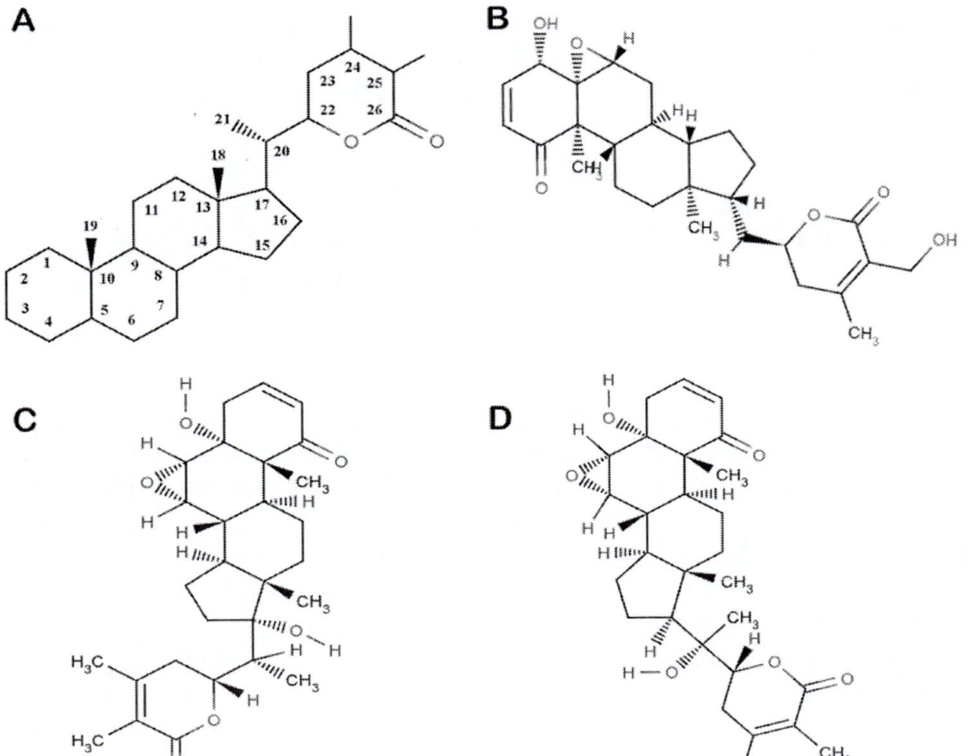

Figure 1. Structures of withanolides: (A) withanolide skeleton (B) withaferin A (C) withanone (D) withanolide A.

Finding the binding mechanisms using wet lab experiments is quite a rigorous approach involving a substantial amount of time. On the other hand, computational tools have become increasingly important in expedition of drug discovery and design processes. Methods from computational chemistry are now used routinely to study drug-receptor complexes in atomic detail and to calculate properties of small-molecule drug candidates.

Homology Modelling, Molecular Docking and MD Simulations

To predict the binding of ligand to the receptors, the presence of three dimensional structure of the receptor is essential. However the slow pace of X-Ray/NMR structure deposition in Protein Data Banks limits the availability of protein 3-D structures. This problem can be addressed to quite an extent by building good quality homology models of the receptor based on the available template structure(s). These structures can then be used to predict the affinity and orientation of the ligand in the binding cavity by using molecular docking. Over the past decade, molecular docking has developed into a method of great use in medicinal chemistry [42,43] and it works well in predicting the structure of the bound ligand in the protein. Recently, it has been shown that molecular docking to homology models of a protein is a good choice when no structure of the target is available [44]. Since molecular docking provides only a static view of protein-ligand interaction, molecular dynamics (MD) simulation can be used to study the interactions in motion. The goal of the MD simulations is to account for protein flexibility and protein/ligand movement that cannot be achieved in the docking simulation.

Probing the Pharmacological Mode of Action of Withanolides

Activation-Suppression of NF-kB [45]

NF-κB is one of the most attractive topics in current biological, biochemical, and pharmacological research, and in the recent years the number of studies focusing on its inhibition/regulation has increased manifolds [46-48]. NF-κB is secreted predominantly in cytoplasm in the form of an inactive complex with IκB inhibitor proteins. Binding to IκB (Inhibitor of kappa B) prevents NF-κB:IκB complex from translocating to the nucleus, thereby maintaining NF-κB in an inactive state. NF-κB signalling is generally considered to occur through NF-κB activation being inititated by stimuli like proinflammatory cytokine TNF (tumor necrosis factor) alpha and bacterial lipopolysaccharide (LPS).

Signalling pathways lead to activation of the beta subunit of the IKK (IκB kinase) complex, which then phosphorylates IκB proteins leading to their degradation and subsequent release of NF-κB. The freed NF-κB dimers translocate to the nucleus where it binds to the target genes. The constitutive activation of NF-κB contributes to multiple cellular outcomes and pathophysiological conditions such as rheumatoid arthritis, asthma, inflammatory bowel disease [49], AIDS [50] and cancer [51]. Thus there lies a huge therapeutic potential beneath inhibition of NF-κB signalling pathway for reducing menace of these chronic ailments [52].

Small ligands (both natural and synthetic) are gaining particular attention in this context. Most recently, WA was shown to potentiate apoptosis of tumor cells by suppression of NF-κB activation [46-48]. Our computational analysis provided a rationalization of the ability of naturally occurring WA to alter the NF-κB signalling pathway. Semi-flexible docking of WA into NEMO suggested non-formation formation of active IKK complex. In addition to this, the large value of binding energy involved in binding of WA to the active NEMO/IKKβ complex obtained by our flexible-docking experiment consolidated the disruption of active IKK complex stability. From this study it can be concluded that WA is a potent anticancer agent as ascertained by its potent NF-κB modulating capability.

Modeling of Human Proteasome and Its Inhibition [53]

UPP is the major proteolytic system in the cytosol and nucleus of all eukaryotic cells [54,55] which regulates cellular events including mitotis, differentiation, signal transduction, apoptosis, and inflammation [56]. UPP controls activation of the transcriptional factor NF-κB (nuclear factor κB), which is a regulatory protein playing central role in a variety of cellular processes, including immune and inflammatory responses, apoptosis, and cellular proliferation. Since the primary interaction of proteasomes occurs with endogenous proteins, the signalling action of transcription factor NF-κB can be blocked by inhibition of proteasomes, thus inhibiting the completion of the cell cycle and mitotic proliferation of cancerous cells and ultimately leading to cell death.

It has been suggested that proteasomal activity is essential for tumour cell proliferation and development of drug resistance. Thus inhibition of proteasomal activity is a promising way to retard or block degradation of specific proteins to correct diverse pathologies [57]. Though quite a number of selective and efficient proteasomal inhibitors exist nowadays, their toxic side effects limit their potential in possible disease treatment [58]. Thus there is an indispensable need for exploration of novel natural products as antitumor drug candidates.

There is an evidence rationalising the proteasome inhibitory action of WA in which WA is shown to inhibit chymotrypsin like activity of a purified rabbit 20S proteasome (IC_{50}=4.5μM) and 26S proteasome in human prostrate cancer cultures (at 5-10μM) and in xenografts (4-8 mg/Kg/day) [59].

Our work supported the mammalian proteasomes inhibiting activity of WA along with elucidation of its possible mode of action.

Since WA is a small herbal molecule, it is expected to provide one of the modest modes of inhibition along with added favours of ease in oral administration and decreased immunogenicity.

The molecular docking results suggested that WA can inhibit the mammalian proteasomes irreversibly and with a high rate through acylation of the N-terminal Thr1 of the β-5 subunit and is supported by the previously reported experimental results [57]. Results from molecular dynamics simulations in water showed that the trajectories of both the native human 20S proteasome and the proteasome complexed with WA were stable over a considerably long time period of 4 ns suggesting the dynamic structural stability of human 20S proteasome/WA complex.

Chaperone/Co-Chaperone Junction Inhibition of Heat Shock Protein (HSP) 90 and Cdc37 [60]

Formation of active Hsp90/Cdc37 complex is one of the essential steps in facilitation of chaperone client interaction, non-assembly of which can lead to prevention of the chaperone-client association resulting in apoptosis of tumor cells. Recently, WA has been shown to potentiate apoptosis of tumor cells by inhibiting Hsp90 in human pancreatic cancer cells [61]. From our flexible docking analysis of WA into active Hsp90/Cdc37 complex in which key interfacing residues of the complex were kept flexible, disruption of the active association complex can be discerned. While docking of WA into segregated Hsp90 leaves the interface residues untouched. Thus the molecular docking analysis of WA into Hsp90 and active Hsp90/Cdc37 complex conducted in this study provides significant evidence in support of the proposed mechanism of chaperone assembly suppression by inhibition or disruption of active Hsp90/Cdc37 complex formation being accounted by non-assembly of the catalytically active Hsp90/Cdc37 complex. Results from the molecular dynamics simulations in water showed that trajectories of the protein complexed with ligand WA were stable over a considerably long time period of 2.6 ns with the energies of the complex being lowered in comparison to the un-docked association complex, suggesting the thermodynamic stability of WA complexed Hsp90/Cdc37.

Dual Inhibition of HSP90/Cdc37 Complex for Supra-Additive Tumor Suppression [62]

Molecular chaperones are required for the correct folding of many proteins inside cells. However, they also play a significant role in induction of cancer diseased states. The role of molecular chaperones in cancer was first characterized with regard to Hsp90 (heat shock protein 90) [63], which is a facilitator of oncogenesis that stabilizes a wide range of over expressed or mutated oncogenic proteins and permits their tumorigenic influence to arise [64,65]. Hsp90 is constitutively expressed at 2 to 10 fold higher levels in tumor cells compared to their normal counterparts [66,67]. The chaperone Hsp90 is involved in regulating the stability and activation state of more than 200 'client' proteins and takes part in the cancer diseased states [68]. The major clientele - protein kinases depend on Hsp90 for their proper folding and functioning. Cdc37, a kinase targeting co-chaperone of Hsp90, mediates the interactions between Hsp90 and protein kinases [69]. Targeting of Cdc37 has the prospect of delivering predominantly kinase-selective molecular responses as compared to the current pharmacologic Hsp90 inhibitors [70]. We carried out a bio-computational study with the aim of exploring the dual inhibition of Hsp90/Cdc37 Chaperone/Co-chaperone association complex by the naturally occurring drug candidates WA and 17-DMAG along with their possible modes of action. Our molecular docking studies reveal that withaferin A in combination with 17-DMAG can act as potent chaperone system inhibitors. The structural and thermodynamic stability of the ligands' bound complex were also observed from molecular dynamics simulations in water. The results suggested a novel supradditive tumor suppressive action mechanism of herbal ligands which can be looked forward for further clinical investigations for possible anticancer drug formulations.

Conclusion

Herbs have been a prime source of highly effective conventional drugs for the treatment of many forms of cancer. In many instances, the actual compound isolated from the plant may not serve as the drug, but leads to the development of potential novel agents. With the development of new technologies, some of the agents which failed earlier clinical studies are now stimulating renewed interest. With the rapid identification of new proteins having significant regulatory effects on tumor cell cycle progression, and their conversion into targets for high throughput screening, molecules isolated from plants and other natural organisms need to be studied at a greater pace for their possible inhibitory action. Understanding the molecular mechanisms of ligand binding is an essential requirement for development and screening of new drugs and their functional analogues. However, the slow and rigorous wet lab experiments limit the progress efficiency in this field dovetailing the need for alternative approaches. Recent developments in computational tools have given an edge by tackling these problems. In the present chapter, applications of bioinformatics approaches have been described to find out the binding mechanisms of prospective natural drugs - withanolides.

References

[1] Jemal A, Siegel R, Xu JQ, Ward E. Cancer Statistics, 2010. *Ca-a Cancer Journal for Clinicians* 2010;60 (5):277-300.

[2] Hoyert DL, Heron MP, Murphy SL, Kung HC. Deaths: final data for 2003. *Natl. Vital. Stat. Rep.* 2006;54 (13):1-120.

[3] Harvey AL. Medicines from nature: are natural products still relevant to drug discovery? *Trends Pharmacol Sci.* 1999;20 (5):196-8.

[4] Holland BK. Prospecting for drugs in ancient texts. *Nature* 1994;369 (6483):702.

[5] Verdine GL. The combinatorial chemistry of nature. *Nature* 1996;384 (6604):11-3.

[6] Patwardhan B, Vaidya ADB, Chorghade M. Ayurveda and natural products drug discovery. *Current Science* 2004;86 (6):789-99.

[7] Koehn FE, Carter GT. The evolving role of natural products in drug discovery. Nat Rev *Drug Discov* 2005;4 (3):206-20.

[8] Gordaliza M. Natural products as leads to anticancer drugs. *Clinical and Translational Oncology* 2007;9 (12):767-76.

[9] Himes RH. Interactions of the catharanthus (Vinca) alkaloids with tubulin and microtubules. *Pharmacol. Ther* 1991;51 (2):257-67.

[10] Leveque D, Jehl F. Molecular pharmacokinetics of catharanthus (vinca) alkaloids. *J. Clin. Pharmacol.* 2007;47 (5):579-88.

[11] Hait WN, Rubin E, Goodin S. Tubulin-targeting agents. Cancer Chemother Biol *Response Modif* 2005;22:35-59.

[12] Meresse P, Dechaux E, Monneret C, Bertounesque E. Etoposide: discovery and medicinal chemistry. *Curr. Med. Chem.* 2004;11 (18):2443-66.

[13] Cragg GM, Newman DJ. Plants as a source of anti-cancer agents. *J. Ethnopharmacol* 2005;100 (1-2):72-9.

[14] Jordan MA, Wilson L. Microtubules as a target for anticancer drugs. *Nature Reviews Cancer* 2004;4 (4):253-65.
[15] Srivastava V, Negi AS, Kumar JK, Gupta MM, Khanuja SP. Plant-based anticancer molecules: a chemical and biological profile of some important leads. *Bioorg. Med. Chem.* 2005;13 (21):5892-908.
[16] Bohlin L, Rosen B. Podophyllotoxin derivatives: Drug discovery and development. *Drug Discovery Today* 1996;1 (8):343-51.
[17] You Y. Podophyllotoxin derivatives: current synthetic approaches for new anticancer agents. *Curr. Pharm Des.* 2005;11 (13):1695-717.
[18] Rowinsky EK, Onetto N, Canetta RM, Arbuck SG. Taxol: the first of the taxanes, an important new class of antitumor agents. *Semin. Oncol* 1992;19 (6):646-62.
[19] Nabholtz JM, Gligorov J. The role of taxanes in the treatment of breast cancer. *Expert Opin Pharmacother* 2005;6 (7):1073-94.
[20] Dubois J. Recent progress in the development of docetaxel and paclitexal analogues. *Expert Opinion on Therapeutic Patents* 2006;16 (11):1481-96.
[21] Ojima I, Geng XD, Wu XY, Qu CX, Borella CP, Xie HS, Wilhelm SD, Leece BA, Bartle LM, Goldmacher VS, Chari RVJ. Tumor-specific novel taxoid-monoclonal antibody conjugates. *Journal of Medicinal Chemistry* 2002;45 (26):5620-3.
[22] Bergstralh DT, Ting JP. Microtubule stabilizing agents: their molecular signaling consequences and the potential for enhancement by drug combination. *Cancer Treat Rev.* 2006;32 (3):166-79.
[23] Sengupta S, Thomas SA. Drug target interaction of tubulin-binding drugs in cancer therapy. *Expert Rev. Anticancer Ther* 2006;6 (10):1433-47.
[24] Malonga H, Neault JF, Diamantoglou S, Tajmir-Riahi HA. Taxol anticancer activity and DNA binding. *Mini Rev. Med. Chem.* 2005;5 (3):307-11.
[25] Kingston DGI, Bane S, Snyder JP. The Taxol pharmacophore and the T-Taxol bridging principle. *Cell Cycle* 2005;4 (2):279-89.
[26] Potmeisel M, Pinedo H. In: *Camptothecins: new anticancer agents.* Boca Raton, Florida: CRC Press, 1995. pp. 149-50.
[27] Hartmann JT, Lipp HP. Camptothecin and podophyllotoxin derivatives: inhibitors of topoisomerase I and II - mechanisms of action, pharmacokinetics and toxicity profile. *Drug Saf* 2006;29 (3):209-30.
[28] Wall ME. Camptothecin and taxol: discovery to clinic. *Med. Res. Rev.* 1998;18 (5):299-314.
[29] Potmesil M. Camptothecins: from bench research to hospital wards. *Cancer Res* 1994;54 (6):1431-9.
[30] Garcia-Carbonero R, Supko JG. Current perspectives on the clinical experience, pharmacology, and continued development of the camptothecins. *Clinical Cancer Research* 2002;8 (3):641-61.
[31] Dennis MJ, Beijnen JH, Grochow LB, van Warmerdam LJ. An overview of the clinical pharmacology of topotecan. *Semin Oncol* 1997;24 (1 Suppl 5):S5-12-S5-8.
[32] Cersosimo RJ. Irinotecan: a new antineoplastic agent for the management of colorectal cancer. *Ann. Pharmacother* 1998;32 (12):1324-33.
[33] Erlichman C, Sargent DJ. New treatment options for colorectal cancer. *N Engl. J. Med.* 2004;351 (4):391-2.

[34] Mishra L, Singh B, Dagenias S. Scientific basis for the therapeutic use of Withania somnifera (ashwagandha): a review. *Altern. Med. Rev.* 2000;5:334-6.

[35] Kapoor LD. *Handbook of Ayurvedic Medicinal Plants*. In. London, UK: CRC Press, 2001. pp. 337-8.

[36] Ray A, Gupta M. Withasteroids, a growing group of naturally occurring steroidal lactones. In: Hertz W, Kerby G, Moore R, Steglich W, Tamm C, editors. *Progress in the chemistry of natural organic products.* New York: Springer-Verlag, 1994. pp. 1-106.

[37] Bandyopadhyay M, Jha S, Tepfer D. Changes in morphological phenotypes and withanolide composition of Ri-transformed roots of Withania somnifera. *Plant Cell Reports* 2007;26 (5):599-609.

[38] Kulkarni S, George B, Mathur R. Protective effect of Withania somnifera root extract on electrographic activity in a lithium pilocarpine model of status epilepticus. *Phytother Res.* 1998;12:451-3.

[39] Bhattacharya A, Ghosal S, Bhattacharya SK. Anti-oxidant effect of Withania somnifera glycowithanolides in chronic footshock stress-induced perturbations of oxidative free radical scavenging enzymes and lipid peroxidation in rat frontal cortex and striatum. *Journal of Ethnopharmacology* 2001;74 (1):1-6.

[40] Bhattacharya SK, Muruganandam AV. Adaptogenic activity of Withania somnifera: an experimental study using a rat model of chronic stress. *Pharmacology Biochemistry and Behavior* 2003;75 (3):547-55.

[41] Chaudhary G, Sharma U, Jagannathan NR, Gupta YK. Evaluation of Withania somnifera in a middle cerebral artery occlusion model of stroke in rats. *Clinical and Experimental Pharmacology and Physiology* 2003;30 (5-6):399-404.

[42] Chen H, Lyne PD, Giordanetto F, Lovell T, Li J. On evaluating molecular-docking methods for pose prediction and enrichment factors. *J. Chem. Inf. Model* 2006;46 (1):401-15.

[43] Cummings MD, DesJarlais RL, Gibbs AC, Mohan V, Jaeger EP. Comparison of automated docking programs as virtual screening tools. *Journal of Medicinal Chemistry* 2005;48 (4):962-76.

[44] Yang JM, Chen YF, Shen TW, Kristal BS, Hsu DF. Consensus scoring criteria for improving enrichment in virtual screening. *J. Chem. Inf. Model* 2005;45 (4):1134-46.

[45] Grover A, Shandilya A, Punetha A, Bisaria VS, Sundar D. Inhibition of the NEMO/IKKβ association complex formation, a novel mechanism associated with the NF-κB activation suppression by Withania somnifera's key metabolite withaferin A. *BMC Genomics* 2010;11 (Suppl 4):S 25.

[46] Oh JH, Kwon TK. Withaferin A inhibits tumor necrosis factor-alpha-induced expression of cell adhesion molecules by inactivation of Akt and NF-kappa B in human pulmonary epithelial cells. *International Immunopharmacology* 2009;9 (5):614-9.

[47] Ichikawa H, Takada Y, Shishodia S, Jayaprakasam B, Nair MG, Aggarwal BB. Withanolides potentiate apoptosis, inhibit invasion, and abolish osteoclastogenesis through suppression of nuclear factor-kappa B (NF-kappa B) activation and NF-kappa B-regulated gene expression. *Molecular Cancer Therapeutics* 2006;5 (6):1434-45.

[48] Kaileh M, Vanden Berghe W, Heyerick A, Horion J, Piette J, Libert C, De Keukeleire D, Essawi T, Haegeman G. Withaferin A strongly elicits I kappa B kinase beta

hyperphosphorylation concomitant with potent inhibition of its kinase activity. *Journal of Biological Chemistry* 2007;282 (7):4253-64.

[49] Barnes PJ. Nuclear factor kappa B. *International Journal of Biochemistry and Cell Biology* 1997;29 (6):867-70.

[50] Hiscott J, Kwon H, Genin P. Hostile takeovers: viral appropriation of the NF-kappa B pathway. *Journal of Clinical Investigation* 2001;107 (2):143-51.

[51] Amit S, Ben-Neriah Y. NF-kappa B activation in cancer: a challenge for ubiquitination- and proteasome-based therapeutic approach. *Seminars in Cancer Biology* 2003;13 (1):15-28.

[52] Karin M, Delhase M. The I kappa B kinase (IKK) and NF-kappa B: key elements of proinflammatory signalling. *Seminars in Immunology* 2000;12 (1):85-98.

[53] Grover A, Shandilya A, Bisaria VS, Sundar D. Probing the anticancer mechanism of prospective herbal drug Withaferin A on mammals: a case study on human and bovine proteasomes. *BMC Genomics* 2010;11 (Suppl 4):S 15.

[54] Lupas A, Zwickl P, Wenzel T, Seemuller E, Baumeister W. Structure and function of the 20S proteasome and of its regulatory complexes. *Cold Spring Harbor Symposia on Quantitative Biology* 1995;60:515-24.

[55] Hershko A, Ciechanover A. The ubiquitin system. *Annual Review of Biochemistry* 1998;67:425-79.

[56] Myung J, Kim KB, Crews CM. The ubiquitin-proteasome pathway and proteasome inhibitors. *Medicinal Research Reviews* 2001;21 (4):245-73.

[57] Mohan R, Hammers H, Bargagna-Mohan P, Zhan X, Herbstritt C, Ruiz A, Zhang L, Hanson A, Conner B, Rougas J, Pribluda V. Withaferin A is a potent inhibitor of angiogenesis. *Angiogenesis* 2004;7 (2):115-22.

[58] Colson K, Doss DS, Swift R, Tariman J, Thomas TE. Bortezomib, a newly approved proteasome inhibitor for the treatment of multiple myeloma: nursing implications. *Clin. J. Oncol. Nurs* 2004;8 (5):473-80.

[59] Yang HJ, Shi GQ, Dou QP. The tumor proteasome is a primary target for the natural anticancer compound withaferin a isolated from "Indian Winter Cherry". *Molecular Pharmacology* 2007;71 (2):426-37.

[60] Grover A, Shandilya A, Agrawal V, Pratik P, Bhasme D, Bisaria VS, Sundar D. Hsp90/Cdc37 Chaperone/co-chaperone complex, a novel junction anticancer target elucidated by the mode of action of herbal drug Withaferin A. *BMC Bioinformatics* 2011;12 (Suppl 1):S30.

[61] Yu YK, Hamza A, Zhang T, Gu MC, Zou P, Newman B, Li YY, Gunatilaka AAL, Zhan CG, Sun DX. Withaferin A targets heat shock protein 90 in pancreatic cancer cells. *Biochemical Pharmacology* 2010;79 (4):542-51.

[62] Grover A, Shandilya A, Agrawal V, Pratik P, Bhasme D, Bisaria VS, Sundar D. Blocking the chaperone kinome pathway: mechanistic insights into a novel dual inhibition approach for supra-additive suppression of malignant tumors. *Biochem. Biophys. Res. Commun.* 2011;404 (1):498-503.

[63] Whitesell L, Lindquist SL. HSP90 and the chaperoning of cancer. Nature Reviews *Cancer* 2005;5 (10):761-72.

[64] Pearl LH. Hsp90 and Cdc37 - a chaperone cancer conspiracy. *Current Opinion in Genetics and Development* 2005;15 (1):55-61.

[65] Calderwood SK, Khaleque MA, Sawyer DB, Ciocca DR. Heat shock proteins in cancer: chaperones of tumorigenesis. *Trends in Biochemical Sciences* 2006;31 (3):164-72.

[66] Ferrarini M, Heltai S, Zocchi MR, Rugarli C. Unusual Expression and Localization of Heat-Shock Proteins in Human Tumor-Cells. *International Journal of Cancer* 1992;51 (4):613-9.

[67] Welch WJ, Feramisco JR. Purification of the Major Mammalian Heat-Shock Proteins. *Journal of Biological Chemistry* 1982;257 (24):4949-59.

[68] Kamal A, Boehm MF, Burrows FJ. Therapeutic and diagnostic implications of Hsp90 activation. *Trends in Molecular Medicine* 2004;10 (6):283-90.

[69] Hunter T, Poon RYC. Cdc37: A protein kinase chaperone? *Trends in Cell Biology* 1997;7 (4):157-61.

[70] Gray PJ, Stevenson MA, Calderwood SK. Targeting Cdc37 inhibits multiple signaling pathways and induces growth arrest in prostate cancer cells. *Cancer Research* 2007;67 (24):11942-50.

In: Bioinformatics
Editor: Renu Tuteja
ISBN 978-1-62100-913-9

Chapter X

Cancer Phylogenetics: Computational Modeling of Tumor Evolution

Felix Bast*
Centre for Biosciences, Central University of Punjab,
Bathinda, Punjab, India

Abstract

The field of phylogenetics is one of the core areas of Bioinformatics which deals with computational methods to infer evolutionary heritage of organisms and genes. While phylogenetics has been extensively utilized in taxonomy and systematics of organisms, it is only very recently that the system started expanding to other fields- most importantly in cancer biology where it profoundly transformed our understanding of clonal evolution. Many of our findings in cancer phylogenetics credit to the fact that the tumor is not merely a collection of transformed cells with random mutation events; rather it is an evolving population.

Many of the facets underpinning modern evolutionary synthesis can be applied to classify cancers and track its progression from initiating somatic mutation to symptomatic neoplasm. It is now widely accepted that all sub-clones within cancer are phylogenetically related and probability of a particular sub-clone progressing into neoplasm depending upon its time of initiation and evolutionary fitness. Computational models of tumor evolution have also contributed in identifying common clades- "cancer sub-types"- associated with particular cancers in different patients that in turn helped in translating our understanding of oncogeny to the development of "targeted therapeutics"- rationally designed drugs that are molecularly targeted to particular sub-types. Advent of next generation ultra-deep genome sequencing technologies has been rapidly transforming the very landscape of cancer phyogenetics.

This chapter introduces the concept of cancer phylogenetics and reviews some of the recent advances in this field. This chapter also summarizes various phylogenetic approaches including distance matrix methods, parsimony, maximum likelihood,

* Email: felix.bast@gmail.com.

Bayesian methods and probabilistic inference that have potential applications in cancer research.

1. Introduction

Bioinformatics deals with use of statistical, mathematical and computational methods for processing biological information. Over the past half a century Bioinformatics have triggered a genomic revolution in which algorithms were designed for applications such as DNA sequencing, genomic sequence analysis and microarray expression profiling that have greatly contributed in our understanding of cancer as a multifactorial evolutionary phenomenon.

Traditionally cancer research has concentrated on the discovery of oncogenes and tumor suppressor genes and it was only very recently that the importance of factors contributing in the disruption of normal cellular differentiation became apparent.

Most cancer therapeutics work only for a small subset of patients and this remains the same problem even for advanced molecularly targeted drugs, for the reasons unknown as of now. This problem is causing such a financial burden in our healthcare system wherein large proportion of population receiving expensive-however ineffective-treatment. Biological characterization of each patient's tumor is needed in order to develop a scenario for delivering right targeted therapeutics to right cancer type[1].

Many cancers consist of heterogeneous population of subclones containing over-expressed, mutated and silenced genes. Identifying an effective molecular target remains a serious problem due to this heterogeneity [2]. Currently there is a substantial interest in the field of targeted therapeutics in which molecularly designed drugs attack a particular abnormality or disease mechanism present in distinct subtype of cancer [3].

Proper identification and characterization of such subtypes are essential for this approach. The field of Bioinformatics has immensely assisted in the logical groupings of target subtypes for a particular drug.It can however be stated that notion of cancer subtype as a static entity in the disease progression is a simplification; progression of neoplasm is rather a dynamic process. We need to understand not only general pathways of disease progression but also specific molecular steps along the pathway in order to maximize our chances of successfully curing it [4].

Resisting subclones evade in treatment targeting overexpressing and mutated genes [5]. Targeting a key "red dot" gene -mutation of which in early oncogenesis leads to dysregulation of molecular pathway involved intumerigenesis of all subclones- seems to be the winning strategy [6, 7].

Examples include HER-2, EGFR, KIT and ABL in breast cancer, non-small cell lung cancer, gastrointestinal stromal tumors and chronic myeogenous leukemia, respectively. Identifying red-dot targets, development and application of drugs targeting these and molecular characterization of each patient's tumor demands development of a bioinformatics framework encompassing multidisciplinary teams focusing translational research [6].

2. Cancer as an Evolutionary Process

Neoplastic progression recapitulates evolutionary processes in many aspects. Cancer can be viewed as a large population of genetically and epigenetically heterogeneous cells competing amongst and against host cells for space and resources[8]. As aberrations on somatic cells have heritable effect on the fitness of clones, expansion or contraction of mutant clones follows some of the well characterized processes of evolution like the natural selection and genetic drift [9-11]. Evolution of the neoplastic clones (clonal evolution) normally selects for increased proliferation and survival, leading ultimately to invasion, metastasis and therapeutic resistance [12, 13]. Although consideration of cancer as an evolutionary and ecological process is not new, nothing much has been done for the application of evolutionary biology and ecology in the field of cancer research and therapeutics.

2.1. Mutations

Variations in populations need to be heritable for the evolution to happen. Degree of variability in the heterogeneous neoplasms in fact demonstrated to predict progression to malignancy; i.e., more genetically diverse the cancer population is, more easily it would progress to malignancy [2, 14, 15]. Genetic heterogeneity in return is a result of genetic instability and is a hall mark of all cancers studied till date[15-18].

While number of mutations clearly increases risk of a somatic cell to get transformed, most of the genetic (mutations) and epigenetic alterations observed in cancer are evolutionarily neutral; i.e., they do not much contribute to the fitness of the population. Most of the non-neutral mutations are thought to be deleterious and therefore act against neoplastic progression [19]. Number and nature of mutations needed for the transformation remain unknown even for the most widely studied cancer retinoblastoma. A thorough characterization of key transforming mutations will contribute in the development of biomarkers and identification of targets for cancer therapy[20-22]. Mutation rates themselves can be taken as a biomarker for neoplastic progression as well as to analyze drug efficacy.

2.2. Genetic Drift in Cancer

Genetic drift is a stochastic evolutionary phenomenon in which allele frequency of a gene changes in a population due to random sampling. It is thought to be a key process contributing in neoplastic transformation and progression.[10, 15, 23, 24] However most of the parameters that are needed to understand role of this process in cancer progression remains to be measured; for example, cell generation time, turn-over numbers and effective population size of cancer clones. Like elsewhere, genetic drift becomes more relevant in cancer clones when population size is low. Evolutionary events like population bottlenecks in which population size gets significantly reduced are demonstrated to be fertile grounds for genetic drift to take place. For example, loss of breast epithelium due to apoptosis in menstrual cycle is a normal process giving rise to the process of population bottleneck [25]. Significant reduction in population size in cancer therapies might also contribute in the process of genetic drift [11]. Cells with fitness disadvantages can get fixed sometimes, especially after the cancer therapy, due to population bottleneck giving rise to the genetic drift. Selectively advantageous mutations and mutations that are evolutionary neutral- however linked to the previous (hitchhiker mutations) - gets fixated ultimately through this stochastic process[26]. As neutral mutations are thought to be occurring at constant rates, it can be utilized as a molecular tumor clock- a representative of time since the initiation of neoplasm [27, 28].

2.3. Natural and Artificial Selections in Cancer

Natural selection involvescertain individuals possessing adaptive traits getting fixed in a population through reproduction and isolation. Heritable variation of reproductive success is an invariable requirement for the natural selection and this happens frequently in neoplasms as the genetic alterations are heritable and confer selective advantages or disadvantages on the cell. Tissues with repeated cycles of wounding and proliferations apparently have the repercussions of natural selection to the max, as these cycles effectively sieves clones having the best survival strategies; i.e., most proliferating[9, 10]. The process by which an adaptive mutation spreading through population is known as selective sweep and this might happen in the case of neoplasm clones, typically ending in the fixation of mutations that increase its fitness[29-31]. Several issues of natural selection in cancer remain to be answered. For example, what makes mutation patterns so different between different types of cancers and different organs? Selective pressures acting upon different cell types/organs may be quite different and factors affecting those needs to be thoroughly understood. It is also interesting to know the fitness effects of mutations in the genes that are not being used in normal cell types.

Interventional therapies in cancer give rise to deliberate artificial selection upon the neoplasm, potentially resulting in the emergence of resistance clones. The Luria-Delbruck experiment demonstrated that genetic mutations do not arise in response to the selection in bacteria; selective pressure merely selects pre-existing mutations [32]. Applying the same principle in the process of carcinogenesis, it can be deduced that the earlier we intervene in cancer the less probable it would be for the emergence of resistant mutant. Cancers that result from relatively smaller number of mutations like retinoblastoma are genetically less heterogeneous and therefore less likely to harbor resistant mutants. Traditionally chemo-therapy has been administered as large pulsed doses at regular intervals. Taking clues from evolutionary biology it might be the case that low-but continuous-doses work better as it generate lesser resistant mutants.

3. Phylogenetics

Phylogenetics- an evolutionary classification system that relies exclusively on the genetic relatedness of the objects -has become an indispensable tool in post-genomic era. While phylogenetics has been extensively used in biological systematics, the system has only begun to get expanded to other fields such as gene finding, comparative genomics, haplotype inference, cancer biology and diagnostics[22, 33-35]. Phylogenetic inferences have a universal assumption that evolution is a process of divergence that can be modeled statistically. While this first-order assumption works in most cases, there are several evolutionary processes that possess certain difficulties. For example, Horizontal Gene Transfer (HGT) and inter specific recombination are documented in many bacterial lineages that results in transfer of genetic material across distantly related species[36]. Accurate evolutionary relationship can be modeled in the presence of these evolutionary processes by a phylogenetic network.

3.1. Species Tree vs. Gene Tree

The basic phylogenetic inference connects species through a bifurcating tree (cladogram). A set of orthologous genes (loci) are sampled from different species and genealogies are estimated to construct this species tree. Internal nodes of these trees represent speciation and other major taxonomic events.On the other hand gene trees represent evolutionary history of individual genes and can provide information such as gene duplication, nucleotide substitution and extinction events [37]. As a gene at a locus in a genome replicates and copies are passed on to daughter cells, branching points are generated in the gene trees. Within a species tree many tangled gene trees can be found due to meiotic recombination. Different genes should form different trees; in fact, even a single locus could have many trees as a result of intragenic recombination [38-40]. In light of this incongruence between species and gene trees, how gene trees can be used to reconstruct species trees? This can be done under the assumption that time intervals between species-branching events are much greater than time intervals between lineage-branching events in each species, and therefore gene and species divergences are likely to be nearly concurrent.

3.2. Character Conflicts in Cladistics Analysis

Genetic relatedness-or similarity (homology)-between specimens are of two fundamental types; ancestral and derived. Ancestral homologies (plesiomorphy) are very misleading and dangerous for any type of phylogenetic inferences [41-43]. For example consider lineages of birds, crocodiles and lizards. Bird lineage has undergone rapid evolution to get wings and other skeletal adaptations for flight. In comparison, lineages of both lizards and crocodiles have evolved more slowly to retain ancestral reptile characteristics such as walking on four legs and scales, and left looking similar. This similarity between reptiles and crocodiles is ancestral homology and it does not provide evidence that they have a Most Recent Common Ancestor (MRCA)-also known as concestor-than either does with birds. Consider number of digits on the legs of frog, dog and horse. Frog and dog have five digits-an ancestral state for all tetrapods (group consisting of reptiles, amphibians, birds and mammals). Horses have only one of the five digits left. Thus inferring this ancestral homology to conclude that frog and dog have more recent common ancestor than either do with horse, we are fundamentally erred. Indeed both horse and dog are mammals and share an MRCA than either does with frog. However if we are studying relatedness of a frog, dog and fish, five digits on foot is no longer an ancestral homology as it was not present in the common ancestor of all three species. This derived homology (synapomorphy) has evolved within the group we are studying and correctly tell us about evolutionary heritage (phylogeny).Derived homologies are reliable evidence that two species share MRCA. Phylogenetics is a quest to define groups of specimens by discovering these derived homologies (phylogenetic characters). Shared derived state among the group of evolving specimens, synapomorphy, is a potential **"biomarker" for that group. This evolutionary definition of biomarker as synapomorphy** offers a natural classification of cancers based on theirontogenicrelatedness [22].

Convergent evolution describes acquisition of same phenotype in unrelated genetic lineages. A classic example is the wing in birds and bats. MRCA for both birds and bats did not have wings; both groups acquired this trait via convergent evolution. The term homoplasy refers to the similarity between two taxa of unrelated ancestry due to convergent evolution. Similar to ancestral homologies, homoplasies are troublesome for the phylogenetic inference [44, 45].

Truly derived homologies cannot conflict and in an ideal scenario character conflicts can be reduced to zero by cladistics analyses. However in reality conflicts can only be resolved to something higher than zero because techniques are all prone to errors. Homoplasies can be mistaken for homologies and convergence can be deceptively exact. One way to deal with conflicting phylogeny is to infer the one supported by majority (majority consensus).

3.3. Coalescent Theory

Tracing the evolutionary tree backwards by discovering MRCAs of the specimens comes under the realms of genealogy [46]. Species genealogies typically utilize one or multiple independent genetic loci to delineate between species that are being analyzed. The coalescent theory deals with modeling genetic drift of a population back in time to investigate genealogy of antecedents. Therefore it is a retrospective model, attempting to trace all alleles of a gene shared by individuals to MRCA [47, 48]. The basic coalescent model is a random variant; takes into consideration of only genetic drift and assumes no natural selection, population structure, recombination and gene flow. However recent advances in this model allow several of the above mentioned extensions to be analyzed concurrently and have potential applications in the field such as disease gene mapping [47, 49, 50].

3.4. Molecular Phylogenetics

As large amount of nucleotide and amino acid sequences are made available through public domain repositories, range of biological topics that it influences through evolutionary consideration expands rapidly. Sequences of DNA and proteins can be used to infer phylogeny the way morphological characters used in the past.While proteins blazed the trail (sequencing Insulin and its phylogeny reconstruction [51]), most of the recent developments in the molecular phylogenetics are concerned with DNA sequences.

Homology/homoplasy distinction, while powerful in deciding which morphological characters are phylogenetically informative, is so much less powerful in the case of molecular phylogenetics, as there are only four distinct states for nucleic acids and 20 for proteins[42]. Functional convergences, such as evolution of wings in species that fly, fuels discovery of phylogenetically informative morphological characters. If a definitive relationship between structure and function is not available, a conclusion is impossible. This is the case for molecules and therefore we have to treat them the way morphologists treat organs with unknown functions. At the same time quantities of evidence presented by molecules are very large in comparison with morphology. Only through molecular phylogenetic studies that we understand information regarding structure and function of unknown proteins from rapidly accumulating genomic sequence data.

4. Modeling Evolution

Nucleotide substitution models, which encode hypotheses on the relative rates of mutations along the DNA sequences, are crucial components of molecular phylogenetic analysis. This is because of the fact that the longer the amount of time two sequences are diverged from the common ancestor (*i.e.*, immediately after coalescence), it is more likely that two or more consecutive mutations occur on any particular nucleotide position. Genetic distances between two diverging sequences increase linearly only for a short time since the divergence. Once separated, probability for two mutations to occur at any particular site becomes higher; therefore simple distance matrix approaches will undercount number of mutation events. One particular problem is the issue of "long-branch attraction" in which two distantly related but convergently evolving sequences can be misinterpreted as closely related.

Mathematical models of sequence evolution include variables that represent features of the evolution, but the numerical values are not known *a priori*. These variables are termed as parameters. There are two statistical approaches in modeling molecular evolution. Empirical models derive values of parameters from pre-computed analysis of large dataset. Parameter values are estimated only once and are applied to all datasets and therefore result in fixed parameter values. In this kind of models, particular data will only have negligible influence. The alternative approach is parametric models where pre-specified parameter values are absent. Parametric models allow the parameter values to be derived from each dataset.

4.1. Models of DNA Evolution

Different parametric approaches were developed to model DNA evolution in which three main nucleic acid parameters were frequently taken; 1. Base frequency, 2.Base exchangeability and 3.Rate heterogeneity.

Base frequency parameters gives frequencies of four nucleotides; A,T,G and C, averaged over all sites in a DNA sequence. Factors such as overall GC contents are known to affect base frequency parameters. These parameters also skew frequency of certain bases during substitutions. Base exchangeability parameters are concerned with relative frequencies of bases to be substituted for one another (A <> T, A <>G, A<>C, T<>G, T<>C and G<>C) minus frequency of individual bases. If two bases are biochemically similar, base exchangeability will be higher, so that similar bases get substituted for one another. For example, A and G are purines and C and T are pyrimidines and therefore frequencies of A<>G and C<>T (purine<>purine or pyrimidine<>pyrimidine; "transition substitutions") are higher than that of the rest of the substitutions (purine<>pyrimidine; “transversion” substitutions). Describing each site’s rate of substitution as random sampling from a continuous probability distribution like ”gamma distribution” is a common approach to model rate heterogeneity and an important step to fit models to data. It is often used together with base exchangeability and base frequency parameters.First DNA substitution model (JC) was based upon base frequency parameters [52]. Two parameters, viz., base transitions and transversions were taken into consideration in subsequently designed Kimura-2 Parameter model -K2P[53], which was also based upon equal base frequency parameters. Felsenstein in 1981 introduced a model (F81) based up on unequal base frequencies in which substitution rate corresponds to equilibrium frequency of the target nucleotide[54]. Hasegawa, Kishino and Yano in 1985 unified last two models into a six parameter base exchangeability model (HKY) that distinguishes between transitions and transversions[55]. In 1986 a Generalized Time Reversible (GTR) model was developed by Simon Tavare in which six substitution rate parameters as well as four equilibrium base frequency parameters were taken into consideration [56]. Taken together with gamma distribution parameter, GTR is one of the most widely used DNA substitution model today.

4.2. Choosing the Right Model

Selection of an appropriate nucleotide substitution model is often critical for any phylogenetic analysis, especially given that under-parameterized or overly restrictive models produce aberrant behavior when assumptions are violated and over-parameterized or complex models are computationally expensive and may be over-fit. A common method to achieve this is by means of the pairwiseLikelihood Ratio Test (LRT) between the models that estimates “goodness of fit” between the model and the data set. One disadvantage in using LRT is that it always estimates higher likelihood scores for complex parameters than the simple. Therefore most of the algorithms implementing LRT, for *e.g.*, MODELTEST [57], are designed such that it selects the simplest model with likelihood scores not worse than more complex models. Another disadvantage for LRT is that as it is a pairwise analysis, order in which models are compared has a major effect on the one that is eventually selected. Newer model tests like Akaike Information Criterion (AIC) or Bayesian Information Criterion (BIC) calculates likelihood estimates on individual model than a pair so that these are independent on the order[58]. These tests also have inbuilt correction factors to penalize over-parameterized (in case of AIC) or complex (in case of BIC) models.

4.3. Complex Models of Evolution

Evolutionary models giving insights into structure and selection are extensively used in studies of amino acid sequence evolution. Although such models are very complex and demand superior computational capabilities, they provide important information about molecular evolution. Codon replacement models have been recently developed in studies of coding DNA sequence evolution, in which both DNA substitutions and evolutionary constraints acting upon protein product are concurrently taken into consideration[59]. For example, by studying relationships between rates of synonymous (nucleic acid substitutions will not change Amino Acid) and non-synonymous (Amino Acid altering) DNA substitutions, models will be able to detect positive selection without *a priori* knowledge of protein structure or function.

Several other models are based in addressing association of heterogeneity of patterns and rates of evolution among sites with structural organization of RNA or protein by simultaneously considering phylogeny and protein structure[60]. Structural and functional properties of proteins were used to model amino acid sequence evolution by analyzing amino acid substitutions at various structural motives [61].

5. Overview of Inferential Methodology

Properties of molecular sequence data such as availability of large amount of characters and recognizability of independent characters have encouraged utilization of statistical models to infer phylogenies. Inference methods will ideally extract maximum amount of information available in the dataset, combine this information with prior knowledge of patterns of sequence evolution and will deal with model parameters whose values are not known *a priori*. Representation of molecular hypotheses about the evolutionary ancestry of the sequences is achieved by phylogenetic trees (phylograms). There are several statistical approaches exist to infer phylogeny from molecular sequences. In this section we will cover the broad two classes of statistical approaches that have dominated in the literature: distance matrix methods and discrete data methods, although neither of them reproduces the evolutionary tree absolutely.

5.1. Distance Matrix Methods

Distance matrix methods measure genetic distances between the sequences and therefore explicitly requires MSA as an input. Pairwise evolutionary distances of the sequences in MSA are calculated and represented in a rooted or unrootedphylogram such that closely related sequences appear in the same interior node. Advantages of the distance matrix methods lie in the analyses being fast and from its ability to model substitution bias to correct multiple mutations. It produces only one tree- seemingly the best bet- however, costing the phylogenetic accuracy at a great deal. Neighbor-Joining (NJ) algorithm is one of the widely implemented distance matrix methods [62]. Unlike other methods, NJ does not assume that the lineages concurrently evolve (molecular clock hypothesis) and therefore produces an unrooted tree. By including known taxa as outgroup, it is possible to root the NJ phylogram, and when done that way, it always produces an ultrametric tree (equal distance from root to the branch tips).

5.2. Discrete Data Methods

Two of the discrete data methods commonly used are Maximum Parsimony (MP) and Maximum Likelihood (ML). The MP, a relatively simple non-parametric statistical method, infers that the best representation of evolutionary relationships is the one that requires minimum number of steps (*i.e.*, nucleic acid substitutions). The input data for MP analysis, known as "characters", are phenotypical or genealogical attributes that are heritable and observed to vary between the taxa. MP produces a number of phylograms with considerable topological variations and therefore an evaluation of all such phylograms is very complicated. A strict consensus phylogram is usually constructed by heuristic approaches that usually involve steepest-descent style minimization mechanisms. Major disadvantage of this method is that the character states are generally noisy to an extent that overly simplistic approach of MP results in erroneous conclusions. Its inability to apply nucleotide substitution models and the common notion of evolution being non-parsimonious, further limits MP's usefulness in phylogenetic inference.

Most of the modern phylogenetic analyses are based on ML, a parametric statistical method, which is reported to be more accurate (*i.e.*, more likely to predict the evolution) and robust (less sensitive to faulty assumptions and models) than other phylogenetic inferences [63]. ML criterion assesses probability of particular mutations by a substitution model and allows varying rates of evolution across both lineages and sites [54]. Highly probable ML phylograms tend to have Interior branches that require minimum number of mutations to construct, and *vice versa*. ML is a preferable method for phylogenetic analysis of distantly related sequences, although it demands greater computational capabilities.

A much faster alternative that is often simultaneously performed with ML in the same data set is Bayesian Inference (BI) - name of which was derived from an 18th century statistician Thomas Bayes. The BI combines prior probabilities of a phylogeny with likelihood of trees to produce posterior probability distribution on phylograms [64]. Because tree topologies and branch lengths are not treated as parameters-as in ML- but as random variables, it is impossible to obtain BI probabilities analytically. Therefore BI probabilities are approximated by numerical simulations like Markov Chain Monte Carlo (MCMC) or Metropolis Coupled MCMC (MCMCMC). These chains explore the posterior probability grids in an integrative manner with model parameters. Trees are then sampled at fixed intervals and a consensus tree is constructed. The proportion of time that the chain visited sampled trees having a particular interior branch of the consensus tree is expressed as Bayesian posterior probabilities (PP, [65]). The computer program Mr.Bayes is often used to estimate the BI [66].

6. Genetic Heterogeneity and Cancer Classification Systems

Cancer staging classifications were traditionally based on the degree of differentiation of the tumor sample as determined histopathologically. This is then statistically analyzed to assign a particular stage. Finding common basis for defining type and subtypes of cancer is possible only if each subtype has unique developmental and maintenance pathways.

In recent times it has become increasingly evident that disruption of normal differentiation is an important component of tumorigenesis [67, 68]. Discovery of genes involved in this disruption have started helping the researchers to map neoplastic differentiation pathways and to classify cancer subtypes according to their maturation status.

One method of studying role of a particular gene in an organism is by monitoring its expression level. Gene expression data have been widely used recently in classifying various cancer sybtypes[69-74].

Expression of a particular gene changes during cell development and level of its expression can be correlated with amount of mRNA it encodes. Expression patterns of many genes in parallel can be monitored by complimentary DNA (cDNA) microarrays[75, 76]. Microarray data shows that each cancer specimen has a unique set of genes that are being expressed at any particular time. Gene expression profiles of these specimens are unique to distinguish it from other specimens. If these values are averaged-as is the case with phonetic clustering based staging methods- **profile's identity** gets distorted.

Principles of phylogenetics were first applied in the field of cancer to aid in the classification. Several researchers have attempted to utilize hierarchical clustering methods that are primarily used in phylogenetics to organize genes discovered by microarray methods into phylogenetic trees based on their expression patterns. One such method is based upon pairwise average linkage cluster analysis [77]. Recently more advanced mathematical and statistical data mining approaches including geometric mixture models and neural networks have been successfully used for the construction of phylograms from microarray data (9-14) to aid in cancer classification [78-81].

7. Phylogenetic Structures in Cancer

Identification and characterization of novel cancer subtypes have reached new heights under the technical framework of molecular phylogenetics. The principle behind cancer phylogenetics is straight forward; cancers are not merely a collection of mutated aberrant cells; rather it is an evolving population. Ongoing evolution of the neoplasm –i.e., patterns of tumor progression- can be tracked through analyzing mutations since the transformation event.

Ultimatelyasubclone with best evolutionary fitness comes to dominate the picture. Development and clinical presentation of symptomatic neoplasm might appear as a linear process macroscopically. However in subcellular level, competing genotypes in heterogeneous subclones and various extinction events become apparent. There are various direct and indirect evidences for such cancer progression models [82].

Some cancer subtypes are observed to be having high prevalence of somatically acquired mutations [83, 84]. Immediately after applying a selective pressure like drugs molecularly targetedto a particular subclone, preexisting drug-resistant clones were observed to be rapidly emerging- a phenomenon reflected in modern evolutionary synthesis [85]. Long latency between initiating mutations and clinical presentation of neoplasm is yet another indirect evidence for these models [86].

More recently multiple genetically related subclones in a case of B-Cell Chronic Lymphocytic Leukemia and clonal interrelationships within were demonstrated by phylogenetic analyses aided by massively parallel *de novo* DNA sequencing methods [87]. These studies ultimately help in the identification of the subpopulations of invasive, drug-resistant and relapsing cells as well as annotating initiating genetic alterations.

7.1. Two Approaches

There are two approaches widely used in Cancer Phylogenetics. Tumor by tumor approach utilizes difference between tissues between tumor populations-inter-tumor heterogeneity. Cell by cell approach utilizes difference between individual cells within tumor-intra-tumor heterogeneity. These two approaches with various methodological avenues for discovering phylogenetic structures in cancer are illustrated in Figure 1. This can be compared with approaches used in phylogeography- a study deals with finding historical processes responsible for contemporary distribution of organisms. Sampling from one microenvironment helps to assess species diversity and distribution within that habitat. At the same time, sampling from a wider geographic area helps to discover inter-habitat heterogeneity and distribution patterns.

Based on the presence or absence of specific mutation events, microarray gene expression measurements and gene copy numbers measured by Comparative Genomic Hybridization (CGH), it was Desper et al. who applied the principles of phylogenetics for the first time in the field of cancer to calculate evolutionary distances for inferring inter-tumor *oncogenetic trees* [35].

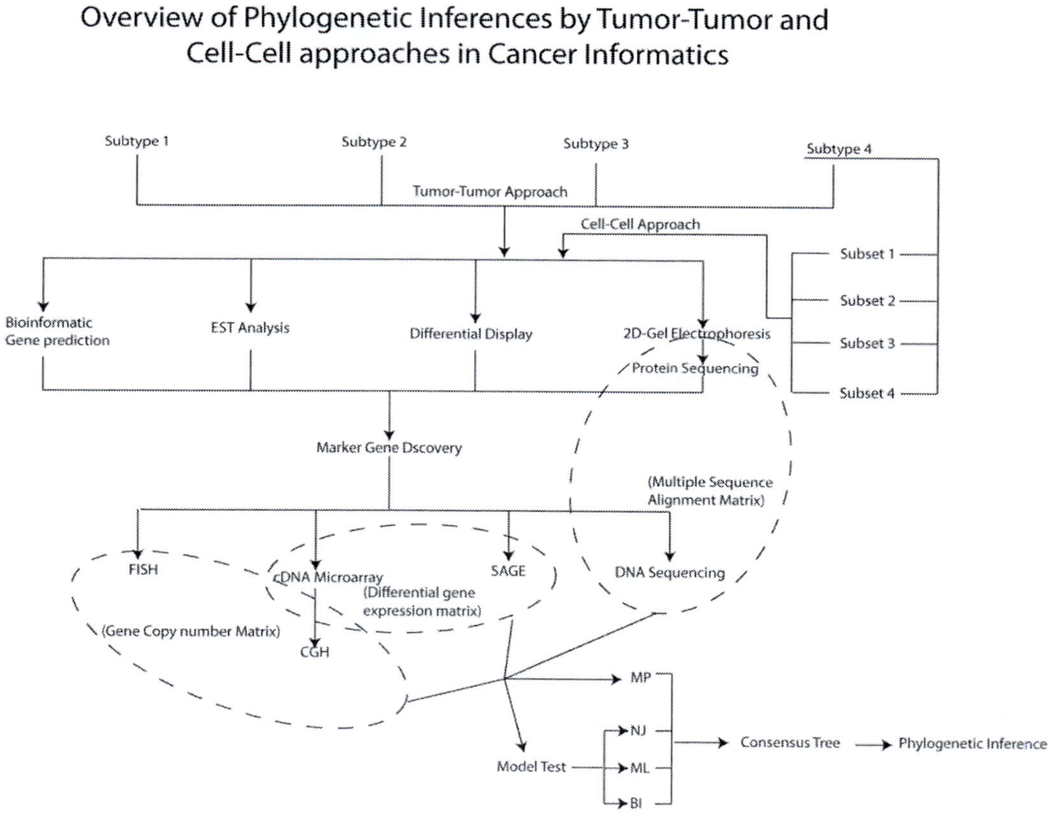

Figure 1. Two approaches and underlying methodological avenues for discovering phylogenetic structures in cancer. EST- Expressed Sequence Tags; SAGE- Serial Analysis of Gene Expression; FISH- Fluorescent In-situ Hybridization; CGH-Comparative Genomic Hybridization; MP-Maximum parsimony; NJ-Neighbor Joining; ML-Maximum Likelihood; BI-Bayesian Inference.

Average disease progression pathways evolving across a patient population were modeled by treating these cancer subtypes as internal nodes in the phylogenetic trees. Molecularly similar tumor subtypes form members of the same clades and distances of each subtype from the normal-state corresponds to the degrees of progression. Since then, many studies have conducted in this field and applied robust phylogenetic inference methods including maximum parsimony, maximum likelihood and Bayesian inferences.This tumor by tumor approach allows assays of many distinct probes per cancer capable of covering an entire transcriptome during expression profiling or copy number changes of a particular gene over an entire genome. However an accurate intracellular picture of cancer progression including existence of transitory cell populations and interrelationship between them cannot be drawn from this approach. It considers each tumor as a particular progression state in the overall disease progression and presumes that one can draw an evolutionary tree connecting all these states. However, tumors do not comprise homogenous populations of cells; it in turn preserves remnants of disease progression. While cell proliferation rates were observed to be gradually progressing as the cancer progresses, cells of the later stages augment, not replace, cells in earlier stages. That is, looking at individual tumors one can draw overall picture of cancer itself (Figure 2).

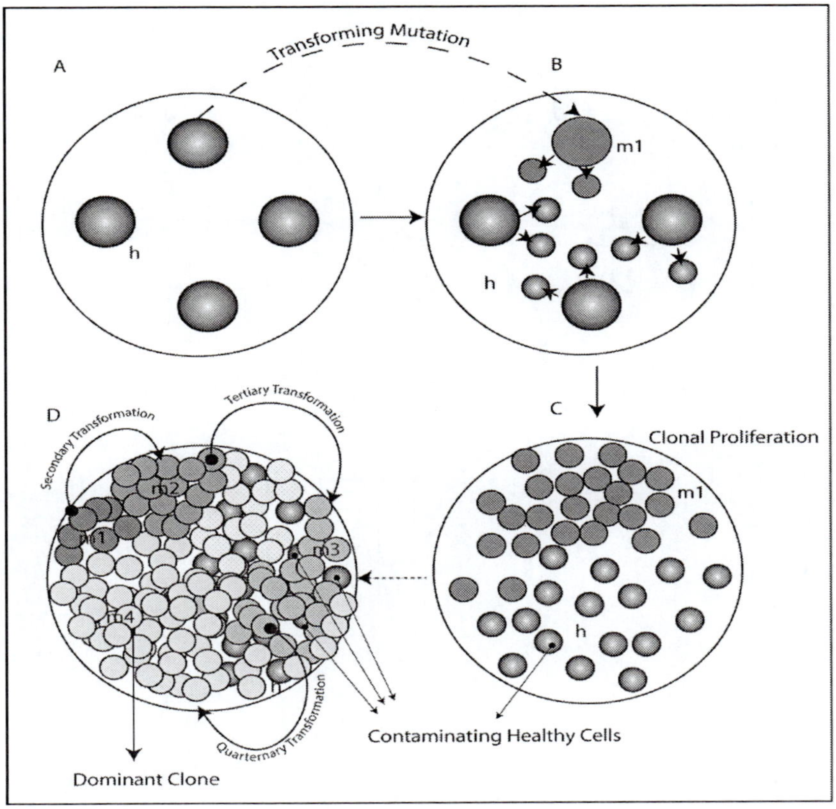

Figure 2. Neoplastic progression ("Chain Reaction") in cancer, leading to intra-tumor genetic heterogeneity. A. Healthy cells (h); B. Initial transformation event giving rise to mutated cell- a neoplasm (m1); C. Proliferation of M_1 neoplasm; D. Fully differentiated neoplasm retains the remnants of complete progression pathways.

This concept is the basis of alternative cell by cell approach which assumes that tumors preserve remnants of earlier cell populations as they develop [88, 89]. Tumors therefore contain transformed cells at various stages of progression as well as healthy "contaminating"cells. Various lines of evidence support this argument. Earlier studies involved gene copy number analyses by Fluorescent In-Situ Hybridization (FISH) that demonstrated single tumors as containing subsets of clones, each possessing one type of genetic alterations such as successive acquisition of a mutation sequence or differing degrees of gene amplification[90]. Evidences from the recent genomic sequencing experiments substantiate this finding that as tumors progress they retain remnants of ancient states along their progression pathways. Primary tumors, in comparison with secondary and other tumors resulted from metastases, shows higher degree of genetic heterogeneity indicating that metastases results from furtherhematopoietic differentiations of primary populations. By determining intra-tumor cellular heterogeneity it is possible to model common progression pathways and metastases.Since only a few previously characterized markers of progression can be used per cell at one time, this approach only gives relatively crude measures of progression states.

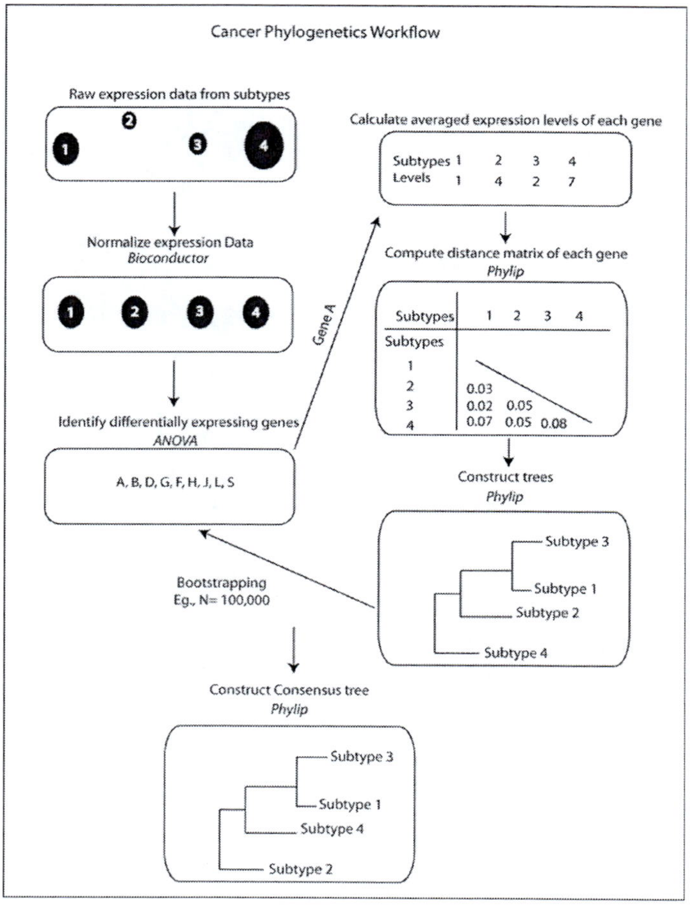

Figure 3. Cancer Phylogenetics workflow for tumor by tumor approach. Software/statistical parameters that can be used in main steps are indicated in italics. Modified from [95].

7.2. Outline of Methodology

Consider that we have microarray expression profile data from specimens that have been classified into different subtypes immuno/histo-pathologically. First step is to normalize the data across various specimens/subtypes. Software such as Bioconductor [91]are frequently used for the normalization. Next step is performing an ANOVA (one-way) to analyze for differentially expressed genes; i.e., a gene whose expression levels in one subtype differs from expression levels in at least one another subtype.Expression levels of each differentially expressed gene (discovered in previous step) is then averaged across all samples of each subtype to calculate average expression levels. Distance matrix is then computed from these averaged expression values, which in turn used to construct phylogenetic tree with software such as Geneious®(www.geneious.com) or Phylip®[92]. This process is then repeated (bootstrapping) with different sets of differently expressed genes to construct various trees with differing topologies [93]. A consensus tree is then constructed from these trees using appropriate algorithm such as Weighted Least Squares [94]. A workflow of this methodology is illustrated in Figure 3, which was successfully implemented for the phylogeny reconstruction of leukemia, breast cancer and liposarcoma subtypes[95].

Conclusion

Phylogeneticsis an important platform for cancer research by providing dynamic, predictive and seamless evolutionary classification and discovery of classes that accurately reveal biological processes and patterns between them. Analysis and interpretation of various phylogenetically informative data generated by modern cancer research will further facilitate our understanding of cancer as an evolutionary phenomenon. Since the advent of massively parallel next generation sequencing methods, huge quantity of omics data emerging at an ever increasing rate offer a unifying paradigm for the computational phylogenetics. Cancer phylogenetics will eventually aid in our quest for understanding differences in chemo-sensitivity among various clones and thereby to the realms of personalized medicine bycreating rationally designed and molecularly targeted drugs delivering right quantityto the right target group.

Future scenarios of cancer phylogenetics might include establishment of phylogenetic trees from genotyping distinct population of cancer cells isolated from spatiotemporally distinct parts of the tumor, including regions of invasion, metastasis and relapse. Single cell expression measurements may prove more valuable for cancer phylogenetics than measurements from gene or chromosome copy number changes. Identification of initial transformation event by phylogenetic inferences will also have potential clinical implications as the presence of residual cells having genetic alterations similar to initial transformation event could predict recurrence. Identification of initial transformation event can also facilitate the discovery of "red-gene" –chemopreventive and therapeutic targeting of which in the initial phases could ultimately prevent development of neoplasm altogether. In addition, oncogenetic trees reveal the directionality of change within a set of specimens, and could be of use in early diagnosis, prognosis, treatment assessment and biomarker identification.

References

[1] Bauvet, F., et al., [Therapeutic consequences of molecular biology advances in oncology]. *Bull Cancer*, 2009. 96(1): p. 59-71.

[2] Cooke, S.L., et al., Genomic analysis of genetic heterogeneity and evolution in high-grade serous ovarian carcinoma. *Oncogene*, 2010. 29(35): p. 4905-13.

[3] Wcislo, G., Molecular underpinnings of the targeted therapy for cancer. *Acta Pol. Pharm,* 2008. 65(6): p. 633-40.

[4] Dinh, P., C. Sotiriou, and M.J. Piccart, The evolution of treatment strategies: aiming at the target. *Breast,* 2007. 16 Suppl 2: p. S10-6.

[5] Turk, D. and G. Szakacs, Relevance of multidrug resistance in the age of targeted therapy. *Curr. Opin. Drug Discov Devel.*, 2009. 12(2): p. 246-52.

[6] Gomez-Lopez, G. and A. Valencia, Bioinformatics and cancer research: building bridges for translational research. *Clin. Transl. Oncol.,* 2008. 10(2): p. 85-95.

[7] Colozza, M., et al., Breast cancer: achievements in adjuvant systemic therapies in the pre-genomic era. *Oncologist,* 2006. 11(2): p. 111-25.

[8] Axelrod, R., D.E. Axelrod, and K.J. Pienta, Evolution of cooperation among tumor cells. *Proc. Natl. Acad. Sci. USA*, 2006. 103(36): p. 13474-9.

[9] Maley, C.C. and B.J. Reid, Natural selection in neoplastic progression of Barrett's esophagus. *Semin. Cancer Biol.*, 2005. 15(6): p. 474-83.

[10] Michor, F., et al., Somatic selection for and against cancer. *J. Theor. Biol.*, 2003. 225(3): p. 377-82.

[11] Merlo, L.M., et al., Cancer as an evolutionary and ecological process. *Nat. Rev. Cancer,* 2006. 6(12): p. 924-35.

[12] Takahashi, K., et al., Clonal and parallel evolution of primary lung cancers and their metastases revealed by molecular dissection of cancer cells. *Clin. Cancer Res.*, 2007. 13(1): p. 111-20.

[13] Stilgenbauer, S., et al., Clonal evolution in chronic lymphocytic leukemia: acquisition of high-risk genomic aberrations associated with unmutated VH, resistance to therapy, and short survival. *Haematologica,* 2007. 92(9): p. 1242-5.

[14] Shackleton, M., et al., Heterogeneity in cancer: cancer stem cells versus clonal evolution. *Cell,* 2009. 138(5): p. 822-9.

[15] Masramon, L., et al., Genetic instability and divergence of clonal populations in colon cancer cells in vitro. *J. Cell Sci.,* 2006. 119(Pt 8): p. 1477-82.

[16] Nowak, M.A., F. Michor, and Y. Iwasa, Genetic instability and clonal expansion. *J.Theor. Biol.*, 2006. 241(1): p. 26-32.

[17] Maley, C.C., et al., Genetic clonal diversity predicts progression to esophageal adenocarcinoma. *Nat. Genet*, 2006. 38(4): p. 468-73.

[18] Breivik, J., The evolutionary origin of genetic instability in cancer development. *Semin.Cancer Biol*, 2005. 15(1): p. 51-60.

[19] Michor, F., et al., Stochastic elimination of cancer cells. *Proc. Biol. Sci.*, 2003. 270(1528): p. 2017-24.

[20] Li, X., et al., Application of biomarkers in cancer risk management: evaluation from stochastic clonal evolutionary and dynamic system optimization points of view. *PLoS Comput. Biol.*, 2011. 7(2): p. e1001087.

[21] Di Pietro, C., et al., The apoptotic machinery as a biological complex system: analysis of its omics and evolution, identification of candidate genes for fourteen major types of cancer, and experimental validation in CML and neuroblastoma. *BMC Med. Genomics*, 2009. 2: p. 20.

[22] Abu-Asab, M., M. Chaouchi, and H. Amri, Evolutionary medicine: A meaningful connection between omics, disease, and treatment. *Proteomics Clin. Appl.*, 2008. 2(2): p. 122-134.

[23] Peng, B., C.I. Amos, and M. Kimmel, Forward-time simulations of human populations with complex diseases. *PLoS Genet*, 2007. 3(3): p. e47.

[24] O'Brien, S.J., A role for molecular genetics in biological conservation. *Proc. Natl.Acad.Sci. USA*, 1994. 91(13): p. 5748-55.

[25] Strassmann, B.I., Menstrual cycling and breast cancer: an evolutionary perspective. *J.Womens Health*, 1999. 8(2): p. 193-202.

[26] Bozic, I., et al., Accumulation of driver and passenger mutations during tumor progression.*Proc. Natl. Acad. Sci. USA*, 2010. 107(43): p. 18545-50.

[27] Shibata, D., Molecular tumour clocks. *Ann. Med.,* 1997. 29(1): p. 5-7.

[28] Shibata, D., et al., Somatic microsatellite mutations as molecular tumor clocks. *Nat.Med.*, 1996. 2(6): p. 676-81.

[29] Maley, C.C., et al., Selectively advantageous mutations and hitchhikers in neoplasms: p16 lesions are selected in Barrett's esophagus. *Cancer Res*, 2004. 64(10): p. 3414-27.

[30] Akey, J.M., et al., Population history and natural selection shape patterns of genetic variation in 132 genes. *PLoS Biol.*, 2004. 2(10): p. e286.

[31] Zhang, J. and H.F. Rosenberg, Diversifying selection of the tumor-growth promoter angiogenin in primate evolution. *Mol. Biol. Evol.*, 2002. 19(4): p. 438-45.

[32] Luria, S.E. and M. Delbruck, Mutations of Bacteria from Virus Sensitivity to Virus Resistance. *Genetics*, 1943. 28(6): p. 491-511.

[33] Schwartz, R. and S.E. Shackney, Applying unmixing to gene expression data for tumor phylogeny inference. *BMC Bioinformatics*, 2010. 11: p. 42.

[34] Shonkwiler, R.W., J.V. Herod, and Springer., Mathematical biologyan introduction with Maple and Matlab, 2009, Springer: New York ; London. p. xiii, 551 p.

[35] Desper, R., et al., Distance-based reconstruction of tree models for oncogenesis. *J.Comput. Biol.*, 2000. 7(6): p. 789-803.

[36] Syvanen, M., Horizontal gene transfer: evidence and possible consequences. *Annu.Rev.Genet.,* 1994. 28: p. 237-61.

[37] Page, R.D. and M.A. Charleston, From gene to organismal phylogeny: reconciled trees and the gene tree/species tree problem. *Mol. Phylogenet Evol*, 1997. 7(2): p. 231-40.

[38] Liu, L. and D.K. Pearl, Species trees from gene trees: reconstructing Bayesian posterior distributions of a species phylogeny using estimated gene tree distributions. *Syst. Biol.,* 2007. 56(3): p. 504-14.

[39] Knowles, L.L. and P.B. Klimov, Estimating phylogenetic relationships despite discordant gene trees across loci: the species tree of a diverse species group of feather mites (Acari: Proctophyllodidae). *Parasitology*, 2011: p. 1-10.

[40] Zhang, L., From Gene Trees to Species Trees II: Species Tree Inference by Minimizing Deep Coalescence Events. *IEEE/ACM Trans Comput Biol Bioinform*, 2011.

[41] Meyer, A., Homology and homoplasy: the retention of genetic programmes. *Novartis Found Symp*, 1999. 222: p. 141-53; discussion 153-7.

[42] Wake, D.B., Homoplasy, homology and the problem of 'sameness' in biology. *Novartis Found Symp*, 1999. 222: p. 24-33; discussion 33-46.
[43] Rendall, D. and A. Di Fiore, Homoplasy, homology, and the perceived special status of behavior in evolution. *J. Hum. Evol.,* 2007. 52(5): p. 504-21.
[44] Yang, M. and G.J. Wyckoff, Detection of selection utilizing molecular phylogenetics: a possible approach. *Genetica,* 2011. 139(5): p. 639-48.
[45] Walsh, D.A. and A.K. Sharma, Molecular phylogenetics: testing evolutionary hypotheses. *Methods Mol. Biol.*, 2009. 502: p. 131-68.
[46] Fu, Y.X. and W.H. Li, Coalescing into the 21st century: An overview and prospects of coalescent theory. *Theor. Popul. Biol.*, 1999. 56(1): p. 1-10.
[47] Crandall, K.A. and A.R. Templeton, Empirical tests of some predictions from coalescent theory with applications to intraspecific phylogeny reconstruction. *Genetics,* 1993. 134(3): p. 959-69.
[48] Rosenberg, N.A. and M. Nordborg, Genealogical trees, coalescent theory and the analysis of genetic polymorphisms. *Nat. Rev. Genet,* 2002. 3(5): p. 380-90.
[49] Tian, J.P. and X.S. Lin, The mutation process in colored coalescent theory. *Bull MathBiol.,* 2009. 71(8): p. 1873-89.
[50] Bulla, I., et al., HIV classification using the coalescent theory. *Bioinformatics*, 2010. 26(11): p. 1409-15.
[51] Sanger, F., Species differences in insulins. *Nature,* 1949. 164(4169): p. 529.
[52] Jukes, T.H. and C.R. Cantor, {*Evolution of protein molecules*}. 1969.
[53] Takahata, N. and M. Kimura, A model of evolutionary base substitutions and its application with special reference to rapid change of pseudogenes. *Genetics*, 1981. 98(3): p. 641-57.
[54] Felsenstein, J., Evolutionary trees from DNA sequences: a maximum likelihood approach. *J. Mol. Evol.*, 1981. 17(6): p. 368-376.
[55] Hasegawa, M., H. Kishino, and T. Yano, Dating of the human-ape splitting by a molecular clock of mitochondrial DNA. *J.Mol. Evol.*, 1985. 22(2): p. 160-174.
[56] Tavaré, S., Some probabilistic and statistical problems in the analysis of DNA sequences. *Some mathematical questions in biology-DNA sequence analysis*, 1986. 17: p. 57-86.
[57] Posada, D. and K.A. Crandall, Modeltest: testing the model of DNA substitution. *Bioinformatics*, 1998. 14(9): p. 817.
[58] Posada, D. and T.R. Buckley, Model selection and model averaging in phylogenetics: advantages of Akaike information criterion and Bayesian approaches over likelihood ratio tests. *Syst. Biol.,* 2004. 53(5): p. 793.
[59] Yang, Z., et al., Codon-substitution models for heterogeneous selection pressure at amino acid sites. *Genetics,* 2000. 155(1): p. 431.
[60] Rzhetsky, A., Estimating substitution rates in ribosomal RNA genes. *Genetics*, 1995. 141(2): p. 771-83.
[61] Goldman, N., J.L. Thorne, and D.T. Jones, Assessing the impact of secondary structure and solvent accessibility on protein evolution. *Genetics*, 1998. 149(1): p. 445-58.
[62] Saitou, N. and M. Nei, The neighbor-joining method: a new method for reconstructing phylogenetic trees. *Mol. Biol. Evol.*, 1987. 4(4): p. 406-25.

[63] Huelsenbeck, J.P., The robustness of two phylogenetic methods: four-taxon simulations reveal a slight superiority of maximum likelihood over neighbor joining. *Mol. Biol.Evol.*, 1995. 12(5): p. 843.

[64] Huelsenbeck, J.P., et al., Bayesian inference of phylogeny and its impact on evolutionary biology. *Science,* 2001. 294(5550): p. 2310.

[65] Yang, Z. and B. Rannala, Bayesian phylogenetic inference using DNA sequences: a Markov Chain Monte Carlo method. *Mol. Biol. Evol.*, 1997. 14(7): p. 717.

[66] Huelsenbeck, J.P. and F. Ronquist, MRBAYES: Bayesian inference of phylogenetic trees. *Bioinformatics,* 2001. 17(8): p. 754-5.

[67] Herynk, M.H., et al., Accelerated mammary maturation and differentiation, and delayed MMTVneu-induced tumorigenesis of K303R mutant ERalpha transgenic mice. *Oncogene*, 2009. 28(36): p. 3177-87.

[68] Stiewe, T., The p53 family in differentiation and tumorigenesis. *Nat. Rev. Cancer*, 2007. 7(3): p. 165-8.

[69] Fu, L.M. and C.S. Fu-Liu, Multi-class cancer subtype classification based on gene expression signatures with reliability analysis. *FEBS Lett*, 2004. 561(1-3): p. 186-90.

[70] Dyrskjot, L., Classification of bladder cancer by microarray expression profiling: towards a general clinical use of microarrays in cancer diagnostics. *Expert Rev. Mol.Diagn,* 2003. 3(5): p. 635-47.

[71] Sotiriou, C., et al., Breast cancer classification and prognosis based on gene expression profiles from a population-based study. *Proc. Natl. Acad. Sci. USA*, 2003. 100(18): p. 10393-8.

[72] Ahr, A., et al., Molecular classification of breast cancer patients by gene expression profiling. *J. Pathol.,* 2001. 195(3): p. 312-20.

[73] Anbazhagan, R., et al., Classification of small cell lung cancer and pulmonary carcinoid by gene expression profiles. *Cancer Res*, 1999. 59(20): p. 5119-22.

[74] Golub, T.R., et al., Molecular classification of cancer: class discovery and class prediction by gene expression monitoring. *Science*, 1999. 286(5439): p. 531-7.

[75] Romualdi, C., et al., Pattern recognition in gene expression profiling using DNA array: a comparative study of different statistical methods applied to cancer classification. *Hum. Mol. Genet*, 2003. 12(8): p. 823-36.

[76] Schena, M., et al., Quantitative monitoring of gene expression patterns with a complementary DNA microarray. *Science,* 1995. 270(5235): p. 467.

[77] Eisen, M.B., et al., Cluster analysis and display of genome-wide expression patterns. *Proc. Natl. Acad. Sci. USA*, 1998. 95(25): p. 14863-8.

[78] Huang, M.L., et al., Usage of Case-Based Reasoning, Neural Network and Adaptive Neuro-Fuzzy Inference System Classification Techniques in Breast Cancer Dataset Classification Diagnosis. *J. Med. Syst.,* 2010.

[79] Wang, S.L., et al., Tumor classification by combining PNN classifier ensemble with neighborhood rough set based gene reduction. *Comput. Biol. Med.,* 2010. 40(2): p. 179-89.

[80] Ubeyli, E.D., Adaptive neuro-fuzzy inference systems for automatic detection of breast cancer. *J. Med. Syst,* 2009. 33(5): p. 353-8.

[81] Kuo, S.J., et al., Classification of benign and malignant breast tumors using neural networks and three-dimensional power Doppler ultrasound. *Ultrasound Obstet Gynecol*, 2008. 32(1): p. 97-102.

[82] Hanahan, D., The hallmarks of cancer. *Cell*, 2000. 100(1): p. 57-70.
[83] Stephens, P.J., et al., Complex landscapes of somatic rearrangement in human breast cancer genomes. *Nature*, 2009. 462(7276): p. 1005-10.
[84] Greenman, C., et al., Patterns of somatic mutation in human cancer genomes. *Nature,* 2007. 446(7132): p. 153-8.
[85] Roche-Lestienne, C., et al., Several types of mutations of the Abl gene can be found in chronic myeloid leukemia patients resistant to STI571, and they can pre-exist to the onset of treatment. *Blood,* 2002. 100(3): p. 1014.
[86] Hong, D., et al., Initiating and cancer-propagating cells in TEL-AML1-associated childhood leukemia. *Science,* 2008. 319(5861): p. 336.
[87] Campbell, P.J., et al., Subclonal phylogenetic structures in cancer revealed by ultra-deep sequencing. *Proc. Natl. Acad. Sci. USA*, 2008. 105(35): p. 13081-6.
[88] Pennington, G., et al., Reconstructing tumor phylogenies from heterogeneous single-cell data. *J. Bioinform. Comput. Biol.*, 2007. 5(2a): p. 407-27.
[89] Pennington, G., et al., Expectation-maximization method for reconstructing tumor phylogenies from single-cell data. *Comput. Syst. Bioinformatics Conf*, 2006: p. 371-80.
[90] Smith, C.A., et al., Correlations among p53, Her-2/neu, and ras overexpression and aneuploidy by multiparameter flow cytometry in human breast cancer: evidence for a common phenotypic evolutionary pattern in infiltrating ductal carcinomas. *Clin. CancerRes.*, 2000. 6(1): p. 112-26.
[91] Gentleman, R.C., et al., Bioconductor: open software development for computational biology and bioinformatics. *Genome Biol.*, 2004. 5(10): p. R80.
[92] Felsenstein, J., {PHYLIP}: phylogenetic inference package, version 3.5 c. 1993.
[93] Felsenstein, J., Confidence limits on phylogenies: an approach using the bootstrap. *Evolution*, 1985. 39(4): p. 783-791.
[94] Desper, R. and O. Gascuel, Theoretical foundation of the balanced minimum evolution method of phylogenetic inference and its relationship to weighted least-squares tree fitting. *Mol. Biol. Evol.*, 2004. 21(3): p. 587.
[95] Riester, M., et al., A differentiation-based phylogeny of cancer subtypes. *PLoS Comput.Biol.*, 2010. 6(5): p. e1000777.

Index

B

C

D

E

F

K

L

M

N

O

P

Q

R

S

T